Hydrothermal Carbonization

Hydrothermal Carbonization

Editor

M. Toufiq Reza

Basel • Beijing • Wuhan • Barcelona • Belgrade • Novi Sad • Cluj • Manchester

Editor
M. Toufiq Reza
Department of Biomedical
and Chemical Engineering
and Sciences,
Florida Institute
of Technology,
Melbourne, FL, USA

Editorial Office
MDPI
St. Alban-Anlage 66
4052 Basel, Switzerland

This is a reprint of articles from the Special Issue published online in the open access journal *Energies* (ISSN 1996-1073) (available at: https://www.mdpi.com/journal/energies/special_issues/hydrothermal_carbonization).

For citation purposes, cite each article independently as indicated on the article page online and as indicated below:

Lastname, A.A.; Lastname, B.B. Article Title. *Journal Name* **Year**, *Volume Number*, Page Range.

ISBN 978-3-0365-8456-0 (Hbk)
ISBN 978-3-0365-8457-7 (PDF)
doi.org/10.3390/books978-3-0365-8457-7

Cover image courtesy of M. Toufiq Reza.

© 2023 by the authors. Articles in this book are Open Access and distributed under the Creative Commons Attribution (CC BY) license. The book as a whole is distributed by MDPI under the terms and conditions of the Creative Commons Attribution-NonCommercial-NoDerivs (CC BY-NC-ND) license.

Contents

About the Editor vii

M. Toufiq Reza
Hydrothermal Carbonization
Reprinted from: *Energies* **2022**, *15*, 5491, doi:10.3390/en15155491 1

Manfredi Picciotto Maniscalco, Maurizio Volpe and Antonio Messineo
Hydrothermal Carbonization as a Valuable Tool for Energy and Environmental Applications: A Review
Reprinted from: *Energies* **2020**, *13*, 4098, doi:10.3390/en13164098 5

Fabio Merzari, Jillian Goldfarb, Gianni Andreottola, Tanja Mimmo, Maurizio Volpe and Luca Fiori
Hydrothermal Carbonization as a Strategy for Sewage Sludge Management: Influence of Process Withdrawal Point on Hydrochar Properties
Reprinted from: *Energies* **2020**, *13*, 2890, doi:10.3390/en13112890 31

Louise Delahaye, John Thomas Hobson, Matthew Peter Rando, Brenna Sweeney, Avery Bernard Brown, Geoffrey Allen Tompsett, et al.
Experimental and Computational Evaluation of Heavy Metal Cation Adsorption for Molecular Design of Hydrothermal Char
Reprinted from: *Energies* **2020**, *13*, 4203, doi:10.3390/en13164203 53

Taina Lühmann and Benjamin Wirth
Sewage Sludge Valorization via Hydrothermal Carbonization: Optimizing Dewaterability and Phosphorus Release
Reprinted from: *Energies* **2020**, *13*, 4417, doi:10.3390/en13174417 77

Kyoung S. Ro, Judy A. Libra and Andrés Alvarez-Murillo
Comparative Studies on Water- and Vapor-Based Hydrothermal Carbonization: Process Analysis
Reprinted from: *Energies* **2020**, *13*, 5733, doi:10.3390/en13215733 93

Md Rifat Hasan, Nepu Saha, Thomas Quaid and M. Toufiq Reza
Formation of Carbon Quantum Dots via Hydrothermal Carbonization: Investigate the Effect of Precursors
Reprinted from: *Energies* **2021**, *14*, 936, doi:10.3390/en14040986 111

Md Tahmid Islam, Nepu Saha, Sergio Hernandez, Jordan Klinger and M. Toufiq Reza
Integration of Air Classification and Hydrothermal Carbonization to Enhance Energy Recovery of Corn Stover
Reprinted from: *Energies* **2021**, *14*, 1397, doi:10.3390/en14051397 121

Gabriel Gerner, Luca Meyer, Rahel Wanner, Thomas Keller and Rolf Krebs
Sewage Sludge Treatment by Hydrothermal Carbonization: Feasibility Study for Sustainable Nutrient Recovery and Fuel Production
Reprinted from: *Energies* **2021**, *14*, 2697, doi:10.3390/en14092697 135

Gabriella Gonnella, Giulia Ischia, Luca Fambri and Luca Fiori
Thermal Analysis and Kinetic Modeling of Pyrolysis and Oxidation of Hydrochars
Reprinted from: *Energies* **2022**, *15*, 950, doi:10.3390/en15030950 147

About the Editor

M. Toufiq Reza

M. Toufiq Reza, Ph.D., is an Assistant Professor of Chemical Engineering at the Department of Biomedical and Chemical Engineering and Sciences of Florida Institute of Technology. He received his M.S. and Ph.D. in Chemical Engineering from the University of Nevada, Reno, in 2011 and 2013, respectively. He has published more than ninety peer-reviewed journal articles, five patents, two book chapters, and more than one hundred oral and poster presentations. Dr. Reza was named as one of recipients of the I&EC Research 2021 Class of Influential Researchers. He has also contributed to the AIChE Journal 2022 Futures Issue. He was the recipient of the 2022 New Holland Young Researcher award from American Society of Agricultural and Biological Engineers (ASABE) and 2022 FASABE Young Researcher Award. He is also listed in the "World Ranking of Top 2% Scientists" in the 2021 and 2022 database published by Stanford University. Dr. Reza leads the Biofuels Research Lab at Florida Tech, with primary research interest on upcycling wastes via hydrothermal carbonization. His current research projects have been funded by the National Science Foundation, U.S. Department of Agriculture, U.S. Environmental Protection Agency, American Chemical Society, Florida Sea Grant, U.S. Agency for International Development, Sugarbush Foundation, and other agencies. Dr. Reza is a member of the American Institute of Chemical Engineers, American Chemical Society, and American Society of Agricultural and Biological Engineers. He is the secretary of PRS-707 Food & Organic Waste Management & Utilization at the ASABE. He is an Associate Editor of *Frontiers in Fuels*, a Section Editor of the *Energies* journal, and an Editorial Board member of the *Biomass Conversion and Biorefinery* journal.

Editorial

Hydrothermal Carbonization

M. Toufiq Reza

Department of Biomedical and Chemical Engineering and Sciences, Florida Institute of Technology, 150 West University Blvd, Melbourne, FL 32901, USA; treza@fit.edu

Over the past decade, hydrothermal carbonization (HTC) has emerged as a promising thermochemical pathway for treating and converting wet wastes into fuel, materials, and chemicals. Many of the earlier studies have been carried out to understand HTC reaction mechanisms and reaction kinetics using model compounds [1,2]. More recently, HTC studies have shifted from model compounds to researching lignocellulosic biomasses [3–5]; however, these works remain focused on solid biofuel production in the form of hydrochar. Perhaps the real potential for HTC lies in wet wastes (such as sewage sludge, agricultural wastes, animal wastes, etc.) and their conversion to fertilizer, high value materials, and sustainable chemicals [1]. However, the kinetics and reaction mechanism of the real wet wastes often deviate from those of the model compounds [6]. Therefore, studying the reaction kinetics of real wet waste is an important step towards revealing the viability of HTC in the commercial process.

This Special Issue titled, "Hydrothermal Carbonization" has invited researchers from around the World to contribute to these issues in the field of HTC research. Including the editorial, ten papers are listed in this Special Issue. Researchers from eight different countries have contributed their research to this Special Issue. The highlights of the Special Issue are as follows:

- Maniscalco et al. [7] reviewed the recent progress made towards using HTC as a valuable tool for energy and environmental applications. The review identified that pristine hydrochar from organic wastes such as food waste are suitable for combustion or soil amendment based on feedstock. However, the review also reported recent progresses with respect to further activating hydrochar for energy storage and pollutant adsorption applications.
- Three articles are reported aspects of HTC related to sewage sludge. Merzari et al. [8] have evaluated HTC systems at various points in wastewater treatment plants to treat various type of sludges (e.g., primary, secondary, and digestate sludge) for different applications. Meanwhile, Luhmann and With have optimized the dewaterability and phosphorus release from sewage sludge during HTC [9]. Sewage sludge is one of the major sources of phosphorus and recovering phosphorus effectively from sewage sludge may enhance the success of HTC. Finally, Gerner et al. [10] have reported the nutrient recovery and fuel production potentials from sewage sludge. The study reveals that the HTC process could be much cheaper than the current practice of sewage sludge treatment in Switzerland; therefore, fuel production and nutrient recovery from sewage sludge are economically viable.
- Ro et al. [11] have compared the HTC process with vapothermal carbonization (VTC). When feedstock does not have an adequate volume of water to create subcritical water conditions, the thermochemical conversion is called VTC. There are distinct product formations for HTC and VTC [12]. This study reveals the minimum amount of water required in the feedstock to be considered conducive to HTC. This fundamental study will clarify future research directions and ensure the practice of HTC.
- Islam et al. [13] reported an approach integrating air classification with HTC. In the study, air classification separated high-ash-fraction (waste) corn stover from the

Citation: Reza, M.T. Hydrothermal Carbonization. *Energies* **2022**, *15*, 5491. https://doi.org/10.3390/en15155491

Received: 19 July 2022
Accepted: 27 July 2022
Published: 29 July 2022

Publisher's Note: MDPI stays neutral with regard to jurisdictional claims in published maps and institutional affiliations.

Copyright: © 2022 by the author. Licensee MDPI, Basel, Switzerland. This article is an open access article distributed under the terms and conditions of the Creative Commons Attribution (CC BY) license (https://creativecommons.org/licenses/by/4.0/).

light ash fraction type [14]. This paper reported that the HTC of waste products can be performed and the hydrochar can be used to enhance the pelletization of clean corn stover.
- Gonnella et al. [15] have studied the thermal analysis and kinetics modelling of both the pyrolysis and oxidation of hydrochar. Although hydrochar has been reported as solid fuel, few sources have reported the combustion of hydrochar.
- Delahaye et al. [16] evaluated the cationic heavy metal adsorption of hydrochar both experimentally and computationally. Confirmed via density functional theory (DFT) simulation, this fundamental study reveals that hydrochar possesses superior binding sites for copper (II) ions compared to ion exchange resins.
- Finally, Hasan et al. [17] reported how various precursors affect the carbon quantum dot (CQD) properties via HTC. CQDs are very high-value materials with a variety of applications, such as bioimaging, photocatalysis, and energy storage. This paper demonstrates the possibility of separating CQDs from HTC process liquid. The utilization of HTC process liquid has always been a concern for HTC commercialization and this paper reveals a potential application for this very liquid.

Overall, HTC has shown tremendous potential toward commercialization. We hope that readers will find this Special Issue informative and inspiring. The editor is thankful to the authors for submitting their HTC research. The *Energies* editors and the reviewers are also acknowledged for reviewing the manuscripts, which have made significant improvements to this Special Issue.

Funding: The work is partially funded by USDA NIFA AFRI grant no. 2019-67019-31594 and grant no. 2021-67022-34487.

Conflicts of Interest: The author declares no conflict of interest.

References

1. Román, S.; Libra, J.; Berge, N.; Sabio, E.; Ro, K.; Li, L.; Ledesma, B.; Álvarez, A.; Bae, S. Hydrothermal Carbonization: Modeling, Final Properties Design and Applications: A Review. *Energies* **2018**, *11*, 216. [CrossRef]
2. Libra, J.A.; Ro, K.S.; Kammann, C.; Funke, A.; Berge, N.D.; Neubauer, Y.; Titirici, M.-M.; Fühner, C.; Bens, O.; Kern, J.; et al. Hydrothermal carbonization of biomass residuals: A comparative review of the chemistry, processes and applications of wet and dry pyrolysis. *Biofuels* **2011**, *2*, 71–106. [CrossRef]
3. Islam, M.T.; Sultana, A.I.; Saha, N.; Klinger, J.L.; Reza, M.T. Pretreatment of Biomass by Selected Type-III Deep Eutectic Solvents and Evaluation of the Pretreatment Effects on Hydrothermal Carbonization. *Ind. Eng. Chem. Res.* **2021**, *60*, 15479–15491. [CrossRef]
4. Reza, M.T.; Yang, X.; Coronella, C.J.; Lin, H.; Hathwaik, U.; Shintani, D.; Neupane, B.P.; Miller, G.C. Hydrothermal Carbonization (HTC) and Pelletization of Two Arid Land Plants Bagasse for Energy Densification. *ACS Sustain. Chem. Eng.* **2015**, *4*, 1106–1114. [CrossRef]
5. Saba, A.; Saha, P.; Reza, M.T. Co-Hydrothermal Carbonization of coal-biomass blend: Influence of temperature on solid fuel properties. *Fuel Process. Technol.* **2017**, *167*, 711–720. [CrossRef]
6. Reza, M.T.; Yan, W.; Uddin, M.H.; Lynam, J.G.; Hoekman, S.K.; Coronella, C.J.; Vasquez, V.R. Reaction kinetics of hydrothermal carbonization of loblolly pine. *Bioresour. Technol.* **2013**, *139*, 161–169. [CrossRef] [PubMed]
7. Maniscalco, M.P.; Volpe, M.; Messineo, A. Hydrothermal Carbonization as a Valuable Tool for Energy and Environmental Applications: A Review. *Energies* **2020**, *13*, 4098. [CrossRef]
8. Merzari, F.; Goldfarb, J.; Andreottola, G.; Mimmo, T.; Volpe, M.; Fiori, L. Hydrothermal Carbonization as a Strategy for Sewage Sludge Management: Influence of Process Withdrawal Point on Hydrochar Properties. *Energies* **2020**, *13*, 2890. [CrossRef]
9. Lühmann, T.; Wirth, B. Sewage Sludge Valorization via Hydrothermal Carbonization: Optimizing Dewaterability and Phosphorus Release. *Energies* **2020**, *13*, 4417. [CrossRef]
10. Gerner, G.; Meyer, L.; Wanner, R.; Keller, T.; Krebs, R. Sewage Sludge Treatment by Hydrothermal Carbonization: Feasibility Study for Sustainable Nutrient Recovery and Fuel Production. *Energies* **2021**, *14*, 2697. [CrossRef]
11. Ro, K.S.; Libra, J.A. Alvarez-Murillo, Comparative Studies on Water- and Vapor-Based Hydrothermal Carbonization: Process Analysis. *Energies* **2020**, *13*, 5733. [CrossRef]
12. Funke, A.; Reebs, F.; Kruse, A. Experimental comparison of hydrothermal and vapothermal carbonization. *Fuel Process. Technol.* **2013**, *115*, 261–269. [CrossRef]
13. Islam, M.T.; Saha, N.; Hernandez, S.; Klinger, J.; Reza, M.T. Integration of Air Classification and Hydrothermal Carbonization to Enhance Energy Recovery of Corn Stover. *Energies* **2021**, *14*, 1397. [CrossRef]

14. Lacey, J.A.; Aston, J.E.; Westover, T.L.; Cherry, R.S.; Thompson, D.N. Removal of introduced inorganic content from chipped forest residues via air classification. *Fuel* **2015**, *160*, 265–273. [CrossRef]
15. Gonnella, G.; Ischia, G.; Fambri, L.; Fiori, L. Thermal Analysis and Kinetic Modeling of Pyrolysis and Oxidation of Hydrochars. *Energies* **2022**, *15*, 950. [CrossRef]
16. Delahaye, L.; Hobson, J.T.; Rando, M.F.; Sweeney, B.; Brown, A.B.; Tompsett, G.A.; Ates, A.; Deskins, N.A.; Timko, M.T. Experimental and Computational Evaluation of Heavy Metal Cation Adsorption for Molecular Design of Hydrothermal Char. *Energies* **2020**, *13*, 4203. [CrossRef]
17. Hasan, M.R.; Saha, N.; Quaid, T.; Reza, M.T. Formation of Carbon Quantum Dots via Hydrothermal Carbonization: Investigate the Effect of Precursors. *Energies* **2021**, *14*, 986. [CrossRef]

Review

Hydrothermal Carbonization as a Valuable Tool for Energy and Environmental Applications: A Review

Manfredi Picciotto Maniscalco, Maurizio Volpe and Antonio Messineo *

Faculty of Engineering and Architecture, Kore University of Enna, Cittadella Universitaria, 94100 Enna, Italy; manfredi.maniscalco@unikore.it (M.P.M.); maurizio.volpe@unikore.it (M.V.)
* Correspondence: antonio.messineo@unikore.it

Received: 6 July 2020; Accepted: 4 August 2020; Published: 7 August 2020

Abstract: Hydrothermal carbonization (HTC) represents an efficient and valuable pre-treatment technology to convert waste biomass into highly dense carbonaceous materials that could be used in a wide range of applications between energy, environment, soil improvement and nutrients recovery fields. HTC converts residual organic materials into a solid high energy dense material (hydrochar) and a liquid residue where the most volatile and oxygenated compounds (mainly furans and organic acids) concentrate during reaction. Pristine hydrochar is mainly used for direct combustion, to generate heat or electricity, but highly porous carbonaceous media for energy storage or for adsorption of pollutants applications can be also obtained through a further activation stage. HTC process can be used to enhance recovery of nutrients as nitrogen and phosphorous in particular and can be used as soil conditioner, to favor plant growth and mitigate desertification of soils. The present review proposes an outlook of the several possible applications of hydrochar produced from any sort of waste biomass sources. For each of the applications proposed, the main operative parameters that mostly affect the hydrochar properties and characteristics are highlighted, in order to match the needs for the specific application.

Keywords: HTC; waste biomass; energy recovery; environmental remediation; nutrients recovery; activated carbon

---

1. Introduction

The increasing need of finding new renewable energy alternatives to fossil fuels together with the need to safely dispose of organic waste has pushed, in the last few years, the investigation for more efficient and reliable technologies for waste biomass energy exploitation and conversion towards valuable materials. Indeed, the constant rise in the global energy consumption together with the parallel decrease in the fossil fuel reservoirs has driven the scientific community toward the research of more sustainable sources [1]. Waste biomass, and in particular the organic fraction of municipal solid waste (OFMSW) and industrial and sewage sludges are among the most studied residual feedstocks to produce energy and valuable carbonaceous materials [2–6].

One of the major drawbacks of choosing waste biomass for energy production, is the need for one or more pre-treatments before use [7]. The low energy density due to the high moisture content of waste biomass, makes it often not suitable for a direct energy conversion by combustion and/or gasification [8–10]. The high moisture in residual biomass raises significantly the operational cost of transportation and energy consumption for drying [11]. However, due to its great abundance, globally availability, carbon neutrality and the necessity to find efficient and valuable technologies to treat and convert it into sustainable energy sources and exploitable materials, waste biomasses has gained increasing attention and investigation [12].

Between different valuable alternative treatments for waste biomass, wet thermochemical technologies for their flexibility and, principle, their low investment and operating costs are receiving increasing attention in the last years [13,14]. In particular, hydrothermal carbonization (HTC), also known as wet pyrolysis, is a thermochemical process performed in the presence of sub-critical water at temperature usually between 180 and 280 °C and autogenous saturated vapor conditions (10–80 bars). Residence time is varied from minutes up to several hours [3,15,16]. At high temperature and pressure, even at subcritical conditions, water undergoes a dramatic properties change acting more as an organic solvent and its increased ion product favors reactions that are typically catalyzed by acids or bases, promoting biomass decomposition through hydrolysis, dehydration and decarboxylation reactions [17,18]. By means of these reactions, it is possible to increase the carbon content of the initial feedstock, by removing most of the more volatile oxygenated compounds (furans and low molecular fatty acids) that are typically moved to the aqueous phase [19]. It is also well documented in the literature that, during HTC, part of the inorganics are moved to the liquid phase [19–21] thus, on the one hand, reducing in the solid residue (hydrochar), minerals and ash formation during combustion, and on the other hand, increasing the high heating values (HHV) [17,22]. This behavior leads to a reduced risk of slugging and fouling phenomena during combustion of hydrochars in boilers if compared to the use of pyrochars [2,23,24]. Hydrochar can be used in numerous applications, as a bio fuel for energy production [1,13,25], as a source of carbon and nutrients in soil application [19,26,27] or even as starting material for further advanced utilization in supercapacitors or a porous matrix for the adsorption of pollutants [28,29]. Moreover, the possible scalability of the process up to the industrial level [30], as well as relatively mild process conditions, make HTC a valid solution for the near future for the transformation of a wide range of raw biomass into valuable materials [13,30]. In Figure 1 is reported a flowchart of hydrochar production from waste biomass and its possible energetic and environmental applications.

Since HTC is performed under wet conditions, this technology can find its best application for biomasses that present high initial moisture content. Indeed, when dealing with direct combustion, a detrimental overall efficiency was shown when the starting biomass presents high moisture content [10]. The main reaction pathways involved in the hydrothermal carbonization process are still under discussion among the scientific community [31]. One possible route was discussed by Jatzwauck and Schumpe [32], according to which the HTC process is essentially divided into three steps. The first one is mainly related to the hydrolyzation of cellulose, hemicellulose and part of the lignin. During the second step, the generated intermediates endure further re-arrangement resulting in the generation of gaseous products. As an alternative to the formation of the gas phase, intermediate products can conglomerate to form secondary char, during the third step. The other mostly discussed route, regarding the evolution of HTC reactions, involves two main pathways: a solid–solid transformation, according to which hydrochar is directly generated from the dehydration of the initial biomass, and a vapor–solid reaction in which the dissolved organics present in the solution, back polymerize to form again a solid material, often named coke or secondary char [33–35].

The present work attempted to review the most relevant research on HTC of the last 5 years, in order to derive the best operative conditions for the production of valuable hydrochar materials. Moreover, whenever possible, the present review was integrated with a further pyrolysis or gasification process in order to evaluate how HTC behaves as a thermo-chemical pre-treatment for the production of porous material or precursor for energy application purposes.

Figure 1. Waste biomass hydrochar potential applications.

2. Effect of HTC Operating Conditions

It is well established that the HTC treatment induces some major transformations inside the biomass particles, in accordance with the severity of the process that is performed. From an operative point of view, the three main parameters that can be varied in a HTC process are temperature, residence time and biomass-to-water ratio. It was found that each of these variables can differently influence the intermediate reaction and the output products in terms of yields and composition [33,36]. The evolution of the main biomass components (lignin, cellulose and hemicellulose) in relation to the reaction severity was recently studied and modelled [37,38]. Heidari et al. [37] found almost negligible interactions between lignin, cellulose and hemicellulose on the final hydrochar properties, while cellulose was mostly affected by reaction severity. Authors also conducted an evaluation on the optimal proportion between biomass components to increase carbon content and calorific value of hydrochar, finding that a cellulose, hemicellulose and lignin content respectively of 40%, 35% and 25%, represents the optimal starting conditions. Another main parameter that can significantly influence the composition of the final hydrochar is the origin of the biomass. It is well known that ligno-cellulosic biomass is mainly composed of lignin, cellulose and hemicellulos, but the extent of their proportion in the overall biomass composition is strictly related to the nature of the material considered. At the same time each of these components undergoes degradation processes at different extent in accordance to the process conditions [39,40]. The first component that is subjected to hydrolyzation is hemicellulose, which starts to degrade at temperatures of 180–200 °C [41], followed by cellulose depolymerization (above 210 °C) [40] and softening of lignin [42]. Complete conversion of hemicellulose for reaction temperature of 180 °C and residence time of 4 h was reported by Olszewski et al., when treating brewer's spent grains [43]. Cellulose was found to be the most difficult component to be degraded, as reported by Titirici [31]. According to the author, cellulose starts to be converted above 210 °C according to two different pathways, as depicted in Figure 2 [31]. In their experiments, Olszewski et al. [43] and Volpe et al. [40] reported a complete cellulose conversion at a reaction temperature of 260 °C.

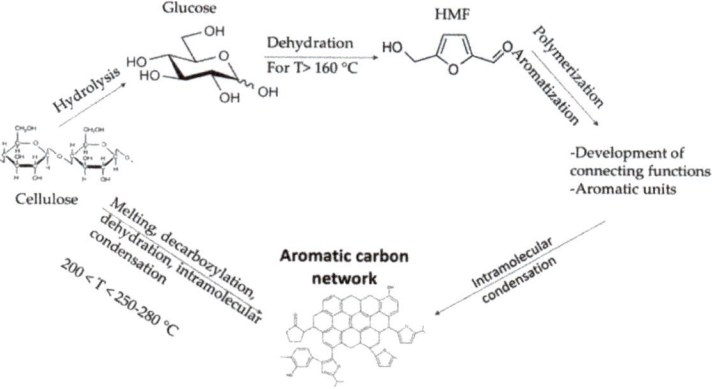

Figure 2. Cellulose conversion pathways, as proposed in Titirici, 2012.

Due to its very complex structure, lignin undergoes decomposition under a wider temperature range with respect to hemicellulose and cellulose [44]. The deterioration starts at around 200 °C, leading to the disruption of highly reactive alkyl-aryl-ether bonds, which further react with other intermediate compounds to form highly stable products [45]. Volpe et al. [19], when performing thermos gravimetric analysis (TGA) on olive mill waste hydrochar, found that while cellulose peak is significantly reduced for HTC temperature of 250 °C, the peak associated to lignin increases [44]. Through the same analysis, authors also confirmed the production of lower molecular compounds and/or tarry products coming from the decomposition of lignin. Indeed, a new peak in TGA curve at around 250 °C was found for the sample produced at 220 and 250 °C. From these considerations, it can be deduced that different reaction temperatures lead to different compositions in terms of cellulose, hemicellulose and lignin. The study conducted by Kim et al. [46] reported the energy retention efficiency (ERE) for each of these components in relation to reaction temperature. Indeed, as previously reported, as HTC temperature is raised, the calorific value increases as well, but the mass yield is reduced. It is therefore of paramount importance to define the best conditions to maximize the ERE, in order to obtain a valuable biochar for energy purposes. The results of this study showed that ERE for lignin and cellulose was maximized, respectively, at 200 °C and 220 °C, while the best temperature for the conversion of biomass into high energetic medium appeared to be 220 °C.

The higher thermal stability of lignocellulosic against non-lignocellulosic biomass was also confirmed by Zhuang et al. [47], testing herb tea waste (lignocellulosic) and penicillin mycelial waste (non-lignocellulosic). The analysis performed on the hydrochars produced showed a superior thermal stability for the tea waste, due to the presence of hemicellulose and lignin which start to degrade above 210 °C, while polysaccharide and protein present in the penicillin mycelial waste were hydrolyzed at temperatures lower than 180 °C. Finally, some considerations should be addressed regarding the influence of pressure, particle size, and the possibility of water recirculation within the system. With respect to the first parameter, the autogenous pressure that is developed inside the reactor maintains the water in the liquid state, favoring its ability to dissolve polar compounds [48]. Higher pressure levels result in smaller hydrochar dimensions and higher pore volume, together with a mild increase in the HHV. Only when pressure is raised above 100 MPa, significant increases in the HHV can be achieved, but the higher cost of the reactor can limit the application [48]. Regarding particle size and water recirculation, Heidari et al. [49] found that when biomass dimension was raised from <0.25 mm to within the range of 1.25–3 mm, a slight increase in the mass yield was observed, but the calorific value was lower. This condition is explained by the more severe conversion that biomass endures since water can more easily penetrate and reach the inner core of the biomass particle. Water recirculation affects almost only the mass yield, increasing it after the first recycle of about 12%, while the contribution to HHV is almost negligible, since it was raised by just 2%.

Proximate and ultimate analysis have been widely used to characterize the hydrochars properties. Proximate analysis is used to get an overall picture of the char composition in terms of fixed carbon (FC), volatile matter (VC), ash and moisture content (M), while the ultimate analysis give the weight percentages of their major elements (C, H, N, S, O). Although not all the reactions involved in HTC treatment are well known, the general evolution pathways of the main biomass elements are fairly well understood. As long as reaction severity is increased, most of the volatile compounds are released, increasing carbon content and heating value. The main reactions occurring are dehydration and de-hydrogenation which imply, respectively, the removal of hydroxyl groups and the rupture of the of the long chain carboxylic acids [50]. Different studies evaluated the effect of residence time and reaction temperature on the final hydrochar yields and properties [4,17,51,52]. The results of the studies assert that the main contribution to biomass degradation and increase in the calorific value is reaction temperature rather than residence time [53]. Lucian et al. [51] reported an increase in the heating value of olive trimming from 22.6 to 27.8 MJ kg^{-1} when reaction temperature was raised from 180 to 250 °C. At the same time, increasing the residence time from 1 to 6 h led to a maximum increase in HHV of +2 MJ kg^{-1}, at the temperature of 220 °C, passing from 24.7 to 26.7 MJ kg^{-1}. The increase in the calorific value is mainly due to the combined effect of the removal of oxygen and volatile compounds and the parallel rise in carbon content. The evolution of the concentration of carbon, oxygen and hydrogen is usually represented through the van Krevelen diagram in H/C and O/C ratios which are plotted. It is fairly stated that an increase in the hydrothermal carbonization temperature boosts dehydration reactions, removing oxygen from the initial biomass structure, resulting in a decrease in the O/C ratio. The evolution of the hydrochar composition is represented in the van Krevelen diagram through a shift toward lower values of O/C, characteristic of highly carbonaceous fuels like lignite or coals. Ulbrich et al. stated that the removal of oxygen for low HTC temperature is only mildly affected by residence time, while in the temperature range of 230–280 °C a more pronounced connection between oxygen reduction and residence time was observed [52]. Figure 3 displays the van Krevelen diagrams for three biomass sources reported in the literature: a sugar rich waste, namely rotten apples and coming from the fruit industry, a lignocellulosic residue (olive trimming) and a protein based biomass (Penicillin mycelial waste) [1,51,54]. It can clearly be seen that each waste endures a specific evolution pathway, in relation to its initial structure and composition. Evidences of the different transformations that biomass endures during HTC, in relation to its origin process parameters, are reported in Table 1. Lignocellulosic material, like olive trimming, presents higher thermal stability than proteins and polysaccharides, resulting in higher mass yield (approx. 50%) and fixed carbon (9%) with respect to fruit and protein-based wastes.

Hydrochar produced from fruit wastes resulted in an increase in fixed carbon as long as reaction temperature is raised, due to the conversion of glucose which starts at around 225 °C [1]. Different research groups showed that mass yield is mainly affected by reaction temperature, rather than residence time. For low reaction temperature, residence time can even have a positive effect on the final mass yield. Chen et al. [55], as well as Lucian et al. [51], reported an increase in mass yield when reaction temperature was set respectively at 190 and 180 °C for a reaction time of 6 h, when compared to mass yields obtained at 1-h reaction time. The increase in hydrochar mass is associated with a re-arrangement and back polymerization of organic compounds from liquid to solid phase [55].

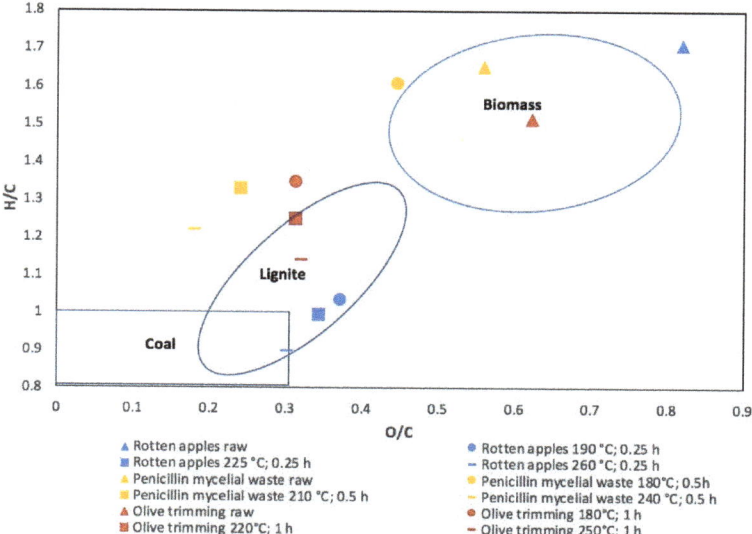

Figure 3. van Krevelen diagram of hydrothermal carbonization (HTC) treatment of rotten apples, Penicillin mycelial and olive trimming.

As previously reported, the last parameter that can be varied when performing HTC treatment is the biomass-to-water ratio (B/W). Volpe et al. [17] found that when biomass is increased in the mix up to a B/W ratio equal to 0.2, the share of secondary char production is increased as well, boosting the HHV of the hydrochar. In terms of ash content, HTC presents a huge advantage over other pre-treatment process such as pyrolysis. In fact, although during "dry" pre-treatment all the inorganics are kept inside the biomass structure, increasing their content in the final product due to the mass loss during the heating, in HTC some of the ash forming elements are removed from the biomass and washed away with the process water [24].

Table 1. Effect of HTC reaction parameters on hydrochar composition, calorific value and mass yield, together with coal properties as reference.

Biomass	FC [%]	VM [%]	C [%]	H [%]	O [%]	HHV$_{raw}$ [MJ/kg]	HTC Temp [°C]	Res. Time [h]	FC$_{HTC}$ [%]	VM$_{HTC}$ [%]	C$_{HTC}$ [%]	H$_{HTC}$ [%]	O$_{HTC}$ [%]	HHV$_{HC}$ [MJ/kg]	MY [%]	Ash$_{HC}$ [%]	Ref.
Pine	13.1	87.8	48.1	6.6	45.1	19.2	220	4.0	30.8	68.8	66.0	5.9	27.5	25.9	76	0.4	[56]
Straw	14.8	80.9	48.2	6.5	40.6	18.2	220	4.0	25.2	72.6	57.9	5.0	34.4	22.4	56	2.2	[56]
Herb tea waste	17.2		45.1	5.9	32.7	19.3	150	0.5	15.3	68.8	54.9	6.9	34.8	19.8	84	15.8	[54]
		69.4					180	0.5	16.0	65.8	58.6	6.9	31.6	20.5	69	18.1	
							210	0.5	16.6	63.4	62.6	6.9	27.6	22.1	63	20.0	
							240	0.5	19.1	58.4	68.4	6.5	22.1	26.3	53	22.5	
Penicillin mycelial waste	14.0	78.3	44.8	6.1	33.4	19.1	150	0.5	14.3	76.9	50.2	7.3	34.4	21.3	71	8.7	[54]
							180	0.5	15.9	73.6	53.3	7.1	31.6	23.6	38	10.5	
							210	0.5	17.1	68.2	64.5	7.1	20.6	27.4	29	14.6	
							240	0.3	10.0	63.8	69.2	7.0	16.5	29.2	23	17.2	
Sewage sludge	3.51	39.9	21.5	3.7	14.2	9.3	150	0.5	3.1	31.6	50.4	8.4	33.5	10.4	79	65.4	[54]
							180	0.5	2.8	24.7	57.1	8.3	27.2	10.2	75	72.5	
							210	0.5	2.4	19.6	66.4	9.4	16.7	8.1	72	77.9	
							240	0.5	2.2	16.8	71.5	9.9	11.0	7.7	68	81	
Brewer spent grains (80% moisture)	16	76.2	51.3	6.9	36.5	22.3	180	2.0	17.8	74.9	60.3	7.1	27.8	26.5	68	4.3	[4]
							200	2.0	20.8	71.9	62.0	7.0	26.4	27.2	64	4.3	
							220	2.0	23.7	69.0	65.8	7.3	22.1	29.2	58	4.3	
							180	4.0	26.7	72.8	62.9	7.0	25.2	27.6	67	4.2	
							200	4.0	23.0	70.5	61.2	7.1	27.2	26.9	63	4.1	
							220	4.0	29.6	66.2	67.1	6.9	21.1	29.3	55	4.2	
Brewer spent grains (90% moisture)	16	76.2	51.2	6.9	36.5	22.3	180	2.0	22.4	74.2	60.2	7.4	28.3	26.7	66	3.4	[4]
							200	2.0	25.1	71.7	62.3	7.2	26.2	27.5	62	3.2	
							220	2.0	28.7	68.0	66.5	7.3	21.9	29.6	52	3.2	
							180	4.0	23.7	73.1	59.9	7.0	29.1	26.2	65	3.1	
							200	4.0	26.5	70.3	63.5	7.2	24.9	28.0	60	3.2	
							220	4.0	31.9	64.8	66.6	6.9	21.8	29.1	51	3.3	
Rotten apples	14.8	83.6	43.5	6.2	47.5	17.5	190	0.25	34.2	65.5	62.4	5.4	30.8	24.8	36	0.3	[1]
							225	0.25	35.9	63.9	64.2	5.3	29.4	25.7	36	0.2	
							260	0.25	37.9	61.6	66.8	5.0	26.8	26.0	39	0.4	

Table 1. Cont.

Biomass	FC [%]	VM [%]	C [%]	H [%]	O [%]	HHV_raw [MJ/kg]	HTC Temp [°C]	Res. Time [h]	FC_HTC [%]	VM_HTC [%]	C_HTC [%]	H_HTC [%]	O_HTC [%]	HHV_HC [MJ/kg]	MY [%]	Ash_HC [%]	Ref.
Grape pomace	11.6	87.5	44.1	6.2	41.9	17.5	190	0.25	28.8	68.6	55.7	5.5	34.5	21.8	38	2.6	[1]
							225	0.25	35.4	62.8	61.4	5.1	29.9	24.5	40	1.7	
							260	0.25	35.0	60.6	64.9	5.0	24.9	24.8	45	4.3	
Peat moss	27.4	65.9	51.1	5.4	42.0	21.3	240	0.25	40.5	52.4	59.5	4.8	34.1	25.2	74	7.1	[57]
								0.5	42.4	51.4	61.5	5.0	31.8	25.3	73	6.2	
								0.75	41.8	51.5	62.1	4.9	31.3	25.8	70	6.8	
Olive trimming	17.6	78.4	48.3	6.1	40	19.8	180	1.0	20.3	76.1	54.3	6.1	34.3	22.6	72	3.5	[51]
							180	3.0	23.2	72.9	56.7	6.2	31.4	23.4	70	3.9	
							180	6.0	21.9	73.8	58.9	5.9	29.2	24.1	73	4.3	
							220	1.0	25.1	70.6	59.4	6.2	28.3	24.7	63	4.3	
							220	3.0	31.4	64.4	63.0	6.3	24.4	26.4	58	4.2	
							220	6.0	28.4	66.9	65.5	6.0	21.8	26.7	57	4.6	
							250	1.0	29.2	66.8	65.3	6.2	22.2	27.8	52	4.0	
							250	3.0	35.7	59.6	68.9	6.3	17.6	29.0	48	4.7	
							250	6.0	39.9	56.3	70.6	5.9	17.5	29.6	50	3.8	
Sub-bituminous Coal	35.1	33.1	60.8	5.7	15.6	28.2										8(*)	[58]

FC = fixed carbon; VM = volatile matter; MY = mass yield. (*) the value of ash for sub-bituminous coal is referred to the raw material and not the hydrothermally carbonized one.

3. HTC of Waste Biomass for Energy Production and Storage

Hydrothermal carbonization has been proven to be an effective tool for waste biomass pre-treatment, to produce highly carbonaceous materials that can be used for energetic applications [59–62]. In the last 5 years, research projects have focused their attention on the possible energetic use of the HTC products as a carbon-rich material for direct combustion or as a precursor for the realization of supercapacitors [63–66]. Prior to any of its further use, hydrochar must be dried to remove all the moisture.

3.1. Direct Combustion

Hydrochar has shown better energetic properties with respect to the raw original biomass. In fact, the increased heating value and the reduced volatile content ensure a better combustion and exploitation of the calorific properties of the biomass. Moreover, the lower ash content due to the removal in the liquid phase of part of the inorganics, reduce fouling and slagging phenomena that can lead to inefficacy and increased maintenance of the boiler [16,67,68].

Numerous studies investigated the possible use of hydrochar from HTC of sludge for direct combustion in a boiler. Most of the researches involve sludge from municipal waste water treatment plants, but recently also industrial sludge, like from a paper mill plant was proposed [69]. Merzari et al. studied the HTC as a strategy for sewage sludge management [70,71]. Peng et al. [72] tested various temperatures and residence times for the pre-treatment of sewage sludge from a waste water treatment plant, before its combustion. Temperature was varied in the range of 180–300 °C, with a 40 °C interval with a residence time of 30 min. while, to evaluate the contribution of reaction time authors kept the temperature fixed at 260 °C for 30, 60, 90, 350 and 480 min. Authors found that 260 °C and a residence time between 30 and 90 min, led to the highest increase in the higher heating value (2–10%). A further prolongation of reaction time up to 360 min. induced only an increment in the ash content up to 69.26%. The ash-related problems are common issues when dealing with sludges from waste water treatment plants. Wang et al. [73] studied the behavior of high-ash municipal sewage sludge when subjected to HTC in the temperature range of 170–350 °C. Raw sludge presented poor energetic properties, due to the very high ash and volatile matter (VM) content and the low fixed carbon amount. Similarly to what was reported previously by Peng et al. [72] and by Chen et al. [74], a mild HTC treatment with temperatures below 260 °C determines the best hydrochar properties for the further combustion stage. Indeed, the reduced ash content, the higher dewaterability and the greater HHV with respect to the other pre-treatment conditions, make this temperature the optimum one at which to process sludges. A valid practice to obtain hydrochar with better combustion performances, is to perform a co-HTC with carbon rich biomass, in order to raise the carbon content and the calorific value [57,75,76]. Zheng et al. [77] tested co-HTC of food waste (FW) and municipal sludges (MS) under different ratios and carbonization temperatures. The share of food waste in the mix was varied between 30, 50 and 70%, while the temperatures tested were 180 °C, 230 °C and 280 °C. The addition of food waste can partly mitigate the drawbacks related to poor combustion properties of sewage sludge hydrochar by raising the HHV and the carbon content, while lowering at the same time the ash amount. As long as the amount of food waste was raised in the feedstock mix, authors reported a drastic increase in the HHV which passed from 9.62 MJ kg^{-1} for the hydrochar prepared at 230 °C from only MS to 19 and 23 MJ kg^{-1} when the amount of FW was, respectively, 50 and 70%. At a temperature of 230 °C, HTC of only food waste produced a good quality solid fuel with a HHV of 31 MJ kg^{-1} and an ash content of 6%. Another perspective on the possible use of food waste as feedstock for HTC conversion was presented by Wang et al. [78] and McGaughy et al. [79]. Wang et al. [78] found a similar trend in the evolution of the calorific value of the hydrochar, with a peak of 31.7 MJ kg^{-1} when FW was treated at 260 °C for 1 h. In addition, fouling and slagging phenomena were reduced. Indeed, as long as reaction temperature increases, the removal of inorganics like Cl, S and N, together with a great share of alkali metals, is enhanced which contribute the most to slagging and fouling problems. However, when the carbonization temperature was raised above 220 °C, authors encountered an increased emission of NOx, which should be avoided during combustion. McGaughy and Reza [79], on the basis of HTC experiments with food waste, proposed a simulation model of a plant processing 1 ton of fresh refuse per day, resulting in a positive net energy balance for the whole process.

Food waste in the form of digestate from biogas plants, was experimented by Cao et al. [80]. The digestion significantly reduced the potential for energy conversion, since it converted a great share of the carbonaceous content to methane and carbon oxides and increased the ash amount with respect to the raw material. HTC treatment did not show significant improvement in terms of combustion potential. Poor energetic properties were also reported in [81] when dealing with the wet fraction of municipal solid waste digestate, but mostly due to the scarce potential of the initial feedstock.

A good potential for solid fuel conversion was proposed in [82,83] for the lignocellulosic material. In [82] authors tested the efficacy of a 220 °C, 90 min HTC treatment on six different biomasses, namely: olive pomace, walnut shell, hazelnut shell, apricot seed, tea stalk and wood sawdust. Each of the biomasses studied, presented better combustion properties with respect to the initial feedstock, in terms of heating value, carbon content and ignition conditions. The best energetic properties were found to be related to the olive pomace with an increase in the HHV of up to 25.6 MJ kg^{-1} and an ash content of 5.5%, while hazelnut husks showed the poorest properties with a HHV of 20.6 MJ kg^{-1} and an ash percentage of approximately 12%. Moreover, all the hydrochar presented higher ignition temperatures with respect to the raw biomasses, which significantly improve the handling and storage properties of the fuels, reducing the risk of self-ignition and combustion. Similar properties of the obtained hydrochar were also reported by Chen et al. when dealing with HTC of sweet potato peels [84].

3.2. Supercapacitors & Batteries Application

The research on possible use of biomass-derived material as source for electrode fabrication has recently increased the attention on the use of HTC pre-treatment. The present review will consider only the methods involving the use of waste biomass as starting material. This consideration is due to the fact that some studies had focused the attention mostly on the activation step, using synthetized glucose or lignin as starting material, while in the present work only the biomass-derived precursors were considered. Agricultural wastes, as well as plant wastes or high lignocellulosic materials were reported as suitable for energy storage applications [85–88]. Application for supercapacitors has been more and more studied in recent years. Supercapacitors can deliver great power output, ensuring at the same time long life and great efficiency after hundreds of cycles [86]. Supercapacitors store charge reversibly through ions immobilization in the electric double layer, on which non-faradaic and pseudocapacitive reaction occurs [87]. Hence, carbon-based material for supercapacitors applications must ensure good electric conductivity, great surface area with possible hierarchical pore distribution, together with a reasonable cost. For such reason, biomass material represents a greatly attractive option in this field [87].

Soybean root was tested by Guo et al. [89], resulting in a great electrode capacitance and cycling stability. Authors performed an acid HTC treatment at 180 °C for 18 h by using a 5% *w/w* H_2SO_4 solution. The obtained hydrochars were further active with KOH at 800 °C for 3 h, under a N_2 atmosphere. The Brunauer–Emmet–Teller (BET) and Scanning Electron Microscope (SEM) analysis showed the formation of a hierarchical interconnected porous structure composed of large macro and meso-pores on the external surface, which further develop into a capillary net of micropore sites. These latter spots are the ones that mainly contribute to ions storage [90]. Cyclic voltammetry (CV) showed great electrical response even at a high scanning rate, reporting an almost rectangular trend of the curve even at 800 mV s^{-1}, as well as a significant capacitance of 328 F g^{-1} for a current density of 1 A g^{-1}.

Tests on bamboo shells were reported by [65], resulting in a lower capacitance with respect to the ones reported by Gou et al. [89], and equal to 209 F g^{-1} at 0.5 A g^{-1}. Similarly in this case, authors performed an acid HTC treatment with 1% *w/w* H_2SO_4 solution and the obtained hydrochars, before being activated in KOH at 600–800 °C, were carbonized with melamine under N_2 at a temperature of 600 °C, with a char-to-melamine ratio of 1:4 *w/w*. To better understand the effect of melamine treatment on the final product, a blank assay was also conducted. The results showed a remarkable increase in surface area and pore development, as well as in the capacitance properties of the hydrochar. Melamine for hydrochar activation was also studied by Sevilla et al. [91], mixing it with $K_2C_2O_4$ and the hydrochar obtained from eucalyptus sawdust produced at 250 °C for 4 h. The use of $K_2C_2O_4$, instead of the more frequently used potassium hydroxide, according to the authors, is due to three main reasons: its lower corrosive potential, the higher obtainable product yield and its less disruptive action on the morphology of chars. The use of $K_2C_2O_4$ instead of KOH can almost double the char production, for the same amount of reactant adopted, while increasing at the same time the electrical properties. Hydrochar from pine cones, hemp waste, tobacco rods, peony pollen and argy worm-wood were successfully tested as precursors for supercapacitors application through KOH activation [63,92–95]. Excellent capacitance properties were found by Zhao et al. [63] when working with hydrothermally carbonized tobacco rods. In this work, authors firstly treated the biomass through HTC at 200 °C for 12 h and then activated it at 800 °C for 1 h with a char-to-KOH ratio of 1:3 in mass. Results showed that the activation stage increased the specific surface area up to 2115 m^2 g^{-1}, raising significantly also the specific capacitance. Indeed, the produced electrodes showed superb electrical properties with a capacitance as high as 286 and 212 F g^{-1}, when the current density was, respectively, 0.5 and 20 A g^{-1}. Higher capacitance was recorded when using malva nut as biomass precursor by Ye et al. [96]. HTC was performed at 200 °C for 18 h and hydrochars were further activated at 700 °C in a ratio of 3:1 with KOH. Capacitance up to 279 and 219 F g^{-1} were achieved for current densities of 1 and 20 A g^{-1}, respectively. Acid HTC with KOH activation resulted in lower performance when pine cones were used as starting biomass [93]. Manyala et al. tested the addition of 0.5 ml of sulfuric acid into 80 ml of deionized (DI) water, as medium for the HTC reaction. The further activation was performed for 1 h at the temperatures of 600, 700, 800 and 900 °C, with a KOH-to-char ratio of 1:1 (w/w). When temperature reached 900 °C a significant reduction in porosity and electrical properties of the char was observed. This condition is probably due to pore walls collapsing at higher temperature, which enlarge their average dimensions, reducing at the same time the amount of micropores which act as a storage site during charging. A wider inspection of the possible activating agent was conducted by Jain et al. [97] on coconut shells. Biomass was firstly treated in autoclave with H_2O_2 or $ZnCl_2$, testing different dosages and temperatures, and further activated under CO_2 at 800 °C for 2 h. Three different HTC conditions were tested in this work:

1. Temperature of 315 °C, 20 min with a $ZnCl_2$-to-shell ratio of 1:1 (w/w);
2. Temperature of 275 °C, 20 min with a $ZnCl_2$-to-shell ratio of 3:1 (w/w);
3. Temperature of 200 °C, 20 min in a 10% H_2O_2 solution with a successive stage at 275 °C for 20 min with a $ZnCl_2$-to-shell ratio of 3:1 (w/w).

Each set of tests was further proposed under a wide range of biomass concentrations in the solution, ranging from 0.05 to 0.66 mg L^{-1}. For the case of $ZnCl_2$-to-shell ratio of 1:1, authors reported a decrease in the specific surface area as long as biomass concentration was raised, passing from 1700 m^2 g^{-1} for a biomass concentration of 0.05 mg L^{-1}, to 1350 m^2 g^{-1} when the concentration was raised to 0.22 mg L^{-1}. The tests using H_2O_2 presented a similar trend, with an increase in surface area from 1750 to 2450 m^2 g^{-1} when the biomass concentration was raised from 0.16 to 0.5 mg L^{-1}, but a further increase up to 0.66 mg L^{-1} led to an almost 20% reduction in specific surface area. In terms of electrical properties, the highest energy density as well as the highest capacitance was found to be related to carbon produced with H_2O_2 and $ZnCl_2$ with a biomass concentration of 0.5 mg L^{-1}, with a capacitance of 246 and 221 F g^{-1} for a current density of 0.25 and 5 A g^{-1}, respectively. Impressive performances were also reported in [95] when testing $ZnCl_2$-activated hydrocar produced from Coca Cola®. Authors found one of the highest capacitances ever recorded for waste precursors, equal to 352.7 F g^{-1} for a current density of 1 A g^{-1}. Table 2 summarizes some of the studies reported above with also some reference values for synthetized materials used for supercapacitor applications.

Table 2. Performances of biomass-derived hydrochar for supercapacitors application.

Biomass	HTC Medium	HTC Temp. [°C]	HTC Res. Time [h]	Activation Agent	Activation Temp. [°C]	Activation Time [h]	Max Capacitance [F g^{-1}]	Ref.
Soybean root	5% H_2SO_4 solution	180	18	KOH	800	3	328 *	[89]
Bamboo shell	1% H_2SO_4 solution	200	24	KOH+ melamine	600–800	1	209 **	[65]
Tobacco rods	H_2O	200	12	KOH	800	1	287 **	[63]
Malva nuts	H_2O	200	18	KOH	700	1.5	279 *	[96]
Pine cones	1% H_2SO_4 solution	160	12	KOH	600–800	1	78 **	[93]
Self-stacked solvated graphene				-	-	-	245 **	[99]
PANI nanowires through electrochemical deposition				-	-	-	818 **	[100]

* Capacitance calculated at 0.5 A g^{-1}. ** Capacitance calculated at 1 A g^{-1}.

Aside from applications for supercapacitors, hydrochar has been successfully used for the production of electrode material in lithium or sodium ion batteries as well as in the vanadium redox ones [101–104]. In [101], authors used corn stalk as substrate for the production of a bio-based lamellar molybdenum disulfide electrode to be used as anode in a lithium ion battery. To obtain the desired material, a thiourea and ammonium molybdate solution was used as reaction medium for the hydrothermal carbonization of cornstalks. Hydrochars were further treated in a furnace at 1000 °C in N_2 atmosphere and then placed into an ammonium chloride solution at 60 °C, to separate the excess of calcium. The obtained material presented a discharge capacity, after 100 cycles, for a current density of 0.1 and 1 A g^{-1} of, respectively, 1129 and 339 mAh g^{-1}. A lithium ion battery anode was also produced from a cellulose-derived carbon nanosphere obtained from corn straw by Yu et al. [102]. A sulfuric acid bath and a following sodium hydroxide bath were used to extract cellulose from corn straw. The obtained products were hydrothermally carbonized at 200 °C for 24, 36, 48 and 60 h before being further carbonized at 600 °C in Ar, to obtain carbon nanospheres. Authors found that if reaction time is too long, it can induce particles agglomeration, reducing the available specific area and, therefore, the storage capacity. The best carbonization time was found to be 36 h, which showed a specific discharge capacity after 100 cycles of 577 mA g^{-1} for a current density of 74 mA g^{-1}. Aside from lithium ion, also applications for sodium ion batteries were studied. In [103], authors tested lath-shaped carbon made through HTC followed by a 800 °C carbonization step produced from peanut shells, to be used as anode in Sodium ion battery. In this case, the discharge capacity was found to be equal to 265 mA g^{-1} at 30 mA g^{-1} after 100 cycles. Moreover, tests for a possible cathode material were performed by Palomares et al. [105], using waste from vine shoots and eucalyptus wood as precursor to realize a sodium vanadium fluorophosphate electrode. Authors found that when hydrochar is further subjected to a flash thermal treatment (700 °C, 10 min in N_2), its electrochemical properties reached a specific capacity of more than twofold that of pristine hydrochar.

Promising results were also achieved in vanadium redox flow batteries when testing activated hydrochar [104] as electrode material.

4. Environmental Remediation

4.1. Adsorption of Contaminants

Waste biomass has been widely experimented for the realization of activated carbons for air or water remediation [106]. Indeed, as already seen for the preparation of supercapacitors, biomass can be transformed into a highly carbonaceous porous media, with superior properties in term of specific area and surface composition. Contaminants are adsorbed within the carbon matrix both through physical trapping on the internal surface or chemical bonding on the charged surface of the porous material. Moreover, depending on the characteristics needed for the final material, the HTC treatment and activation step can be further tailored by adjusting the operative parameters.

Two studies involving the use of digestate material as precursor for the realization of activated carbon for the adsorption of contaminants were found in the literature [107,108]. The research conducted by Bernardo et al. [108] on the use of hydrochar from digestate of municipal solid waste, showed a promising perspective for its application on the removal of phosphate from waste water. Digestate was hydrothermally carbonized at 250 °C for 1 h, in acid (with addition of H_2SO_4) or native conditions, before being activated at 600 °C for 2 h with a KOH-to-hydrochar ratio of 3:1 in mass. Despite the reduction in specific surface area when biomass is subjected to acid treatment, the two sets of chars present the same adsorption capacity of 12 mg of orthophosphate for each gram of carbon. Results indicate that morphological properties do not play a significant role in orthophosphate adsorption. According to the authors, the presence of Al^{3+}, Ca^{2+}, $Fe^{2+/3+}$ and Mg^{2+} ions in the acid hydrochar, can boost the formation of mineral complexes with phosphate ions, reducing its content. Similar behaviors were also reported in [109–111], when operating with bio-chars for the removal of phosphate from aqueous streams.

Hydrochars were also successfully used to remove CO_2 from gaseous streams by using agro-industrial waste or Coca Cola® [98,107,112]. Bio chars from agro-industrial waste were produced at different temperatures (190–250 °C), carbonization times (3 and 6 h) and pH levels (5 and 7), in order to evaluate the optimal reaction conditions [107]. Activated carbons were then realized by mixing hydrocarbons (HCs) with KOH in a mass ratio of 1:4 and then heated up to 600 °C for 2 h. Activation considerably boosted surface area and micro-meso porosity of the material. When testing the adsorption capacity, authors found a much greater affinity toward CO_2 than CH_4, with an adsorption capacity of 8.8 mol_{CO2} kg^{-1}, for the carbon produced at 250 °C for 6 h at a pH of 5. Tests on the adsorption of CO_2 were also performed with activated carbons realized from the HTC of garlic peel and Coca Cola® [98,112]. Hydrochar from garlic peels [112] were activated under different activation temperatures (600, 700 and 800 °C) and KOH-to-char ratios (0:1; 2:1 and 4:1) in order to investigate their effect on porosity development and adsorption capacity. Tests revealed that an increase in activation temperature and KOH amount led to a rise in specific surface area and pore volume. However, CO_2 adsorption is mainly driven by micropores availability, resulting in lower adsorption capacity as long as temperature and potassium hydroxide increases. Indeed, too severe activation conditions can induce pore enlargement as well as pore walls collapse, reducing the possible sites for CO_2 immobilization. Superior performances were achieved when using Coca Cola® as precursor in the HTC treatment [98]. The work conducted by Boyjoo et al. showed outstanding capacity when Coca Cola® was hydrothermally carbonized at 200 °C for 4 h and further activated either with $ZnCl_2$ or KOH. This latter condition induced the highest increment in the adsorption ability, leading to an adsorption capacity of 5.22 mmol g^{-1} at 25 °C and 1 Atm, which is one of the highest ever recorded for biomass precursors.

Bernardo et al. produced activated carbon by hydrothermal carbonization of digested sludge and tested their activity toward phosphorous adsorption. They demonstrated that the high porosity together with a high concentration of cations as in Ca, Al, Fe etc., favored phosphates removal from wastewater [108].

Experiments for heavy metals removal were also conducted starting from agro-industrial wastes as reported in [113–115]. In [114], authors tested the removal capacity of Antimony (III) and Cadmium (II) on pyro-char and hydrochar realized from animal manure. The results of the adsorption tests showed higher yields when pyrolytic chars were used, as well as an increase in the removal capacity as long as pH was raised from 3 to 6. The total adsorption potentials for Antimony were 2.24–3.98 and 4.44–16.28 mg g^{-1}, respectively, for hydrochar and pyrochar, while higher capacities were reported for the case of Cadmium, achieving a removal of 19.80–27.18 and 33.48–81.32 mg g^{-1}. Aside from Cadmium removal, promising results on the adsorption of heavy metals contaminants such as lead (Pb) [116,117], copper (Cu) [118–120], Zinc (Zn) [55,113,115] as well as pharmaceutical and chemical waste were obtained in different researches [106,121–124]. For copper removal, different studies proved that surface charge, pH and adsorbent dosage were the main parameters affecting the adsorption capacity, while for lead superior efficiency (>99.5%) was achieved with Ni/Fe-doped hydrochar [125]. Additionally, organic compounds like methylene blue, methyl orange and Congo red were removed from contaminated solutions through the adsorption on hydrochar produced from biomass waste [126–132]. Up to 655.7 mg of methylene blue for each gram of hydrochar was absorbed when bamboo sawdust was carbonized into 1 M hydrochloric acid solution, before being further treated with NaOH for 1 h. Acid modification led to 100%–200% increase in the adsorption capacity when the solution was kept in the pH range of 10–12 [127].

The removal of Congo red and 2-naptol was studied by Li et al. in two following works, using bamboo sawdust as starting material [133,134]. Both the two pollutants can have toxic effects on both aquatic life and human organisms and, therefore, need to be removed from the water stream. Moreover, due to their recalcitrant behavior to biological and thermal treatment, the use of adsorbent is more and more studied. In both the studies published by the authors, they found that pore development and the presence of oxygen functional groups on the surface can positively influence the adsorption capacity of the activated carbon. HTC reaction was conducted according to three main procedures: with pure water, with acid medium (HNO_3, H_2SO_4 and H_3PO_4) or through a two-stage operation with a first hydrothermal carbonization in a 5% weight NaOH solution followed by an acid one in HNO_3, H_2SO_4 or H_3PO_4. The two-stage process was used to perform a first delignification in order to favor the penetration of the acid inside the material and enhance pore development. The results showed that, among all, the best adsorption performances were achieved in a two-stage process, with H_2SO_4 or H_3PO_4 as acid medium and were equal to 90.51 and 72.93 mg g^{-1}, respectively, for Congo red and 2-naphtol.

Adsorption of crystal violet ($C_{25}H_{30}IN_3$) and malachite green was also investigated in biomass-derived hydrochar in [135–137]. For both contaminants, pH represents a key factor for adsorption since it is responsible for the surface charge modification of the hydrochars and the ionic charge modification of the contaminants. It was found that for crystal violet, the optimum pH was 10, while for malachite green it was found to be 7. Food leftovers were also transformed in adsorbent material to remove rare earth ions [138] and toluene [139] from contaminated streams.

Magnetic modified hydrochar, produced from different biomasses through the addition of iron compounds in the reaction media, were also used to remove tetracycline, roxarsone and persistent free radicals from waste water [140–142]. The magnetic properties of the obtained material can reduce the whole purification process time, by shortening the time needed to remove the adsorbent from the solution and saving costs of operation. In Table 3, some of the presented works are reviewed, according to the starting biomass, HTC condition and contaminant removed.

Table 3. Biomass type and HTC condition for removal of contaminants from waste waters.

Biomass	HTC Reaction Conditions	HTC Modifying Agent	Contaminant Removed	Reference
Digestate	250 °C, 1 h	H_2SO_4	Orthophosphate	[108]
Coca Cola®	200 °C, 4 h	KOH, $ZnCl_2$	CO_2	[98]
Animal manure	250 °C	-	Anthimony (III), Cadmium (II)	[114]
Bamboo sawdust	200 °C, 24 h	HCl + NaOH	Methylene blue	[127]

To sum up, the last five years of studies on the possible use of hydrochar material as precursor for the realization of the adsorption of different contaminants, showed interesting potential for the further development of substrates with enhanced capacities that could represent, in the near future, cheaper and more environmentally friendly options with respect to artificially synthetized materials.

4.2. Phosphorous Recovery

Waste biomass as OFMSW and sewage and agro-industrial sludge contain large amounts of phosphorous. Phosphorous (P) is a "critical raw material" for Europe [143] meaning that this element is of strategical importance but reserves of P rock are limited and mostly concentrated in Morocco, USA and China. As stated above, European legislation is pushing and financing for an increase in recovery of P from waste streams [144]. Hydrothermal carbonation of waste biomass, and in particular of sewage sludge containing up to 4% by weight on a dry basis, has demonstrated the possibility of recovering up to 95% of the total phosphorous content [70,145–147]. It has been demonstrated that during HTC, the phosphorous element segregates into the solid phase and treatment of the recovered hydrochar by an acidic leaching leads to the removal of P which moves to the aqueous phase. Phosphorous element in sewage sludge could be found in different forms depending on the reaction conditions [148]. P can be then precipitated by alkalization of the aqueous solution by adding CaO [145] or sodium hydroxide solution [146]. The addition of magnesium chloride and ammonium to the acidic phosphorous solution prior to alkalization led to the precipitation of struvite directly usable as solid fertilizer [146].

4.3. Soil Conditioning

Char addition in soil has been widely recognized as an good practice to favor plant growth, enhancing soil properties in terms of available nutrients, soil porosity and water retention capacity [149–153]. In terms of plant growth, positive outcomes regarding roots nodulation and biomass production were reported in [154,155]. Moreover, different studies demonstrated also the positive effect of hydrochar on the retention of pesticides, allowing an optimized utilization and reducing the possible risk of leakage and contamination of ground waters [156–158]. The fast mineralization of the carbon contained in the char leads to a short-term release of nutrients in the soil, reducing the carbon content by 30–40% within the first 12–19 months [159–161].

The nature of the biomass used to produce the biochar, as well as the operating conditions, can significantly influence the final char composition and, therefore, the possible interaction with the soil [162,163]. A possible drawback of hydrochar addition can be represented by the release of organic compounds that can have toxic effects on soil. To minimize this effect, simple washing, aerobic or low temperature thermal treatment had been reported as further options to reduce the phytotoxic effect of pristine hydrochar [164,165]. Particular attention must be paid also when using HC from sewage sludges, due to the high content and release of heavy metal in the soil [166].

Tests to control ammonia volatilization and optimize nitrogen use for plant growth were proposed in [167–169]. Chu et al. tested the addition of algae-derived hydrochar to increase yield and mitigate N_2O and NH_3 emission [167]. Conversely to the main goal of the work, results showed a significant increase in grain yield up to approximately 25%, together with a rise in N_2O emission and ammonia volatilization. The increased released of nitrogen compounds could be due to the low C/N ratio of the biomass and hydrochar, which could further prompt microbial activities, facilitating nitrogen gasification. Similar results related to emission of N-containing compounds were reported by Subedi et al. [169], as well as by Andert and Mumme [170] when adding pyro or hydro char into soil. On the same aspect, good results with high nitrogen retention, together with a significant decrease in ammonia volatilization was reported in [168] when using acid-modified hydrochar obtained from sewage sludge. Magnesium citrate or magnesium citrate with sulfuric acid were tested as potential additives in the reaction media for the HTC process. In the author's opinion, the acidity of the chars could reduce the ammonia volatilization, while the high porosity of the carbons can favor the adsorption of NH_4^+. Moreover, the presence of surface functional groups rich in anion exchange sites can further reduce the nitrification rate, increasing N retention.

Base-modified hydrochars were also successfully tested on the removal of heavy metals like Lead (Pb) and Cadmium (Cd) [171,172]. In [171], authors evaluated the possible addition of 0, 10, 20 and 30% (w/w) of lime in the solution, before the HTC treatment. The obtained hydrochars were mixed with the contaminated soil in a proportion equal to 1, 2.5 and 5% by weight to evaluate the effect on leaching and contaminants immobilization. Results of the trials showed a remarkable increase in the removal of contaminants as long as the share of hydrochar was raised, achieving an improvement in the immobilization efficiency of 95.1% (Pb) and 64.4% (Cd), with respect to pristine hydrochar. Indeed, as long as hydrochar amount and lime content were raised, pH of soil shifted toward basic values, increasing electrostatic interaction between positively charged metals ions and the anions in hydrochar. Moreover, the formation of precipitates like metal (hydr)oxide or carbonate at high soil pH helped to further reduce the amount of Pb and Cd ions in the samples. Copper was also successfully removed from contaminated soil by Xia et al. [172], by using amino-functionalized hydrochar derived from pinewood sawdust. Authors reported an efficiency in Cu removal from the soil equal to 96.2%, together with a reduction of 98.1% in Cu amount in the leachate.

5. Hydrochar for Biorefinery

HTC products were recently proposed, from lab scale up to pilot plant unit, as precursor for the production of many organic compounds that were traditionally synthetized from fossil fuel-derived hydrocarbons [173–176]. Most of the efforts are focused on the production of ethanol, furfural compounds and glucose. During HTC, biomass endures significant modifications leading to the production of a vast amount of different chemical components, both in solid or liquid forms. Ethanol was successfully produced from the cellulose extracted from agave and sugarcane bagasse hydrochar in [177] and [178], achieving a production up to 145 L of ethanol from 1 ton of raw sugarcane bagasse. In order to maximize cellulose production, low temperature HTC (180–190 °C) is preferred. The obtained hydrochar is further treated trough enzymatic hydrolysis and fermentation to obtain ethanol. Through the enzymatic hydrolysis of both liquid and solid products of the HTC, glucose was produced from palm oil waste by Zakaria et al. [179].

Furfural and 5-hydroxymethylfurfural (HMF) were also produced from Miscanthus, corncob wastes and hemp [180–183]. HMF was successfully generated from both hydrolyzed cellulose or hemicellulose contained in the original biomass. In both cases, acid treatment through H_2SO_4 or catalytic conversion were necessary to produce HMF.

6. Conclusions

HTC represents a valid pre-treatment technology to convert waste biomass into new valuable products which could find applications in a wide range of fields, from energy production to environment remediation and soil conditioning. Since HTC operates in aqueous medium, it is perfectly suitable to convert wet biomass, without any drying step. The most common waste biomass materials include those produced by agro-industries, municipalities (organic fraction of municipal solid waste and sewage sludge) but also from forestry and paper mill industries. The removal of part of the volatile compounds and ashes increases the combustion properties of the hydrochar, making them suitable for direct combustion in energy production applications. Chemical and physical activation, by using KOH or CO_2, respectively, were successfully proposed to increase the surface area and internal porosity of the material. By modifying the morphology and chemical surface, activated hydrochar can be successfully used for the production of energy storage devices for the immobilization of pollutants in gaseous or liquid streams. Pores development as well as elemental and surface composition can be tailored by changing the carbonization and activation parameters. Finally, the addition in soil can induce important benefits on pant growth, increasing water retention and reducing at the same time possible leaching of contaminants into ground water.

Recently, some researchers focused their attention on how to mitigate the energetic impact of the hydrothermal carbonization process by coupling it with renewable energy production systems. Indeed, biochar can represent a valuable product to increase the efficiency of co-digestion systems, with respect to its pristine composition. HC produced from coffee spent, rice straw or microalgae, as well as the liquid phase produced from the hydrothermal carbonization, induced higher biogas production as well as an increased methane content [184–188]. In such a way, the energy content of the produced biogas can be used to run the HTC reactor, ensuring cleaner combustion with respect to the hydrochar itself. Another option that has recently been studied is to couple the HTC reactor with a solar concentration system which can satisfy the thermal load. Linear parabolic and parabolic disc concentrators ensured a carbonization temperature up to 200 and 250 °C respectively, ensuring heating conditions of traditional systems [189,190].

Nevertheless, there are still some mechanisms in the HTC process that need further study to be completely understood, in order to gain a stronger control over the process parameters, their influence on the final product properties and thus their possible applications. Depending on the HTC operating conditions and the nature of the feedstock, the technology still presents some limitations for its full development. Between them, the correct process liquid treatment and management often present challenges due to the presence of toxic hydrocarbons (such as polycyclic aromatic hydrocarbons (PHAs) and high heavy metals concentrations. The presence of toxic compounds severely limits the possibility to use such residue (as for example in agriculture as a possible source of nutrients) increasing the overall costs of the HTC process. Investigations related to toxic hydrocarbons removal and heavy metals recovery represent a necessary step for further development of the HTC technology.

Author Contributions: Conceptualization, M.P.M., M.V. and A.M.; Writing-Original Draft Preparation, M.P.M.; Writing-Review & Editing, M.P.M., M.V. and A.M.; Supervision & final review, A.M. All authors have read and agreed to the published version of the manuscript.

Funding: This research received no external funding.

Conflicts of Interest: The authors declare no conflict of interest.

References

1. Zhang, B.; Heidari, M.; Regmi, B.; Salaudeen, S.; Arku, P.; Thimmannagari, M.; Dutta, A. Hydrothermal carbonization of fruit wastes: A promising technique for generating hydrochar. *Energies* **2018**, *11*, 2022. [CrossRef]
2. Volpe, M.; Fiori, L.; Volpe, R.; Messineo, A. Upgrading of olive tree trimmings residue as biofuel by hydrothermal carbonization and torrefaction: A comparative study. In *Chemical Engineering Transactions*; AIDIC: Milan, Italy, 2016; Volume 50, pp. 113–118. ISBN 9788895608419.
3. Lucian, M.; Volpe, M.; Gao, L.; Piro, G.; Goldfarb, J.L.; Fiori, L. Impact of hydrothermal carbonization conditions on the formation of hydrochars and secondary chars from the organic fraction of municipal solid waste. *Fuel* **2018**, *233*, 257–268. [CrossRef]
4. Arauzo, P.J.; Olszewski, M.P.; Kruse, A. Hydrothermal carbonization brewer's spent grains with the focus on improving the degradation of the feedstock. *Energies* **2018**, *11*, 3226. [CrossRef]
5. Ferrentino, R.; Ceccato, R.; Marchetti, V.; Andreottola, G. Sewage Sludge Hydrochar: An Option for Removal of Methylene Blue from Wastewater. *Appl. Sci.* **2020**, *10*, 3445. [CrossRef]
6. Gopu, C.; Gao, L.; Volpe, M.; Fiori, L.; Goldfarb, J.L. Valorizing municipal solid waste: Waste to energy and activated carbons for water treatment via pyrolysis. *J. Anal. Appl. Pyrolysis* **2018**, *133*, 48–58. [CrossRef]
7. Cavalaglio, G.; Coccia, V.; Cotana, F.; Gelosia, M.; Nicolini, A.; Petrozzi, A. Energy from poultry waste: An Aspen Plus-based approach to the thermo-chemical processes. *Waste Manag.* **2018**, *73*, 496–503. [CrossRef] [PubMed]
8. Prins, M.J.; Ptasinski, K.J.; Janssen, F.J.J.G. More efficient biomass gasification via torrefaction. *Energy* **2006**, *31*, 3458–3470. [CrossRef]
9. Messineo, A.; Ciulla, G.; Messineo, S.; Volpe, M.; Volpe, R. Evaluation of equilibrium moisture content in ligno-cellulosic residues of olive culture. *ARPN J. Eng. Appl. Sci.* **2014**, *9*, 5–11.
10. Heidari, M.; Salaudeen, S.; Norouzi, O.; Acharya, B.; Dutta, A. Numerical comparison of a combined hydrothermal carbonization and anaerobic digestion system with direct combustion of biomass for power production. *Processes* **2020**, *8*, 43. [CrossRef]
11. Li, H.-Y.; Tsai, G.-L.; Chao, S.-M.; Yen, Y.-F. Measurement of thermal and hydraulic performance of a plate-fin heat sink with a shield. *Exp. Therm. Fluid Sci.* **2012**, *42*, 71–78. [CrossRef]

12. Ingrao, C.; Bacenetti, J.; Adamczyk, J.; Ferrante, V.; Messineo, A.; Huisingh, D. Investigating energy and environmental issues of agro-biogas derived energy systems: A comprehensive review of Life Cycle Assessments. *Renew. Energy* **2019**, *136*, 296–307. [CrossRef]
13. Lucian, M.; Fiori, L. Hydrothermal carbonization of waste biomass: Process design, modeling, energy efficiency and cost analysis. *Energies* **2017**, *10*, 211. [CrossRef]
14. Munir, M.T.; Mansouri, S.S.; Udugama, I.A.; Baroutian, S.; Gernaey, K.V.; Young, B.R. Resource recovery from organic solid waste using hydrothermal processing: Opportunities and challenges. *Renew. Sustain. Energy Rev.* **2018**, *96*, 64–75. [CrossRef]
15. Álvarez-Murillo, A.; Román, S.; Ledesma, B.; Sabio, E. Study of variables in energy densification of olive stone by hydrothermal carbonization. *J. Anal. Appl. Pyrolysis* **2015**, *113*, 307–314. [CrossRef]
16. Mäkelä, M.; Benavente, V.; Fullana, A. Hydrothermal carbonization of lignocellulosic biomass: Effect of process conditions on hydrochar properties. *Appl. Energy* **2015**, *155*, 576–584. [CrossRef]
17. Volpe, M.; Goldfarb, J.L.; Fiori, L. Hydrothermal carbonization of Opuntia ficus-indica cladodes: Role of process parameters on hydrochar properties. *Bioresour. Technol.* **2018**, *247*, 310–318. [CrossRef]
18. Kruse, A.; Funke, A.; Titirici, M.-M. Hydrothermal conversion of biomass to fuels and energetic materials. *Curr. Opin. Chem. Biol.* **2013**, *17*, 515–521. [CrossRef]
19. Volpe, M.; Wüst, D.; Merzari, F.; Lucian, M.; Andreottola, G.; Kruse, A.; Fiori, L. One stage olive mill waste streams valorisation via hydrothermal carbonisation. *Waste Manag.* **2018**, *80*, 224–234. [CrossRef]
20. Reza, M.; Lynam, J.G.; Helal Uddin, M.; Coronella, C.J. Hydrothermal carbonization: Fate of inorganics. *Biomass Bioenergy* **2013**, *49*, 86–94. [CrossRef]
21. Reza, M.T.; Emerson, R.; Uddin, M.H.; Gresham, G.; Coronella, C.J. Ash reduction of corn stover by mild hydrothermal preprocessing. *Biomass Convers. Biorefinery* **2015**, *5*, 21–31. [CrossRef]
22. Gao, L.; Volpe, M.; Lucian, M.; Fiori, L.; Goldfarb, J.L. Does Hydrothermal Carbonization as a Biomass Pretreatment Reduce Fuel Segregation of Coal-Biomass Blends During Oxidation? *Energy Convers. Manag.* **2019**, *181*, 93–104. [CrossRef]
23. Smith, A.M.; Singh, S.; Ross, A.B. Fate of inorganic material during hydrothermal carbonisation of biomass: Influence of feedstock on combustion behaviour of hydrochar. *Fuel* **2016**, *169*, 135–145. [CrossRef]
24. Liu, Z.; Balasubramanian, R. Upgrading of waste biomass by hydrothermal carbonization (HTC) and low temperature pyrolysis (LTP): A comparative evaluation. *Appl. Energy* **2014**, *114*, 857–864. [CrossRef]
25. Unrean, P.; Lai Fui, B.C.; Rianawati, E.; Acda, M. Comparative techno-economic assessment and environmental impacts of rice husk-to-fuel conversion technologies. *Energy* **2018**, *151*, 581–593. [CrossRef]
26. Kruse, A.; Koch, F.; Stelzl, K.; Wüst, D.; Zeller, M. Fate of Nitrogen during Hydrothermal Carbonization. *Energy Fuels* **2016**, *30*, 8037–8042. [CrossRef]
27. Rodriguez Correa, C.; Otto, T.; Kruse, A. Influence of the biomass components on the pore formation of activated carbon. *Biomass Bioenergy* **2017**, *97*, 53–64. [CrossRef]
28. Liu, F.; Gao, Y.; Zhang, C.; Huang, H.; Yan, C.; Chu, X.; Xu, Z.; Wang, Z.; Zhang, H.; Xiao, X.; et al. Highly microporous carbon with nitrogen-doping derived from natural biowaste for high-performance flexible solid-state supercapacitor. *J. Colloid Interface Sci.* **2019**, *548*, 322–332. [CrossRef]
29. Yu, X.; Liu, S.; Lin, G.; Yang, Y.; Zhang, S.; Zhao, H.; Zheng, C.; Gao, X. KOH-activated hydrochar with engineered porosity as sustainable adsorbent for volatile organic compounds. *Colloids Surfaces A Physicochem. Eng. Asp.* **2020**, *588*, 124372. [CrossRef]
30. Hitzl, M.; Corma, A.; Pomares, F.; Renz, M. The hydrothermal carbonization (HTC) plant as a decentral biorefinery for wet biomass. *Catal. Today* **2015**, *257*, 154–159. [CrossRef]
31. Titirici, M.M. Hydrothermal Carbons: Synthesis, Characterization, and Applications. In *Novel Carbon Adsorbents*; Elsevier: Amsterdam, The Netherlands, 2012; ISBN 9780080977447.
32. Jatzwauck, M.; Schumpe, A. Kinetics of hydrothermal carbonization (HTC) of soft rush. *Biomass Bioenergy* **2015**, *75*, 94–100. [CrossRef]
33. Volpe, M.; Fiori, L. From olive waste to solid biofuel through hydrothermal carbonisation: The role of temperature and solid load on secondary char formation and hydrochar energy properties. *J. Anal. Appl. Pyrolysis* **2017**, *124*, 63–72. [CrossRef]
34. Knežević, D.; Van Swaaij, W.; Kersten, S. Hydrothermal conversion of biomass. II. conversion of wood, pyrolysis oil, and glucose in hot compressed water. *Ind. Eng. Chem. Res.* **2010**, *49*, 104–112. [CrossRef]

35. Kruse, A.; Dahmen, N. Water—A magic solvent for biomass conversion. *J. Supercrit. Fluids* **2015**, *96*, 36–45. [CrossRef]
36. Sabio, E.; Álvarez-Murillo, A.; Román, S.; Ledesma, B. Conversion of tomato-peel waste into solid fuel by hydrothermal carbonization: Influence of the processing variables. *Waste Manag.* **2016**, *47*, 122–132. [CrossRef]
37. Heidari, M.; Norouzi, O.; Salaudeen, S.; Acharya, B.; Dutta, A. Prediction of Hydrothermal Carbonization with Respect to the Biomass Components and Severity Factor. *Energy Fuels* **2019**, *33*, 9916–9924. [CrossRef]
38. Borrero-López, A.M.; Masson, E.; Celzard, A.; Fierro, V. Modelling the reactions of cellulose, hemicellulose and lignin submitted to hydrothermal treatment. *Ind. Crops Prod.* **2018**, *124*, 919–930. [CrossRef]
39. Mäkelä, M.; Volpe, M.; Volpe, R.; Fiori, L.; Dahl, O. Spatially resolved spectral determination of polysaccharides in hydrothermally carbonized biomass. *Green Chem.* **2018**, *20*, 1114–1120. [CrossRef]
40. Volpe, M.; Messineo, A.; Mäkelä, M.; Barr, M.R.; Volpe, R.; Corrado, C.; Fiori, L. Reactivity of cellulose during hydrothermal carbonization of lignocellulosic biomass. *Fuel Process. Technol.* **2020**, *206*, 106456. [CrossRef]
41. Libra, J.A.; Ro, K.S.; Kammann, C.; Funke, A.; Berge, N.D.; Neubauer, Y.; Titirici, M.M.; Fühner, C.; Bens, O.; Kern, J.; et al. Hydrothermal carbonization of biomass residuals: A comparative review of the chemistry, processes and applications of wet and dry pyrolysis. *Biofuels* **2011**, *2*, 71–106. [CrossRef]
42. Sevilla, M.; Fuertes, A.B. The production of carbon materials by hydrothermal carbonization of cellulose. *Carbon N. Y.* **2009**, *47*, 2281–2289. [CrossRef]
43. Olszewski, M.P.; Nicolae, S.A.; Arauzo, P.J.; Titirici, M.M.; Kruse, A. Wet and dry? Influence of hydrothermal carbonization on the pyrolysis of spent grains. *J. Clean. Prod.* **2020**, *260*, 121101. [CrossRef]
44. Kruse, A.; Zevaco, T.A. Properties of hydrochar as function of feedstock, reaction conditions and post-treatment. *Energies* **2018**, *11*, 674. [CrossRef]
45. Islam, M.A.; Asif, M.; Hameed, B.H. Pyrolysis kinetics of raw and hydrothermally carbonized Karanj (*Pongamia pinnata*) fruit hulls via thermogravimetric analysis. *Bioresour. Technol.* **2015**, *179*, 227–233. [CrossRef] [PubMed]
46. Kim, D.; Lee, K.; Park, K.Y. Upgrading the characteristics of biochar from cellulose, lignin, and xylan for solid biofuel production from biomass by hydrothermal carbonization. *J. Ind. Eng. Chem.* **2016**, *42*, 95–100. [CrossRef]
47. Zhuang, X.; Zhan, H.; Song, Y.; Yin, X.; Wu, C. Structure-reactivity relationships of biowaste-derived hydrochar on subsequent pyrolysis and gasification performance. *Energy Convers. Manag.* **2019**, *199*, 112014. [CrossRef]
48. Heidari, M.; Dutta, A.; Acharya, B.; Mahmud, S. A review of the current knowledge and challenges of hydrothermal carbonization for biomass conversion. *J. Energy Inst.* **2019**, *92*, 1779–1799. [CrossRef]
49. Heidari, M.; Salaudeen, S.; Dutta, A.; Acharya, B. Effects of Process Water Recycling and Particle Sizes on Hydrothermal Carbonization of Biomass. *Energy Fuels* **2018**. [CrossRef]
50. Li, J.; Zhao, P.; Li, T.; Lei, M.; Yan, W.; Ge, S. Pyrolysis behavior of hydrochar from hydrothermal carbonization of pinewood sawdust. *J. Anal. Appl. Pyrolysis* **2020**, *146*, 104771. [CrossRef]
51. Lucian, M.; Volpe, M.; Fiori, L. Hydrothermal carbonization kinetics of lignocellulosic agro-wastes: Experimental data and modeling. *Energies* **2019**, *12*, 516. [CrossRef]
52. Ulbrich, M.; Preßl, D.; Fendt, S.; Gaderer, M.; Spliethoff, H. Impact of HTC reaction conditions on the hydrochar properties and CO_2 gasification properties of spent grains. *Fuel Process. Technol.* **2017**, *167*, 663–669. [CrossRef]
53. Basso, D.; Patuzzi, F.; Castello, D.; Baratieri, M.; Rada, E.C.; Weiss-Hortala, E.; Fiori, L. Agro-industrial waste to solid biofuel through hydrothermal carbonization. *Waste Manag.* **2016**, *47*, 114–121. [CrossRef] [PubMed]
54. Zhuang, X.; Song, Y.; Zhan, H.; Yin, X.; Wu, C. Gasification performance of biowaste-derived hydrochar: The properties of products and the conversion processes. *Fuel* **2020**, *260*, 116320. [CrossRef]
55. Chen, X.; Lin, Q.; He, R.; Zhao, X.; Li, G. Hydrochar production from watermelon peel by hydrothermal carbonization. *Bioresour. Technol.* **2017**, *241*, 236–243. [CrossRef] [PubMed]
56. Magdziarz, A.; Wilk, M.; Wądrzyk, M. Pyrolysis of hydrochar derived from biomass—Experimental investigation. *Fuel* **2020**, *167*, 117246. [CrossRef]
57. Roy, P.; Dutta, A.; Gallant, J. Hydrothermal carbonization of peat moss and herbaceous biomass (*Miscanthus*): A potential route for bioenergy. *Energies* **2018**, *11*, 2794. [CrossRef]

58. Khatami, R.; Stivers, C.; Joshi, K.; Levendis, Y.A.; Sarofim, A.F. Combustion behavior of single particles from three different coal ranks and from sugar cane bagasse in O_2/N_2 and O_2/CO_2 atmospheres. *Combust. Flame* **2012**, *159*, 1253–1271. [CrossRef]
59. Wilk, M.; Magdziarz, A.; Kalemba-Rec, I.; Szymańska-Chargot, M. Upgrading of green waste into carbon-rich solid biofuel by hydrothermal carbonization: The effect of process parameters on hydrochar derived from acacia. *Energy* **2020**, *202*, 117717. [CrossRef]
60. Babinszki, B.; Jakab, E.; Sebestyén, Z.; Blazsó, M.; Berényi, B.; Kumar, J.; Krishna, B.B.; Bhaskar, T.; Czégény, Z. Comparison of hydrothermal carbonization and torrefaction of azolla biomass: Analysis of the solid products. *J. Anal. Appl. Pyrolysis* **2020**, 104844. [CrossRef]
61. Sharma, H.B.; Sarmah, A.K.; Dubey, B. Hydrothermal carbonization of renewable waste biomass for solid biofuel production: A discussion on process mechanism, the influence of process parameters, environmental performance and fuel properties of hydrochar. *Renew. Sustain. Energy Rev.* **2020**, *123*, 109761. [CrossRef]
62. Zhang, X.; Gao, B.; Zhao, S.; Wu, P.; Han, L.; Liu, X. Optimization of a "coal-like" pelletization technique based on the sustainable biomass fuel of hydrothermal carbonization of wheat straw. *J. Clean. Prod.* **2020**, *242*, 118426. [CrossRef]
63. Peng, C.; Zhai, Y.; Zhu, Y.; Xu, B.; Wang, T.; Li, C.; Zeng, G. Production of char from sewage sludge employing hydrothermal carbonization: Char properties, combustion behavior and thermal characteristics. *Fuel* **2016**, *176*, 110–118. [CrossRef]
64. Gunarathne, D.S.; Mueller, A.; Fleck, S.; Kolb, T.; Chmielewski, J.K.; Yang, W.; Blasiak, W. Gasification characteristics of hydrothermal carbonized biomass in an updraft pilot-scale gasifier. *Energy Fuels* **2014**, *28*, 1992–2002. [CrossRef]
65. Huang, G.; Wang, Y.; Zhang, T.; Wu, X.; Cai, J. High-performance hierarchical N-doped porous carbons from hydrothermally carbonized bamboo shoot shells for symmetric supercapacitors. *J. Taiwan Inst. Chem. Eng.* **2019**, *96*, 672–680. [CrossRef]
66. Saari, J.; Sermyagina, E.; Kaikko, J.; Vakkilainen, E.; Sergeev, V. Integration of hydrothermal carbonization and a CHP plant: Part 2—Operational and economic analysis. *Energy* **2016**, *113*, 574–585. [CrossRef]
67. He, C.; Wang, K.; Yang, Y.; Wang, J.Y. Utilization of sewage-sludge-derived hydrochars toward efficient cocombustion with different-rank coals: Effects of subcritical water conversion and blending scenarios. *Energy Fuels* **2014**, *28*, 6140–6150. [CrossRef]
68. Reza, M.T.; Andert, J.; Wirth, B.; Busch, D.; Pielert, J.; Lynam, J.G.; Mumme, J. Hydrothermal Carbonization of Biomass for Energy and Crop Production. *Appl. Bioenergy* **2014**. [CrossRef]
69. Saha, N.; Saba, A.; Saha, P.; McGaughy, K.; Franqui-Villanueva, D.; Orts, W.J.; Hart-Cooper, W.M.; Toufiq Reza, M. Hydrothermal carbonization of various paper mill sludges: An observation of solid fuel properties. *Energies* **2019**, *125*, 858. [CrossRef]
70. Volpe, M.; Fiori, L.; Merzari, F.; Messineo, A.; Andreottola, G. Hydrothermal carbonization as an efficient tool for sewage sludge valorization and phosphorous recovery. *Chem. Eng. Trans.* **2020**, *80*, 199–204. [CrossRef]
71. Merzari, F.; Goldfarb, J.; Andreottola, G.; Mimmo, T.; Volpe, M.; Fiori, L. Hydrothermal carbonization as a strategy for sewage sludge management: Influence of process withdrawal point on hydrochar properties. *Energies* **2020**, *13*, 2890. [CrossRef]
72. Wang, R.; Wang, C.; Zhao, Z.; Jia, J.; Jin, Q. Energy recovery from high-ash municipal sewage sludge by hydrothermal carbonization: Fuel characteristics of biosolid products. *Energy* **2019**, *186*, 115848. [CrossRef]
73. Chen, C.; Liu, G.; An, Q.; Lin, L.; Shang, Y.; Wan, C. From wasted sludge to valuable biochar by low temperature hydrothermal carbonization treatment: Insight into the surface characteristics. *J. Clean. Prod.* **2020**, *263*, 121600. [CrossRef]
74. Park, K.Y.; Lee, K.; Kim, D. Characterized hydrochar of algal biomass for producing solid fuel through hydrothermal carbonization. *Bioresour. Technol.* **2018**, *258*, 119–124. [CrossRef] [PubMed]
75. Saba, A.; McGaughy, K.; Toufiq Reza, M. Techno-economic assessment of co-hydrothermal carbonization of a coal-Miscanthus blend. *Energies* **2019**, *12*, 630. [CrossRef]
76. Zheng, C.; Ma, X.; Yao, Z.; Chen, X. The properties and combustion behaviors of hydrochars derived from co-hydrothermal carbonization of sewage sludge and food waste. *Bioresour. Technol.* **2019**, *258*, 121347. [CrossRef] [PubMed]

77. Wang, T.; Zhai, Y.; Zhu, Y.; Gan, X.; Zheng, L.; Peng, C.; Wang, B.; Li, C.; Zeng, G. Evaluation of the clean characteristics and combustion behavior of hydrochar derived from food waste towards solid biofuel production. *Bioresour. Technol.* **2018**, *266*, 275–283. [CrossRef]
78. McGaughy, K.; Toufiq Reza, M. Hydrothermal carbonization of food waste: Simplified process simulation model based on experimental results. *Biomass Convers. Biorefinery* **2018**, *8*, 233–292. [CrossRef]
79. Cao, Z.; Jung, D.; Olszewski, M.P.; Arauzo, P.J.; Kruse, A. Hydrothermal carbonization of biogas digestate: Effect of digestate origin and process conditions. *Waste Manag.* **2019**, *100*, 138–150. [CrossRef]
80. Pawlak-Kruczek, H.; Sieradzka, M.; Mlonka-Mędrala, A.; Baranowski, M.; Serafin-Tkaczuk, M.; Magdziarz, A.; Niedźwiecki, Ł. Structural and energetic properties of hydrochars obtained from agricultural and municipal solid waste digestates. In Proceedings of the 32nd International Conference on Efficiency, Cost, Optimization, Simulation and Environmental Impact of Energy Systems (ECOS 2019), Wrocław, Poland, 23–28 June 2019.
81. Başakçılardan Kabakcı, S.; Baran, S.S. Hydrothermal carbonization of various lignocellulosics: Fuel characteristics of hydrochars and surface characteristics of activated hydrochars. *Waste Manag.* **2019**, *100*, 259–268. [CrossRef]
82. Yang, W.; Wang, H.; Zhang, M.; Zhu, J.; Zhou, J.; Wu, S. Fuel properties and combustion kinetics of hydrochar prepared by hydrothermal carbonization of bamboo. *Bioresour. Technol.* **2016**, *205*, 199–204. [CrossRef]
83. Chen, X.; Ma, X.; Peng, X.; Lin, Y.; Yao, Z. Conversion of sweet potato waste to solid fuel via hydrothermal carbonization. *Bioresour. Technol.* **2018**, *249*, 900–907. [CrossRef]
84. Celiktas, M.S.; Alptekin, F.M. Conversion of model biomass to carbon-based material with high conductivity by using carbonization. *Energy* **2019**, *188*, 116089. [CrossRef]
85. Sinan, N.; Unur, E. Hydrothermal conversion of lignocellulosic biomass into high-value energy storage materials. *J. Energy Chem.* **2017**, *26*, 783–789. [CrossRef]
86. Wang, C.; Wu, D.; Wang, H.; Gao, Z.; Xu, F.; Jiang, K. Biomass derived nitrogen-doped hierarchical porous carbon sheets for supercapacitors with high performance. *J. Colloid Interface Sci.* **2018**, *523*, 133–143. [CrossRef] [PubMed]
87. Antero, R.V.P.; Alves, A.C.F.; de Oliveira, S.B.; Ojala, S.A.; Brum, S.S. Challenges and alternatives for the adequacy of hydrothermal carbonization of lignocellulosic biomass in cleaner production systems: A review. *J. Clean. Prod.* **2020**, *252*, 119899. [CrossRef]
88. Guo, N.; Luo, W.; Guo, R.; Qiu, D.; Zhao, Z.; Wang, L.; Jia, D.; Guo, J. Interconnected and hierarchical porous carbon derived from soybean root for ultrahigh rate supercapacitors. *J. Alloys Compd.* **2020**, *834*, 155115. [CrossRef]
89. Maniscalco, M.P.; Corrado, C.; Volpe, R.; Messineo, A. Evaluation of the optimal activation parameters for almond shell bio-char production for capacitive deionization. *Bioresour. Technol. Reports* **2020**, *11*, 100435. [CrossRef]
90. Sevilla, M.; Ferrero, G.A.; Fuertes, A.B. Beyond KOH activation for the synthesis of superactivated carbons from hydrochar. *Carbon N. Y.* **2017**, *114*, 50–58. [CrossRef]
91. Liu, Y.; An, Z.; Wu, M.; Yuan, A.; Zhao, H.; Zhang, J.; Xu, J. Peony pollen derived nitrogen-doped activated carbon for supercapacitor application. *Chinese Chem. Lett.* **2019**, *31*, 1644–1647. [CrossRef]
92. Manyala, N.; Bello, A.; Barzegar, F.; Khaleed, A.A.; Momodu, D.Y.; Dangbegnon, J.K. Coniferous pine biomass: A novel insight into sustainable carbon materials for supercapacitors electrode. *Mater. Chem. Phys.* **2016**, *182*, 139–147. [CrossRef]
93. Zhao, Y.Q.; Lu, M.; Tao, P.Y.; Zhang, Y.J.; Gong, X.T.; Yang, Z.; Zhang, G.Q.; Li, H.L. Hierarchically porous and heteroatom doped carbon derived from tobacco rods for supercapacitors. *J. Power Sources* **2016**, *307*, 391–400. [CrossRef]
94. Sun, W.; Lipka, S.M.; Swartz, C.; Williams, D.; Yang, F. Hemp-derived activated carbons for supercapacitors. *Carbon N. Y.* **2016**, *103*, 181–192. [CrossRef]
95. Dai, C.; Wan, J.; Yang, J.; Qu, S.; Jin, T.; Ma, F.; Shao, J. H_3PO_4 solution hydrothermal carbonization combined with KOH activation to prepare argy wormwood-based porous carbon for high-performance supercapacitors. *Appl. Surf. Sci.* **2018**, *444*, 105–117. [CrossRef]
96. Ye, R.; Cai, J.; Pan, Y.; Qiao, X.; Sun, W. Microporous carbon from malva nut for supercapacitors: Effects of primary carbonizations on structures and performances. *Diam. Relat. Mater.* **2020**, *105*, 107816. [CrossRef]

97. Jain, A.; Xu, C.; Jayaraman, S.; Balasubramanian, R.; Lee, J.Y.; Srinivasan, M.P. Mesoporous activated carbons with enhanced porosity by optimal hydrothermal pre-treatment of biomass for supercapacitor applications. *Microporous Mesoporous Mater.* **2015**, *218*, 55–61. [CrossRef]
98. Boyjoo, Y.; Cheng, Y.; Zhong, H.; Tian, H.; Pan, J.; Pareek, V.K.; Jiang, S.P.; Lamonier, J.F.; Jaroniec, M.; Liu, J. From waste Coca Cola® to activated carbons with impressive capabilities for CO_2 adsorption and supercapacitors. *Carbon N. Y.* **2017**, *116*, 490–499. [CrossRef]
99. Du, P.; Hu, X.; Yi, C.; Liu, H.C.; Liu, P.; Zhang, H.L.; Gong, X. Self-powered electronics by integration of flexible solid-state graphene-based supercapacitors with high performance perovskite hybrid solar cells. *Adv. Funct. Mater.* **2015**, *25*, 2420–2427. [CrossRef]
100. Gupta, V.; Miura, N. Electrochemically deposited polyaniline nanowire's network a high-performance electrode material for redox supercapacitor. *Electrochem. Solid-State Lett.* **2005**, *8*, A630–A632. [CrossRef]
101. Zhao, G.; Cheng, Y.; Sun, P.; Ma, W.; Hao, S.; Wang, X.; Xu, X.; Xu, Q.; Liu, M. Biocarbon based template synthesis of uniform lamellar MoS_2 nanoflowers with excellent energy storage performance in lithium-ion battery and supercapacitors. *Electrochim. Acta* **2020**, *331*, 135262. [CrossRef]
102. Yu, K.; Wang, J.; Song, K.; Wang, X.; Liang, C.; Dou, Y. Hydrothermal synthesis of cellulose-derived carbon nanospheres from corn straw as anode materials for lithium ion batteries. *Nanomaterials* **2019**, *9*, 93. [CrossRef]
103. Ren, X.; Xu, S.D.; Liu, S.; Chen, L.; Zhang, D.; Qiu, L. Lath-shaped biomass derived hard carbon as anode materials with super rate capability for sodium-ion batteries. *J. Electroanal. Chem.* **2019**, *841*, 63–72. [CrossRef]
104. Maharjan, M.; Wai, N.; Veksha, A.; Giannis, A.; Lim, T.M.; Lisak, G. Sal wood sawdust derived highly mesoporous carbon as prospective electrode material for vanadium redox flow batteries. *J. Electroanal. Chem.* **2019**, *834*, 94–100. [CrossRef]
105. Palomares, V.; Blas, M.; Serras, P.; Iturrondobeitia, A.; Peña, A.; Lopez-Urionabarrenechea, A.; Lezama, L.; Rojo, T. Waste Biomass as in Situ Carbon Source for Sodium Vanadium Fluorophosphate/C Cathodes for Na-Ion Batteries. *ACS Sustain. Chem. Eng.* **2018**, *6*, 16386–16398. [CrossRef]
106. Rodríguez Correa, C.; Ngamying, C.; Klank, D.; Kruse, A. Investigation of the textural and adsorption properties of activated carbon from HTC and pyrolysis carbonizates. *Biomass Convers. Biorefinery* **2018**, *8*, 317–328. [CrossRef]
107. Rodriguez Correa, C.; Bernardo, M.; Ribeiro, R.P.P.L.; Esteves, I.A.A.C.; Kruse, A. Evaluation of hydrothermal carbonization as a preliminary step for the production of functional materials from biogas digestate. *J. Anal. Appl. Pyrolysis* **2017**, *124*, 461–474. [CrossRef]
108. Bernardo, M.; Correa, C.R.; Ringelspacher, Y.; Becker, G.C.; Lapa, N.; Fonseca, I.; Esteves, I.A.A.C.; Kruse, A. Porous carbons derived from hydrothermally treated biogas digestate. *Waste Manag.* **2020**, *105*, 170–179. [CrossRef] [PubMed]
109. Vikrant, K.; Kim, K.H.; Ok, Y.S.; Tsang, D.C.W.; Tsang, Y.F.; Giri, B.S.; Singh, R.S. Engineered/designer biochar for the removal of phosphate in water and wastewater. *Sci. Total Environ.* **2018**, *616*, 1242–1260. [CrossRef]
110. Takaya, C.A.; Fletcher, L.A.; Singh, S.; Anyikude, K.U.; Ross, A.B. Phosphate and ammonium sorption capacity of biochar and hydrochar from different wastes. *Chemosphere* **2016**, *145*, 518–527. [CrossRef]
111. Shepherd, J.G.; Joseph, S.; Sohi, S.P.; Heal, K.V. Biochar and enhanced phosphate capture: Mapping mechanisms to functional properties. *Chemosphere* **2017**, *179*, 57–74. [CrossRef]
112. Huang, G.G.; Liu, Y.F.; Wu, X.X.; Cai, J.J. Activated carbons prepared by the KOH activation of a hydrochar from garlic peel and their CO_2 adsorption performance. *Xinxing Tan Cailiao/New Carbon Mater.* **2019**, *34*, 247–257. [CrossRef]
113. Chen, Y.; Chen, J.; Chen, S.; Tian, K.; Jiang, H. Ultra-high capacity and selective immobilization of Pb through crystal growth of hydroxypyromorphite on amino-functionalized hydrochar. *J. Mater. Chem. A* **2015**, *3*, 9843–9850. [CrossRef]
114. Han, L.; Sun, H.; Ro, K.S.; Sun, K.; Libra, J.A.; Xing, B. Removal of antimony (III) and cadmium (II) from aqueous solution using animal manure-derived hydrochars and pyrochars. *Bioresour. Technol.* **2017**, *234*, 77–85. [CrossRef] [PubMed]
115. Sun, K.; Tang, J.; Gong, Y.; Zhang, H. Characterization of potassium hydroxide (KOH) modified hydrochars from different feedstocks for enhanced removal of heavy metals from water. *Environ. Sci. Pollut. Res.* **2015**, *22*, 16640–16651. [CrossRef] [PubMed]

116. Zhou, N.; Chen, H.; Xi, J.; Yao, D.; Zhou, Z.; Tian, Y.; Lu, X. Biochars with excellent Pb(II) adsorption property produced from fresh and dehydrated banana peels via hydrothermal carbonization. *Bioresour. Technol.* **2017**, *232*, 204–210. [CrossRef] [PubMed]
117. Petrović, J.T.; Stojanović, M.D.; Milojković, J.V.; Petrović, M.S.; Šoštarić, T.D.; Laušević, M.D.; Mihajlović, M.L. Alkali modified hydrochar of grape pomace as a perspective adsorbent of Pb^{2+} from aqueous solution. *J. Environ. Manage.* **2016**, *182*, 292–300. [CrossRef] [PubMed]
118. Semercioz, A.S.; Göğüş, F.; Çelekli, A.; Bozkurt, H. Development of carbonaceous material from grapefruit peel with microwave implemented-low temperature hydrothermal carbonization technique for the adsorption of Cu (II). *J. Clean. Prod.* **2017**, *165*, 599–610. [CrossRef]
119. Deng, J.; Li, X.; Wei, X.; Liu, Y.; Liang, J.; Tang, N.; Song, B.; Chen, X.; Cheng, X. Sulfamic acid modified hydrochar derived from sawdust for removal of benzotriazole and Cu(II) from aqueous solution: Adsorption behavior and mechanism. *Bioresour. Technol.* **2019**, *290*, 121765. [CrossRef]
120. Zuo, X.J.; Liu, Z.; Chen, M.D. Effect of H_2O_2 concentrations on copper removal using the modified hydrothermal biochar. *Bioresour. Technol.* **2016**, *207*, 262–267. [CrossRef]
121. Fernandez, M.E.; Ledesma, B.; Román, S.; Bonelli, P.R.; Cukierman, A.L. Development and characterization of activated hydrochars from orange peels as potential adsorbents for emerging organic contaminants. *Bioresour. Technol.* **2015**, *183*, 221–228. [CrossRef]
122. Lima, H.H.C.; Maniezzo, R.S.; Kupfer, V.L.; Guilherme, M.R.; Moises, M.P.; Arroyo, P.A.; Rinaldi, A.W. Hydrochars based on cigarette butts as a recycled material for the adsorption of pollutants. *J. Environ. Chem. Eng.* **2018**, *6*, 7054–7061. [CrossRef]
123. Khataee, A.; Kayan, B.; Kalderis, D.; Karimi, A.; Akay, S.; Konsolakis, M. Ultrasound-assisted removal of Acid Red 17 using nanosized Fe_3O_4-loaded coffee waste hydrochar. *Ultrason. Sonochem.* **2017**, *35*, 72–80. [CrossRef]
124. Khoshbouy, R.; Takahashi, F.; Yoshikawa, K. Preparation of high surface area sludge-based activated hydrochar via hydrothermal carbonization and application in the removal of basic dye. *Environ. Res.* **2019**, *175*, 457–467. [CrossRef] [PubMed]
125. Tang, Z.; Deng, Y.; Luo, T.; Xu, Y.; Zhu, N. Enhanced removal of Pb(II) by supported nanoscale Ni/Fe on hydrochar derived from biogas residues. *Chem. Eng. J.* **2016**, *292*, 224–232. [CrossRef]
126. Buapeth, P.; Watcharin, W.; Dechtrirat, D.; Chuenchom, L. Carbon Adsorbents from Sugarcane Bagasse Prepared through Hydrothermal Carbonization for Adsorption of Methylene Blue: Effect of Heat Treatment on Adsorption Efficiency. *IOP Conf. Ser. Mater. Sci. Eng.* **2019**, *515*, 12003. [CrossRef]
127. Qian, W.C.; Luo, X.P.; Wang, X.; Guo, M.; Li, B. Removal of methylene blue from aqueous solution by modified bamboo hydrochar. *Ecotoxicol. Environ. Saf.* **2018**, *157*, 300–306. [CrossRef]
128. Islam, M.A.; Benhouria, A.; Asif, M.; Hameed, B.H. Methylene blue adsorption on factory-rejected tea activated carbon prepared by conjunction of hydrothermal carbonization and sodium hydroxide activation processes. *J. Taiwan Inst. Chem. Eng.* **2015**, *52*, 57–64. [CrossRef]
129. Fang, J.; Gao, B.; Chen, J.; Zimmerman, A.R. Hydrochars derived from plant biomass under various conditions: Characterization and potential applications and impacts. *Chem. Eng. J.* **2015**, *267*, 253–259. [CrossRef]
130. Wu, J.; Yang, J.; Feng, P.; Huang, G.; Xu, C.; Lin, B. High-efficiency removal of dyes from wastewater by fully recycling litchi peel biochar. *Chemosphere* **2020**, *246*, 125734. [CrossRef]
131. Zheng, H.; Sun, Q.; Li, Y.; Du, Q. Biosorbents prepared from pomelo peel by hydrothermal technique and its adsorption properties for congo red. *Mater. Res. Express* **2020**, *7*, 45505. [CrossRef]
132. Li, B.; Wang, Q.; Guo, J.Z.; Huan, W.W.; Liu, L. Sorption of methyl orange from aqueous solution by protonated amine modified hydrochar. *Bioresour. Technol.* **2018**, *268*, 454–459. [CrossRef]
133. Li, Y.; Meas, A.; Shan, S.; Yang, R.; Gai, X. Production and optimization of bamboo hydrochars for adsorption of Congo red and 2-naphthol. *Bioresour. Technol.* **2016**, *207*, 379–386. [CrossRef]
134. Li, Y.; Meas, A.; Shan, S.; Yang, R.; Gai, X.; Wang, H.; Tsend, N. Hydrochars from bamboo sawdust through acid assisted and two-stage hydrothermal carbonization for removal of two organics from aqueous solution. *Bioresour. Technol.* **2018**, *261*, 257–264. [CrossRef] [PubMed]
135. Wei, J.; Liu, Y.; Li, J.; Yu, H.; Peng, Y. Removal of organic contaminant by municipal sewage sludge-derived hydrochar: Kinetics, thermodynamics and mechanisms. *Water Sci. Technol.* **2018**, *78*, 947–956. [CrossRef] [PubMed]

136. Zhang, H.; Zhang, F.; Huang, Q. Highly effective removal of malachite green from aqueous solution by hydrochar derived from phycocyanin-extracted algal bloom residues through hydrothermal carbonization. *RSC Adv.* **2017**, *7*, 5790–5799. [CrossRef]
137. Hammud, H.H.; Shmait, A.; Hourani, N. Removal of Malachite Green from water using hydrothermally carbonized pine needles. *RSC Adv.* **2015**, *5*, 7909–7920. [CrossRef]
138. Feng, Y.; Sun, H.; Han, L.; Xue, L.; Chen, Y.; Yang, L.; Xing, B. Fabrication of hydrochar based on food waste (FWHTC) and its application in aqueous solution rare earth ions adsorptive removal: Process, mechanisms and disposal methodology. *J. Clean. Prod.* **2019**, *212*, 1423–1433. [CrossRef]
139. Xiao, K.; Liu, H.; Li, Y.; Yang, G.; Wang, Y.; Yao, H. Excellent performance of porous carbon from urea-assisted hydrochar of orange peel for toluene and iodine adsorption. *Chem. Eng. J.* **2020**, *382*, 122997. [CrossRef]
140. Rattanachueskul, N.; Saning, A.; Kaowphong, S.; Chumha, N.; Chuenchom, L. Magnetic carbon composites with a hierarchical structure for adsorption of tetracycline, prepared from sugarcane bagasse via hydrothermal carbonization coupled with simple heat treatment process. *Bioresour. Technol.* **2017**, *226*, 164–172. [CrossRef]
141. Zhu, X.; Qian, F.; Liu, Y.; Matera, D.; Wu, G.; Zhang, S.; Chen, J. Controllable synthesis of magnetic carbon composites with high porosity and strong acid resistance from hydrochar for efficient removal of organic pollutants: An overlooked influence. *Carbon N. Y.* **2016**, *99*, 338–347. [CrossRef]
142. Yu, J.; Zhu, Z.; Zhang, H.; Chen, T.; Qiu, Y.; Xu, Z.; Yin, D. Efficient removal of several estrogens in water by Fe-hydrochar composite and related interactive effect mechanism of H_2O_2 and iron with persistent free radicals from hydrochar of pinewood. *Sci. Total Environ.* **2019**, *658*, 1013–1022. [CrossRef]
143. European Committee. *EU Report on Critical Raw Materials List*; EC: Brussels, Belgium, 2017.
144. Ava Green Chemistry Development GMBH. *Sewage sludge reuse Phosphate recovery with an innovative HTC technology (HTCycle)*; Ava Green Chemistry Development GMBH: Munich, Germany, 2015.
145. Marin-Batista, J.D.; Mohedano, A.F.; Rodriguez, J.J.; de la Rubia, M.A. Energy and Phosphorous Recovery Through Hydrothermal Carbonization of Digested Sewage Sludge. *Waste Manag.* **2020**, *105*, 566–574. [CrossRef]
146. Becker, G.C.; Wüst, D.; Köhler, H.; Lautenbach, A.; Kruse, A. Novel approach of phosphate-reclamation as struvite from sewage sludge by utilising hydrothermal carbonization. *J. Environ. Manage.* **2019**, *238*, 119–125. [CrossRef] [PubMed]
147. Bhatt, D.; Shrestha, A.; Dahal, R.K.; Acharya, B.; Basu, P.; MacEwen, R. Hydrothermal carbonization of biosolids from Waste water treatment plant. *Energies* **2018**, *11*, 2286. [CrossRef]
148. Shi, Y.; Luo, G.; Rao, Y.; Chen, H.; Zhang, S. Hydrothermal conversion of dewatered sewage sludge: Focusing on the transformation mechanism and recovery of phosphorus. *Chemosphere* **2019**, *228*, 619–628. [CrossRef] [PubMed]
149. Bento, L.R.; Castro, A.J.R.; Moreira, A.B.; Ferreira, O.P.; Bisinoti, M.C.; Melo, C.A. Release of nutrients and organic carbon in different soil types from hydrochar obtained using sugarcane bagasse and vinasse. *Geoderma* **2019**, *334*, 24–32. [CrossRef]
150. Ji, M.; Sang, W.; Tsang, D.C.W.; Usman, M.; Zhang, S.; Luo, G. Molecular and microbial insights towards understanding the effects of hydrochar on methane emission from paddy soil. *Sci. Total Environ.* **2020**, *714*, 136769. [CrossRef]
151. Eibisch, N.; Durner, W.; Bechtold, M.; Fuß, R.; Mikutta, R.; Woche, S.K.; Helfrich, M. Does water repellency of pyrochars and hydrochars counter their positive effects on soil hydraulic properties? *Geoderma* **2015**, *245*, 31–39. [CrossRef]
152. Zhang, S.; Zhu, X.; Zhou, S.; Shang, H.; Luo, J.; Tsang, D.C.W. Hydrothermal carbonization for hydrochar production and its application. In *Biochar from Biomass and Waste: Fundamentals and Applications*; Elsevier: Amsterdam, The Netherlands, 2018; ISBN 9780128117293.
153. Abel, S.; Peters, A.; Trinks, S.; Schonsky, H.; Facklam, M.; Wessolek, G. Impact of biochar and hydrochar addition on water retention and water repellency of sandy soil. *Geoderma* **2013**, *202*, 183–191. [CrossRef]
154. Scheifele, M.; Hobi, A.; Buegger, F.; Gattinger, A.; Schulin, R.; Boller, T.; Mäder, P. Impact of pyrochar and hydrochar on soybean (*Glycine max* L.) root nodulation and biological nitrogen fixation. *Zeitschrift fur Pflanzenernahrung und Bodenkunde* **2017**, *180*, 199–211. [CrossRef]
155. George, E.; Ventura, M.; Panzacchi, P.; Scandellari, F.; Tonon, G. Can hydrochar and pyrochar affect nitrogen uptake and biomass allocation in poplars? *Zeitschrift fur Pflanzenernahrung und Bodenkunde* **2017**, *180*, 178–186. [CrossRef]

156. Eibisch, N.; Schroll, R.; Fuß, R. Effect of pyrochar and hydrochar amendments on the mineralization of the herbicide isoproturon in an agricultural soil. *Chemosphere* **2015**, *134*, 528–535. [CrossRef]
157. Eibisch, N.; Schroll, R.; Fuß, R.; Mikutta, R.; Helfrich, M.; Flessa, H. Pyrochars and hydrochars differently alter the sorption of the herbicide isoproturon in an agricultural soil. *Chemosphere* **2015**, *119*, 155–162. [CrossRef] [PubMed]
158. Bera, T.; Purakayastha, T.J.; Patra, A.K.; Datta, S.C. Comparative analysis of physicochemical, nutrient, and spectral properties of agricultural residue biochars as influenced by pyrolysis temperatures. *J. Mater. Cycles Waste Manag.* **2018**. [CrossRef]
159. Gronwald, M.; Vos, C.; Helfrich, M.; Don, A. Stability of pyrochar and hydrochar in agricultural soil—A new field incubation method. *Geoderma* **2016**, *284*, 85–92. [CrossRef]
160. Malghani, S.; Jüschke, E.; Baumert, J.; Thuille, A.; Antonietti, M.; Trumbore, S.; Gleixner, G. Carbon sequestration potential of hydrothermal carbonization char (hydrochar) in two contrasting soils; results of a 1-year field study. *Biol. Fertil. Soils* **2015**, *51*, 123–134. [CrossRef]
161. De Jager, M.; Röhrdanz, M.; Giani, L. The influence of hydrochar from biogas digestate on soil improvement and plant growth aspects. *Biochar* **2020**, *2*, 177–194. [CrossRef]
162. Yuan, H.; Lu, T.; Wang, Y.; Chen, Y.; Lei, T. Sewage sludge biochar: Nutrient composition and its effect on the leaching of soil nutrients. *Geoderma* **2016**, *267*, 17–23. [CrossRef]
163. Riedel, T.; Hennessy, P.; Iden, S.C.; Koschinsky, A. Leaching of soil-derived major and trace elements in an arable topsoil after the addition of biochar. *Eur. J. Soil Sci.* **2015**, *66*, 823–834. [CrossRef]
164. Hitzl, M.; Mendez, A.; Owsianiak, M.; Renz, M. Making hydrochar suitable for agricultural soil: A thermal treatment to remove organic phytotoxic compounds. *J. Environ. Chem. Eng.* **2018**, *6*, 7029–7034. [CrossRef]
165. Busch, D.; Stark, A.; Kammann, C.I.; Glaser, B. Genotoxic and phytotoxic risk assessment of fresh and treated hydrochar from hydrothermal carbonization compared to biochar from pyrolysis. *Ecotoxicol. Environ. Saf.* **2013**, *97*, 59–66. [CrossRef]
166. Yue, Y.; Yao, Y.; Lin, Q.; Li, G.; Zhao, X. The change of heavy metals fractions during hydrochar decomposition in soils amended with different municipal sewage sludge hydrochars. *J. Soils Sediments* **2017**, *17*, 763–770. [CrossRef]
167. Chu, Q.; Xue, L.; Cheng, Y.; Liu, Y.; Feng, Y.; Yu, S.; Meng, L.; Pan, G.; Hou, P.; Duan, J.; et al. Microalgae-derived hydrochar application on rice paddy soil: Higher rice yield but increased gaseous nitrogen loss. *Sci. Total Environ.* **2020**, *717*, 137127. [CrossRef] [PubMed]
168. Chu, Q.; Xue, L.; Singh, B.P.; Yu, S.; Müller, K.; Wang, H.; Feng, Y.; Pan, G.; Zheng, X.; Yang, L. Sewage sludge-derived hydrochar that inhibits ammonia volatilization, improves soil nitrogen retention and rice nitrogen utilization. *Chemosphere* **2020**, *245*, 125558. [CrossRef] [PubMed]
169. Subedi, R.; Kammann, C.; Pelissetti, S.; Taupe, N.; Bertora, C.; Monaco, S.; Grignani, C. Does soil amended with biochar and hydrochar reduce ammonia emissions following the application of pig slurry? *Eur. J. Soil Sci.* **2015**, *66*, 1044–1053. [CrossRef]
170. Andert, J.; Mumme, J. Impact of pyrolysis and hydrothermal biochar on gas-emitting activity of soil microorganisms and bacterial and archaeal community composition. *Appl. Soil Ecol.* **2015**, *96*, 225–239. [CrossRef]
171. Xia, Y.; Liu, H.; Guo, Y.; Liu, Z.; Jiao, W. Immobilization of heavy metals in contaminated soils by modified hydrochar: Efficiency, risk assessment and potential mechanisms. *Sci. Total Environ.* **2019**, *685*, 1201–1208. [CrossRef] [PubMed]
172. Xia, Y.; Luo, H.; Li, D.; Chen, Z.; Yang, S.; Liu, Z.; Yang, T.; Gai, C. Efficient immobilization of toxic heavy metals in multi-contaminated agricultural soils by amino-functionalized hydrochar: Performance, plant responses and immobilization mechanisms. *Environ. Pollut.* **2020**, *261*, 114217. [CrossRef]
173. Du, F.L.; Du, Q.S.; Dai, J.; Tang, P.D.; Li, Y.M.; Long, S.Y.; Xie, N.Z.; Wang, Q.Y.; Huang, R.B. A comparative study for the organic byproducts from hydrothermal carbonizations of sugarcane bagasse and its bio-refined components cellulose and lignin. *PLoS ONE* **2018**, *13*, e0197188. [CrossRef]
174. Chang, M.Y.; Huang, W.J. Hydrothermal biorefinery of spent agricultural biomass into value-added bio-nutrient solution: Comparison between greenhouse and field cropping data. *Ind. Crops Prod.* **2018**, *126*, 186–189. [CrossRef]

175. Ruiz, H.A.; Conrad, M.; Sun, S.N.; Sanchez, A.; Rocha, G.J.M.; Romaní, A.; Castro, E.; Torres, A.; Rodríguez-Jasso, R.M.; Andrade, L.P.; et al. Engineering aspects of hydrothermal pretreatment: From batch to continuous operation, scale-up and pilot reactor under biorefinery concept. *Bioresour. Technol.* **2020**, *299*, 122685. [CrossRef]
176. Ashraf, M.T.; Schmidt, J.E. Process simulation and economic assessment of hydrothermal pretreatment and enzymatic hydrolysis of multi-feedstock lignocellulose—Separate vs combined processing. *Bioresour. Technol.* **2018**, *249*, 835–843. [CrossRef]
177. Aguilar, D.L.; Rodríguez-Jasso, R.M.; Zanuso, E.; de Rodríguez, D.J.; Amaya-Delgado, L.; Sanchez, A.; Ruiz, H.A. Scale-up and evaluation of hydrothermal pretreatment in isothermal and non-isothermal regimen for bioethanol production using agave bagasse. *Bioresour. Technol.* **2018**, *263*, 112–119. [CrossRef] [PubMed]
178. Nascimento, V.M.; Rossell, C.E.V.; de Moraes Rocha, G.J. Scale-up hydrothermal pretreatment of sugarcane bagasse and straw for second-generation ethanol production. In *Hydrothermal Processing in Biorefineries: Production of Bioethanol and High Added-Value Compounds of Second and Third Generation Biomass*; Springer: New York, NY, USA, 2017; ISBN 9783319564579.
179. Zakaria, M.R.; Hirata, S.; Hassan, M.A. Hydrothermal pretreatment enhanced enzymatic hydrolysis and glucose production from oil palm biomass. *Bioresour. Technol.* **2015**, *176*, 142–148. [CrossRef] [PubMed]
180. Rivas, S.; Vila, C.; Alonso, J.L.; Santos, V.; Parajó, J.C.; Leahy, J.J. Biorefinery processes for the valorization of Miscanthus polysaccharides: From constituent sugars to platform chemicals. *Ind. Crops Prod.* **2019**, *134*, 309–317. [CrossRef]
181. Deng, A.; Ren, J.; Li, H.; Peng, F.; Sun, R. Corncob lignocellulose for the production of furfural by hydrothermal pretreatment and heterogeneous catalytic process. *RSC Adv.* **2015**, *5*, 60264–60272. [CrossRef]
182. Li, H.; Wang, X.; Liu, C.; Ren, J.; Zhao, X.; Sun, R.; Wu, A. An efficient pretreatment for the selectively hydrothermal conversion of corncob into furfural: The combined mixed ball milling and ultrasonic pretreatments. *Ind. Crops Prod.* **2016**, *94*, 721–728. [CrossRef]
183. Paze, A.; Brazdausks, P.; Rizhikovs, J.; Puke, M.; Tupciauskas, R.; Andzs, M.; Meile, K.; Vedernikovs, N. Changes in the Polysaccharide Complex of Lignocellulose after Catalytic Hydrothermal Pre-Treatment Process of Hemp (*Cannabis Sativa* L.) Shives. In Proceedings of the 23th European Biomass Conference and Exhibition, Vienna, Austria, 1–4 June 2015; pp. 1063–1069. [CrossRef]
184. Codignole Luz, F.; Volpe, M.; Fiori, L.; Manni, A.; Cordiner, S.; Mulone, V.; Rocco, V. Spent coffee enhanced biomethane potential via an integrated hydrothermal carbonization-anaerobic digestion process. *Bioresour. Technol.* **2018**, *256*, 102–109. [CrossRef]
185. Passos, F.; Ferrer, I. Influence of hydrothermal pretreatment on microalgal biomass anaerobic digestion and bioenergy production. *Water Res.* **2015**, *68*, 364–373. [CrossRef]
186. Weide, T.; Brügging, E.; Wetter, C. Anaerobic and aerobic degradation of wastewater from hydrothermal carbonization (HTC) in a continuous, three-stage and semi-industrial system. *J. Environ. Chem. Eng.* **2019**, *7*, 102912. [CrossRef]
187. He, L.; Huang, H.; Zhang, Z.; Lei, Z.; Lin, B. Le Energy Recovery from Rice Straw through Hydrothermal Pretreatment and Subsequent Biomethane Production. *Energy Fuels* **2017**, *31*, 10850–10857. [CrossRef]
188. Lucian, M.; Volpe, M.; Merzari, F.; Wüst, D.; Kruse, A.; Andreottola, G.; Fiori, L. Hydrothermal carbonization coupled with anaerobic digestion for the valorization of the organic fraction of municipal solid waste. *Bioresour. Technol.* **2020**, *314*, 123734. [CrossRef]
189. Ischia, G.; Orlandi, M.; Fendrich, M.A.; Bettonte, M.; Merzari, F.; Miotello, A.; Fiori, L. Realization of a solar hydrothermal carbonization reactor: A zero-energy technology for waste biomass valorization. *J. Environ. Manage.* **2020**, *259*, 110067. [CrossRef] [PubMed]
190. Xiao, C.; Liao, Q.; Fu, Q.; Huang, Y.; Chen, H.; Zhang, H.; Xia, A.; Zhu, X.; Reungsang, A.; Liu, Z. A solar-driven continuous hydrothermal pretreatment system for biomethane production from microalgae biomass. *Appl. Energy* **2019**, *236*, 1011–1018. [CrossRef]

© 2020 by the authors. Licensee MDPI, Basel, Switzerland. This article is an open access article distributed under the terms and conditions of the Creative Commons Attribution (CC BY) license (http://creativecommons.org/licenses/by/4.0/).

Article

Hydrothermal Carbonization as a Strategy for Sewage Sludge Management: Influence of Process Withdrawal Point on Hydrochar Properties

Fabio Merzari [1], Jillian Goldfarb [2], Gianni Andreottola [1], Tanja Mimmo [3], Maurizio Volpe [1,4] and Luca Fiori [1,*]

[1] Department of Civil, Environmental and Mechanical Engineering, University of Trento, via Mesiano 77, 38123 Trento, Italy; fabio.merzari@unitn.it (F.M.); gianni.andreottola@unitn.it (G.A.); maurizio.volpe@unikore.it (M.V.)
[2] Department of Biological and Environmental Engineering, Cornell University, 226 Riley-Robb Hall, Ithaca, NY 14853, USA; jlg459@cornell.edu
[3] Faculty of Science and Technology, Free University of Bolzano, Piazza Università 5, 39100 Bolzano, Italy; tanja.mimmo@unibz.it
[4] Faculty of Engineering and Architecture, University of Enna Kore, Cittadella Universitaria, 94100 Enna, Italy
* Correspondence: luca.fiori@unitn.it; Tel.: +39-0461-282692

Received: 29 April 2020; Accepted: 1 June 2020; Published: 5 June 2020

Abstract: Conventional activated sludge systems, still widely used to treat wastewater, produce large amounts of solid waste that is commonly landfilled or incinerated. This study addresses the potential use of Hydrothermal Carbonization (HTC) to valorize sewage sludge residues examining the properties of hydrochars depending on HTC process conditions and sewage sludge withdrawal point. With increasing HTC severity (process residence time and temperature), solid yield, total Chemical Oxygen Demand (COD) and solid pH decrease while ash content increases. Hydrochars produced from primary (thickened) and secondary (digested and dewatered) sludge show peculiar distinct properties. Hydrochars produced from thickened sludge show good fuel properties in terms of Higher Heating Value (HHV) and reduced ash content. However, relatively high volatile matter and O:C and H:C ratios result in thermal reactivity significantly higher than typical coals. Both series of carbonized secondary sludges show neutral pH, low COD, enhanced phosphorous content and low heavy metals concentration: as a whole, they show properties compatible with their use as soil amendments.

Keywords: sewage sludge; hydrothermal carbonization; hydrochar; solid biofuel; soil amendment

1. Introduction

In 1991, the European Union (EU) Directive 91/271/EEC set new benchmarks for the collection, treatment and monitoring of wastewater in urban areas [1]. In 2000, the EU produced over 10 million dry tons of sewage sludge (latest available official EU data [2]). Since then, production has steadily increased, increasing burdens on municipal wastewater treatment systems [3]. In 2017, the global sewage sludge production reached approximately 45 million of dry tons per year [4]. In Italy alone, over 1100 kilotons of dry sludge matter were produced in 2010, with almost 40% going to landfills and less than 30% used in agricultural applications [2]. In 2016, the European Commission reported that the 1991 directive was successful in terms of current improvements seen in EU water quality despite an increasing population growth. However, the European Commission also underlined the need to develop "innovative solutions to increase resource efficiency, such as solutions for energy recovery, nutrient recovery, and processing to marketable products and water re-use" [5]. Beyond the EU, global urbanization and the growth of the middle class, combined with stricter environmental regulations, have forced municipalities to re-examine their sludge management practices—transitioning away from traditional disposal via incineration, landfilling, or discharge to oceans/waterways—to favor beneficial reuse [6,7].

Sludge is a complex, heterogeneous mixture comprised of organic compounds such as proteins, peptides, polysaccharides, phenols, aliphatic, aromatic and furan compounds, as well as inorganic materials such as nutrients (phosphorous, potassium, nitrogen), silica, and heavy metals [8] and pathogens and other microbiological pollutants [9]. In a typical wastewater treatment process, primary sludge is characterized as the sludge following mechanical processing (screening, grit removal, sedimentation) containing between 93 and 99.5 wt% water, with a high content of suspended and dissolved organics. Secondary sludge (also known as Waste Activated Sludge, WAS) follows from biological treatment and contains a high amount of microbial cells with a total solids concentration ranging between 0.8 and 1.2 wt% depending on the type of process used [10].

Anaerobic Digestion (AD) is the most widely used sludge management technique. AD converts the organic solids to biogas (predominantly CH_4 and CO_2) via hydrolysis, acidogenesis, acetogenesis, and methanogenesis [11,12]. Despite its popularity, a considerable amount of solid remains after AD; as little as 20–30 wt% of the total organic matter is mineralized [9]. While these solids were once thought to be environmentally benign, their use into the soil may well be a potentially large source of greenhouse gas emissions and point pollutant sources for mercury, lead, cadmium, and copper on arable land, contributing to environmental acidification [13] and posing a toxicological risk in terms of pathogens present [9]. The microbiological processes occurring on land-applied sludge (mainly anaerobic decomposition, nitrification, and denitrification) lead to considerable emissions of methane and nitrous oxide, as well as ammonia and nitrate. While land application of sludge offsets the use of industrial nitrogen-containing fertilizers and represents a considerable net reduction of N_2O, NH_3, and NO_3^- emissions, CH_4 emissions are still estimated to be 6.3 kgton^{-1} of applied sludge [13]. As such, direct land application of secondary sludge may not be the optimal nutrient recovery pathway and may well represent a waste of a renewable energy source.

Hydrothermal Carbonization (HTC) is a process to concentrate the carbon in a given biomass, occurring in water at elevated temperatures in the range of 160–280 °C [14], above saturated vapor pressure, where water's dielectric constant decreases so drastically that it catalyzes the carbonization of biomass while acting as an organic solvent [15,16]. HTC proceeds via a series of mechanisms, including hydrolysis, dehydration, decarboxylation, decarbonylation, and demethanation. HTC is performed in 80–95 vol% water [17–20], making it an ideal processing pathway for wet biomasses [21,22] such as sewage sludge. As well recognized in the literature, HTC not only leads to reduced volume and energy densification of the solid residue, considerably improving its dewaterability [23,24], but also significantly increases its solid fuel properties while stabilizing and disinfecting the sludge [25,26]. Interestingly, HTC of sludge enhances $NO + NH_3$ reactions during solid combustion, significantly reducing NOx emissions across combustion modes [27].

While multiple studies focused on nutrients recovery such as phosphorous [28–30] or probed the impact of processing conditions on the solid and energy yields of hydrochars produced from the HTC of sludge from one withdrawal point in the wastewater treatment (WWT) process [23,31–39], the impact of upstream WWT processes on hydrochars is often not part of the experimental design. Thus, a primary goal of the present work was to determine the most viable point in the WWT process to employ HTC for sludge treatment from a solid fuel production and potential nutrient recovery standpoint. There are two sets of variables of interest in the present work: (1) feedstock withdrawal point from the WWT process; (2) hydrothermal reaction conditions on the composition and oxidation properties of resulting hydrochars. While some research has been done to quantify the environmental and economic benefits of using HTC to treat sewage sludge, understanding the impact of both feedstock and processing conditions on resulting hydrochars is critical to enable better optimization of this waste-to-energy conversion pathway [40].

2. Materials and Methods

Three kinds of samples were collected at different process points from the municipal Wastewater Treatment Plant (WWTP) of Trento, Italy, which uses a Conventional Activated Sludge (CAS) and Membrane BioReactor (MBR) system (more information about CAS/MBR systems in recent prior works in the field: [41,42]). The primary sludge extracted from the primary settler is sent to a static thickener and mixed with secondary sludge coming from the MBR system (Pipe 1 of Figure 1). The thickened sludge exiting the static thickener (Pipe 3) is the first sample taken for the present work. The secondary sludge coming from the CAS system is sent to a dynamic thickener (Pipe 2). After thickening, both streams are sent to the anaerobic digester. Anaerobic digestion lasts 21 days at 33 °C, after which the digested sludge (Pipe 4) is sent to a dewatering system. The effluent from Pipe 4 is the second sample taken for this work. To the digested sludge from Pipe 4 polyelectrolyte is added; this stream is sent to the dewatering system (centrifuge) where it reaches about 25 wt% solids content to enable solid handling. This dewatered sludge after the centrifuge (Pipe 5) is the third sample used in this work. Each of the three samples taken was characterized and stored in a fridge at 4 °C until use.

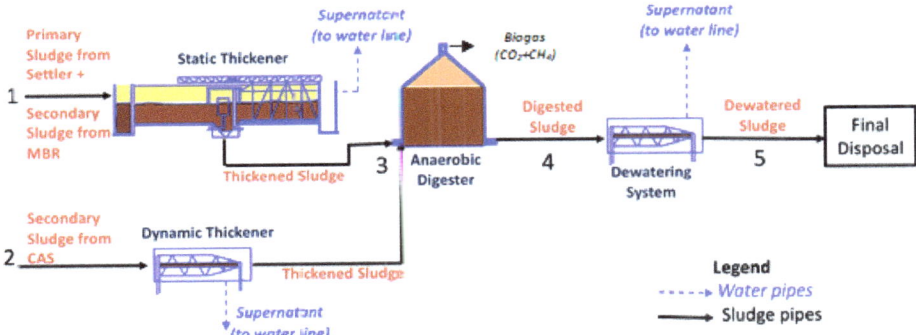

Figure 1. Schematic of wastewater treatment plant in Trento, Italy, identifying three sludge samples (Thickened—Pipe 3; Digested—Pipe 4; Dewatered—Pipe 5) used in the present work.

2.1. Feedstock Characterization

The three feedstocks used were characterized in accordance with accepted environmental practices. All measurements were conducted in triplicate and the average and standard deviation of each data is reported. The total solids content of each sludge was measured by drying in an oven at 105 °C for at least 8 h until constant weight was reached.

The pH of the raw and carbonized samples was measured using a Profi-Line pH 3310 (WTW, Milan, Italy) portable pH-meter by placing 1 g of solids in 20 g of deionized water, shaking for at least 90 min, allowing the mixture to settle for 15 min, and then reading the pH. The total Chemical Oxygen Demand (COD) was measured using a closed reflux titration method [43] using potassium dichromate digestion solution, sulfuric acid reagent, ferroin indicator solution, and a standard ferrous ammonium sulphate titrant according to standard procedures. The same method was used for soluble COD following filtration of the sample through a 0.45 μm filter [43]. Organic nitrogen in the trinegative state was measured via the semi-micro Kjeldahl method. Ammonium nitrogen was measured [43] by first buffering the sample at pH 9.5 with a borate buffer (to decrease hydrolysis), then distilling in a solution of boric acid, and determining the concentration via acid titration with H_2SO_4. To measure the total phosphorous, samples were first digested in H_2SO_4, forming molybdophosphoric acid, which was then reduced by stannous chloride to molybdenum blue. The concentration was measured photometrically at 690 nm and compared against a calibration curve [43].

Ultimate analysis to determine elemental composition of C, H, N, S, and O (by difference) was conducted on a LECO 628 analyzer (LECO, Moenchengladbach, Germany) equipped with Sulphur module for CHN (ASTM D-5373 standard method) and S (ASTM D-1552 standard method) determination. Proximate analyses were done on a LECO Thermogravimetric Analyzer TGA 701 (LECO Corporation, St. Joseph, MI, USA). Samples were heated at 20 °C min^{-1} to 105 °C in air and held until constant weight (< ±0.05%) to provide a dry baseline. They were subsequently heated at 16 °C min^{-1} from 105 °C to 900 °C in nitrogen with a hold time of 7 min, where the mass loss was attributed to Volatile Matter (VM). Finally, samples were held at 800 °C in air to oxidize the Fixed Carbon (FC) until the mass change stay within ±0.5% by weight. Mass remaining after this was considered to be ash (inorganic matter) content. The Higher Heating Value, HHV, was measured using an IKA 200C isoperibolic calorimeter (IKA-Werke GmbH, Staufen, Germany) according to the CEN/TS 14918 standard.

Inductively coupled plasma-optical emission spectroscopy (Arcos Ametek, Spectro, Germany) was used to determine the inorganic concentration of the sludge samples. Briefly, samples were oven-dried at 105 °C until constant weight and then acid-digested in concentrated nitric acid (650 mL^{-1}; Carlo Erba, Milano, Italy) using a single reaction chamber microwave digestion system (UltraWAVE, Milestone Inc., Sheldon CT USA) and Teflon-lined vials to prevent interference. Elements were quantified using certified multi-element standards (CPI International).

2.2. Hydrothermal Carbonization and Product Analysis

Hydrothermal carbonization typically occurs between 180 and 250 °C under autogenous pressure (up to 50 bar) but below the critical point [18]. The present work utilized a 50 mL stainless steel batch HTC reactor rated to withstand 300 °C and 140 bar, with temperature and pressure monitoring and temperature control, as previously described [44,45]. The thickened and digested sludges were used as-received. The reactor was loaded with 35.0 mL ± 0.1 mL biomass, which maintained a biomass (dry biomass)-to-water ratio of about 0.03:1. The dewatered sludge had a solid content of 25 wt% and needed to be diluted to ensure that the biomass was fully submerged. The reactor was filled with 20.00 g ± 0.01 g of dewatered sludge and 15.00 g ± 0.01 g of deionized water to cover the sludge, resulting in a biomass-to-water ratio of 0.17:1. Prior to each run, the reactor was sealed and purged with nitrogen gas; then it was heated up to the desired reaction temperature (190 °C, 220 °C and 250 °C) and held at the set point for the desired reaction time (30 min and 60 min). At least three experimental runs for each of the temperature/time combinations were performed for each of the three sludge samples.

After the reaction time, the reactor was cooled by placing a cold (−25 °C) stainless steel disk under its bottom and by blowing compressed air into its outer walls. The reactor was cooled to ambient temperature in less than 15 min, at which point the produced gas was measured by flowing it into a graduate cylinder filled with water [45]. As reported in the literature, where the CO_2 content is always greater than 90 vol.%, the produced gas was assumed to be comprised entirely of CO_2 [45,46]. The gas yield was estimated according to the ideal gas law under the assumption of standard temperature and pressure as:

$$Y_{gas} = \frac{Mass_{CO_2}}{Mass_{Sludge, dry}} \tag{1}$$

The liquid and solid HTC products were filtered through a pre-dried and weighed piece of cellulose filter paper. The filter paper was then put in the oven overnight at 105 °C and weighed to calculate the solids produced. The solid yield of the hydrochar, $Y_{hydrochar}$, was calculated as:

$$Y_{hydrochar} = \frac{Mass_{hydrochar, dry}}{Mass_{Sludge, dry}} \tag{2}$$

The liquid yield was computed as the complement to 1 of the gas and solid yields.

The solid hydrochar was characterized according to the same methods described in Section 2.1. for Higher Heating Value (HHV), proximate, and ultimate analyses. The hydrochar's relative solid reactivity was measured using a Mettler-Toledo Thermogravimetric Analyzer–Differential Scanning Calorimeter (TGA-DSC-1, Mettler-Toledo LLC, Columbus, Ohio, USA) in an oxidative atmosphere. The TGA-DSC was calibrated with NIST-traceable gold, indium, and aluminum and the mass was measured to ±0.1 µg and temperature to ±0.1 °C. Approximately 10 mg of sample was loaded into a 70 µL alumina crucible. Samples were heated at 20 °C min^{-1} up to 110 °C in air flowing at 50 mL min^{-1} and held for 30 min to drive off any residual moisture. They were subsequently heated at 20 °C min^{-1} up to 950 °C and held for 30 min to oxidize all material. The mass fraction of sample converted (X) at any time, t, was calculated as:

$$X = \frac{m_i - m_t}{m_i - m_f} \qquad (3)$$

where m_i is the initial mass, m_t is the mass at any time, t, and m_f is the final mass after the hold at 950 °C. Derivative thermogravimetric (DTG) curves were plotted as dX/dt (s^{-1}) versus temperature. Differential scanning calorimeter (DSC) data was normalized as heat flow per sample mass at any given instant (m_t). DTG curves are compared to those from an in-house sample of Illinois No. 6 coal, a high volatile bituminous coal from the Illinois #6 (Herrin) seam from the Argonne Premium Coal Bank [47]. The coal sample is well characterized in the literature and is often used as a standard on which to compare solid fuel oxidation [47–49].

Dewaterability—and the improvement due to HTC was determined by measuring the Capillary Suction Time (CST) required for water to be separated from sludge across a filter paper (Whatman 17 CHR, VWR International, Milan, Italy) using a Triton Electronics Ltd. capillary suction timer type 304B according to standard methods [43]. CST provides a quantitative assessment of how readily sludge releases water.

The liquid phase remaining after hydrothermal carbonization was characterized by measuring pH, COD, organic nitrogen, ammonia nitrogen, and phosphorous as described above. Measurements of Readily Biodegradable COD (RBCOD) were performed following the procedure described in literature [50]. To measure NH$_4^+$ nitrogen and soluble COD, samples were screened through a 0.45 µm filter [43].

3. Results

To assess the optimal point to withdraw sludge from the wastewater treatment process for hydrothermal carbonization in terms of resulting hydrochar properties, three samples were pulled from various points along the process: Thickened, Digested, and Dewatered (Figure 1). The feedstocks characteristics are presented in Tables 1 and 2. These three samples were subjected to hydrothermal carbonization at three temperatures (190 °C, 220 °C, 250 °C) and two residence times (30 min, 60 min) each, producing a total of 18 hydrochar samples for analysis.

Looking at the data of the different raw sludges, it is clear that they differ substantially. Thickened sludge contains about 46 wt% elemental carbon and 15 wt% ash. In digested sludge, the carbon decreases to about 26 wt%, and the ash content increases to 45 wt%, due to stabilization during anaerobic digestion. Dewatered sludge contains 36 wt% carbon and about 28 wt% ash. Thus, even if dewatering is a mechanical process, it greatly modifies the sludge characteristics. The supernatant from sludge dewatering has high concentrations of inorganic compounds such as N-NH$_4^+$, P compounds, CaCO$_3$, Mg, K, Na, and other minerals that contribute to the ash content [51]. The dewatering unit washes away these inorganics strongly decreasing the ash content of the dewatered sludge: this reflects on an increase in elemental C, H, N, and O, and also in FC and VM. VM variation is extremely significant, passing from a value of 50 wt% in the digested sludge to about 66 wt% in the dewatered sludge. These differences are also due to the fact that the digested sludge, immediately upstream of the dewatering operation, is chemically conditioned with organic polyelectrolyte (1–10 g/kg dry solids [52]) that is quickly adsorbed on the sludge particles.

Table 1. Characteristics of raw thickened sludge, raw digested sludge, and raw dewatered sludge and products of hydrothermal carbonization of thickened, digested, and dewatered sludge.

Property	HTC Temp (°C)	Thickened Sludge Raw	Thickened 30 min HTC	Thickened 60 min HTC	Digested Sludge Raw	Digested 30 min HTC	Digested 60 min HTC	Dewatered Sludge Raw	Dewatered 30 min HTC	Dewatered 60 min HTC
Moisture Content (wt%)	Raw	97.1 ± 0.1			97.1 ± 0.3			74.9 ± 0.1		
Total Solids (g/L)	Raw	29.1 ± 0.4			29.0 ± 2.3			n.a.		
Fixed + Volatile Solids (g/L)	Raw	26.7 ± 0.4			15.4 ± 0.9			n.a.		
Total COD (g/L)	Raw	28.1 ± 1.4			18.0 ± 0.9			193.2 ± 0.3 [a]		
	190		14.1 ± 2.0	15.3 ± 0.7		9.1 ± 0.2	9.3 ± 0.4		55.3 ± 1.4	56.3 ± 2.7
	220		15.4 ± 1.4	15.1 ± 1.1		11.3 ± 1.2	9.9 ± 0.1		56.7 ± 8.2	54.1 ± 0.3
	250		18.4 ± 0.7	18.4 ± 1.8		10.4 ± 0.7	8.8 ± 0.4		64.9 ± 8.2	57.5 ± 0.3
Soluble COD (g/L)	Raw	2.3 ± 0.1			0.4 ± 0.1			47.2 ± 2.4 [a]		
	190		10.5 ± 1.0	13.4 ± 0.4		6.4 ± 0.3	6.7 ± 0.5		49.8 ± 0.4	55.1 ± 2.8
	220		11.9 ± 1.5	13.5 ± 0.5		6.7 ± 0.3	6.7 ± 0.5		44.8 ± 3.4	45.7 ± 5.5
	250		15.3 ± 0.9	13.7 ± 0.1		8.3 ± 0.8	5.7 ± 0.2		57.5 ± 9.8	46.8 ± 0.6
pH	Raw	7.21 ± 0.1			8.42 ± 0.01			7.0 ± 0.0		
	190		6.0 ± 0.0	5.3 ± 0.0		7.1 ± 0.2	6.8 ± 0.0		6.0 ± 0.0	6.0 ± 0.0
	220		6.0 ± 0.0	6.1 ± 0.0		6.8 ± 0.0	6.2 ± 0.0		6.0 ± 0.0	5.9 ± 0.0
	250		5.8 ± 0.1	6.0 ± 0.0		7.3 ± 0.6	6.9 ± 0.6		6.0 ± 0.0	5.5 ± 0.0
Total Phosphorous in Liquid (mg/L)	Raw	175.00 ± 9.00			187.00 ± 9.00			10.00 ± 1.00 [a]		
	190		32.5 ± 0.1	38.4 ± 0.1		19.6 ± 0.3	22.7 ± 0.1		0.3 ± 0.0	0.4 ± 0.0
	220		20.2 ± 0.1	16.4 ± 0.2		19.8 ± 0.3	17.8 ± 0.1		0.2 ± 0.0	0.2 ± 0.0
	250		16.1 ± 0.1	11.1 ± 0.2		19.3 ± 0.3	12.2 ± 0.1		0.2 ± 0.0	0.2 ± 0.0
Total Phosphorous in Solid (mg/g)	Raw	5.22 ± 0.10			5.81 ± 0.13			9.22 ± 0.33		
	190		6.2 ± 0.7	6.8 ± 0.4		7.2 ± 1.4	7.4 ± 0.7		9.2 ± 0.5	9.4 ± 0.4
	220		7.8 ± 0.3	9.4 ± 0.2		7.5 ± 1.5	7.5 ± 0.5		10.5 ± 0.2	10.6 ± 0.2
	250		9.9 ± 0.5	10.2 ± 4.6		7.4 ± 1.2	6.7 ± 0.9		10.8 ± 0.8	10.9 ± 0.9
Organic Nitrogen in Liquid (g/L)	Raw	0.470 ± 0.230			0.880 ± 0.010			11.023 ± 0.551 [a]		
	190		0.1 ± 0.0	0.1 ± 0.0		1.3 ± 0.0	1.2 ± 0.0		4.5 ± 1.0	4.1 ± 0.2
	220		0.1 ± 0.0	0.1 ± 0.0		1.4 ± 0.0	0.9 ± 0.0		2.4 ± 0.6	1.5 ± 0.1
	250		0.1 ± 0.0	0.1 ± 0.0		1.3 ± 0.0	1.5 ± 0.0		1.7 ± 0.3	1.4 ± 0.7
Ammonia Nitrogen in Liquid (g/L)	Raw	0.490 ± 0.010			0.880 ± 0.010			0.598 ± 0.030 [a]		
	190		0.4 ± 0.0	0.6 ± 0.0		0.7 ± 0.0	0.7 ± 0.0		2.7 ± 0.1	3.8 ± 0.2
	220		0.4 ± 0.0	0.5 ± 0.0		0.8 ± 0.0	1.0 ± 0.0		4.4 ± 0.3	4.6 ± 0.6
	250		0.6 ± 0.0	0.6 ± 0.0		1.0 ± 0.0	0.8 ± 0.0		6.9 ± 0.4	6.5 ± 0.4

ar as received, n.a.—not applicable. [a] Based on 1 g of dry biomass dissolved in one liter of distilled water, HTC—Hydrothermal Carbonization.

Table 2. Products yields and analysis of solid fuel characteristics of raw feedstocks and hydrochars from thickened, digested, and dewatered sludge.

Property	HTC Temp (°C)	Thickened Sludge 30 min HTC	Thickened Sludge 60 min HTC	Digested Sludge 30 min HTC	Digested Sludge 60 min HTC	Dewatered Sludge 30 min HTC	Dewatered Sludge 60 min HTC
Process Yields (dry basis)							
Solid Yield (wt%)	190	77.2 ± 4.7	67.0 ± 0.0	82.8 ± 11.0	81.5 ± 8.9	88.2 ± 2.0	85.6 ± 0.0
	220	60.6 ± 8.8	57.2 ± 3.4	64.6 ± 3.4	70.6 ± 6.4	75.1 ± 1.0	75.2 ± 0.3
	250	49.4 ± 4.8	52.0 ± 6.2	78.1 ± 6.2	64.9 ± 7.8	67.6 ± 2.5	66.8 ± 0.2
Gas Yield (wt%)	190	2.5 ± 0.0	2.7 ± 0.2	3.7 ± 0.9	4.7 ± 0.3	2.1 ± 1.4	2.6 ± 0.2
	220	4.1 ± 0.7	4.2 ± 0.4	5.1 ± 1.3	5.6 ± 1.0	3.7 ± 0.1	4.3 ± 0.6
	250	6.3 ± 1.6	7.2 ± 1.3	6.1 ± 2.1	7.6 ± 1.9	5.1 ± 0.2	5.9 ± 0.1
Liquid Yield (wt%)	190	20.3 ± 2.4	34.9 ± 5.4	43.2 ± 4.9	43.1 ± 4.6	45.2 ± 1.7	44.1 ± 0.1
	220	35.3 ± 4.7	30.7 ± 1.9	34.8 ± 3.9	38.1 ± 2.2	39.4 ± 0.6	39.7 ± 0.4
	250	44.3 ± 3.2	29.6 ± 3.8	42.1 ± 5.0	36.3 ± 2.8	36.4 ± 1.3	36.4 ± 0.2
Ultimate Analysis (dry basis)							
C (wt%)	Raw	45.96 ± 0.20	45.96 ± 0.20	25.60 ± 0.33	25.60 ± 0.33	35.91 ± 0.25	35.91 ± 0.25
	190	41.56 ± 0.94	46.11 ± 0.39	19.22 ± 1.37	14.16 ± 2.57	36.61 ± 0.02	35.07 ± 0.23
	220	44.86 ± 0.51	43.15 ± 0.19	11.70 ± 0.64	10.21 ± 0.58	35.19 ± 0.17	35.75 ± 0.31
	250	41.68 ± 0.29	41.21 ± 0.52	12.51 ± 0.26	12.02 ± 0.05	35.30 ± 0.18	35.57 ± 0.08
H (wt%)	Raw	6.57 ± 0.02	6.57 ± 0.02	3.96 ± 0.06	3.96 ± 0.06	5.42 ± 0.00	5.42 ± 0.00
	190	6.24 ± 0.08	6.46 ± 0.06	2.63 ± 0.16	2.00 ± 0.29	4.92 ± 0.02	4.61 ± 0.03
	220	5.95 ± 0.01	5.69 ± 0.02	1.61 ± 0.08	1.37 ± 0.07	4.32 ± 0.01	4.36 ± 0.03
	250	5.03 ± 0.05	5.01 ± 0.03	1.65 ± 0.01	1.56 ± 0.02	4.11 ± 0.00	4.05 ± 0.01
N (wt%)	Raw	4.26 ± 0.00	4.26 ± 0.00	3.59 ± 0.10	3.59 ± 0.10	5.81 ± 0.02	5.81 ± 0.02
	190	2.23 ± 0.07	2.10 ± 0.07	1.58 ± 0.07	1.07 ± 0.17	4.27 ± 0.06	3.95 ± 0.01
	220	1.87 ± 0.24	1.86 ± 0.07	0.80 ± 0.08	0.66 ± 0.01	3.48 ± 0.02	3.45 ± 0.02
	250	1.89 ± 0.05	1.99 ± 0.04	0.70 ± 0.00	0.69 ± 0.02	3.16 ± 0.04	3.12 ± 0.02
O (wt%)	Raw	27.61 ± 4.86	27.61 ± 4.86	21.16 ± 0.79	21.16 ± 0.79	23.52 ± 0.52	23.52 ± 0.52
	190	28.52 ± 6.69	25.97 ± 4.63	18.43 ± 3.50	15.67 ± 6.89	18.55 ± 0.17	19.33 ± 0.58
	220	25.51 ± 6.34	26.43 ± 5.06	12.35 ± 1.03	18.93 ± 1.10	16.53 ± 0.33	15.34 ± 0.49
	250	22.47 ± 6.44	21.59 ± 3.83	11.72 ± 2.00	7.17 ± 2.53	14.07 ± 0.44	14.12 ± 2.18

Table 2. Cont.

Property	HTC Temp (°C)	Thickened Sludge 30 min HTC	±	Thickened Sludge 60 min HTC	±	Digested Sludge 30 min HTC	±	Digested Sludge 60 min HTC	±	Dewatered Sludge 30 min HTC	±	Dewatered Sludge 60 min HTC	±
Proximate Analysis (dry basis)													
Fixed Carbon (wt%)	Raw	12.56	±	1.43	12.56	±	1.43	4.76	±	3.78	5.90	±	0.62	5.90	±	0.62
	190	4.25	±	2.50	4.58	±	1.75	1.70	±	0.09	2.06	±	0.34	5.37	±	0.23	9.94	±	6.93
	220	5.37	±	4.00	5.23	±	3.97	1.26	±	0.07	0.13	±	0.01	4.04	±	2.06	9.83	±	5.58
	250	5.31	±	5.85	6.12	±	5.93	0.62	±	0.05	1.38	±	0.09	6.27	±	0.61	7.02	±	0.03
Volatile Matter (wt%)	Raw	72.48	±	3.19	72.48	±	3.19	50.25	±	3.51	50.25	±	3.51	65.67	±	0.50	65.67	±	0.50
	190	70.84	±	6.65	71.03	±	5.36	41.09	±	3.22	30.84	±	4.20	58.98	±	0.29	53.02	±	6.61
	220	66.05	±	5.56	63.57	±	7.81	26.17	±	1.59	31.99	±	0.90	55.48	±	2.19	49.07	±	5.72
	250	60.06	±	2.19	56.70	±	3.95	25.97	±	1.68	21.65	±	4.68	50.37	±	0.39	49.85	±	2.03
Ash (Inorganic) (wt%)	Raw	14.96	±	2.31	14.96	±	2.31	44.99	±	3.65	44.99	±	3.65	28.43	±	0.56	28.43	±	0.56
	190	24.91	±	4.57	24.40	±	3.56	57.21	±	1.61	67.10	±	2.27	35.66	±	0.26	37.05	±	6.77
	220	28.58	±	4.78	31.20	±	5.89	72.57	±	0.80	67.88	±	0.45	40.48	±	2.13	41.10	±	5.65
	250	34.63	±	4.02	37.18	±	4.94	73.42	±	0.86	76.97	±	2.34	43.36	±	0.50	43.14	±	1.03
Combustion Analysis													
HHV (MJ/kg)	Raw	20.50	±	0.14	20.50	±	0.14	10.66	±	1.79	10.66	±	1.79	16.02	±	0.09	16.02	±	0.09
	190	19.45	±	0.64	20.71	±	0.03	9.27	±	1.73	7.97	±	1.77	16.30	±	0.19	15.96	±	0.07
	220	20.06	±	0.90	18.72	±	0.03	8.96	±	1.77	7.86	±	0.39	15.70	±	0.09	15.47	±	0.37
	250	18.66	±	0.26	19.17	±	0.61	8.59	±	1.78	9.37	±	1.39	15.98	±	0.11	15.33	±	0.26

3.1. Hydrochar Properties Depend on Sludge Withdrawal Point and Carbonization Conditions

Solid hydrochar yields and properties are strongly influenced by both sewage sludge feedstock and HTC temperature and residence time, as shown in Table 2 (see also Supplementary Materials Figure S1). In general, as the harshness of carbonization increases, the solid hydrochar yield decreases, as is the case with biomasses across the literature [53]. The one exception to this general trend was the digested sludge sample carbonized at 220 °C for 30 min, which had a lower solid yield than its 250 °C and 190 °C counterparts. This anomaly was observed for multiple experimental runs. Such behavior could be attributed to potential re-condensing of tarry materials onto the hydrochar matrix, increasing the observed "solid" yield at 250 °C. Such materials were previously identified on the surface of heterogeneous biomasses that can be maximized at carbonization conditions specific to each biomass [17]. As the solid hydrochar is collected on a 10–100 µm filter, any particles smaller than this size range would be considered part of the liquid yield, which may also explain this anomaly.

As a result of the carbonization process, which releases organic compounds from the solid matrix and forces them into the water phase while generating CO_2, the total COD present substantially decreased upon carbonization (at any time, temperature) of the dewatered sludge sample, with a similar behavior noted for the thickened and digested samples (the effect on these two samples is less pronounced given the lower initial concentration in the raw samples) as shown in Table 1 and Figure S2 in Supplementary Materials. Conversely, soluble COD increased with increasing severity, most notably for the thickened and digested samples, with a significantly higher soluble COD that is maintained throughout carbonization for the dewatered samples. Such behavior has been described in the literature for other biomasses, including sludge, secondary sludge, wood waste, and dairy waste [54]. Given the nature of the carbonization process, where water acts as both a solvent and transport medium, simultaneously drawing organics into the water phase and generating CO_2, this COD mass balance suggests that indeed the insoluble fraction is being converted to CO_2 (and lesser amounts of CO and other non-condensable gases) while the soluble portion is retained in the process water.

The decrease in hydrochar pH for increasingly severe carbonization of the thickened sludge is to be expected; Volatile Fatty Acids (VFAs) present in primary sludge and formed via hydrolysis of triacylglycerols [55], especially acetic acid, are known to remain stable in the liquid phase, only breaking down at higher carbonization temperatures [54]. It was found here that the pH for the thickened sludge dropped from 7.2 (raw) to 6.0 for both the 190 °C and 220 °C conditions and to 5.8 for the 250 °C condition of the 30 min HTC samples and to 5.3, 6.1, and 6.0 for the 190 °C, 220 °C, and 250 °C 60 min samples, respectively (Table 1; plotted in Figure S3 in Supplementary Materials). This suggests that carbonization is able to release some VFAs from the primary matrix, but that the overall acidic nature of the hydrochar increases slightly. It should be noted that these measurements were taken in the same manner used to measure biochar pH, by equilibrating the hydrochar with deionized water and reading the pH of the liquid following settling. Anecdotally, when the pH of the process water is measured immediately following carbonization, the pH increases slightly upon severe carbonization, which would indicate hydrolysis of released VFAs occurs in the process water. For the digested samples, carbonization at 250 °C yields hydrochars with slightly higher pH than their 190 °C and 220 °C counterparts, suggesting a reduced acid content of the char. For the dewatered secondary sludge, the pH decreases with increasing carbonization but is overall lowest to begin with, suggesting that the acid content decreased during anaerobic digestion.

The proposed loss of semi-volatile compounds from the sludge matrices upon carbonization is further supported by the proximate analysis (Figure 2; Table 2; data of Table 2 plotted in Figure S4 in Supplementary Materials). The sludge treatment of WWTP results in a decrease in fixed carbon and volatile matter content upon digestion, with a slight increase in both upon dewatering. Changes in proximate analysis as a function of carbonization conditions versus the raw sludge withdrawn at each point are given in Figure 2. Figure 2 reports the percentage change in value of the various variables (FC, VM, Ash, all data on a dry basis) due to HTC, in respect to the raw biomasses. As compared to the raw sludge sample from each withdrawal point, the volatile matter content decreased with increasingly harsh carbonization. The fixed carbon content also decreases upon HTC, shifting the balance of the hydrochar composition to the inorganic (loosely termed "ash" phase). The change in proximate analysis appears to be more heavily temperature dependent than time dependent for the thickened samples. In this case, the ash content increased by 50 wt%, 100 wt%, and 150 wt% over the raw thickened sample as the carbonization temperature increased from 190 °C to 220 °C to 250 °C, respectively. This increase in inorganic content is offset more so by decreases in fixed carbon than volatile matter. For the digested samples, the increase in ash content—ranging from 25 wt% to 50 wt% as temperature increases—is offset by decreases in both VM and FC, with a slightly higher decrease in FC than VM across HTC conditions. The behavior of the dewatered samples is somewhat different. While for HTC at 30 min there is a slight increase in ash (especially at 220 °C and 250 °C carbonization), the 220 °C 30 min sample is offset more by changes in FC, whereas the 250 °C 60 min sample's higher ash content is due to a decrease in VM. The dewatered 60 min HTC samples show even further divergent behavior; for all temperatures, the FC and ash contents increase, offset by a decrease in VM content. For the 190 °C and 220 °C samples, the change in FC weight percent is actually greater than that of the ash, whereas the 250 °C sample has a larger increase in ash than fixed carbon. This is likely due to the preparation of the samples. Both thickened and digested samples were used as received. The dewatered samples had too low of a moisture content to ensure all solids were submerged in the reactor, such that deionized water was added to the samples prior to carbonization. It is possible that some of the inorganics migrated into the water phase (to establish a concentration equilibrium), and thus, the relative change in ash content was lower for these samples.

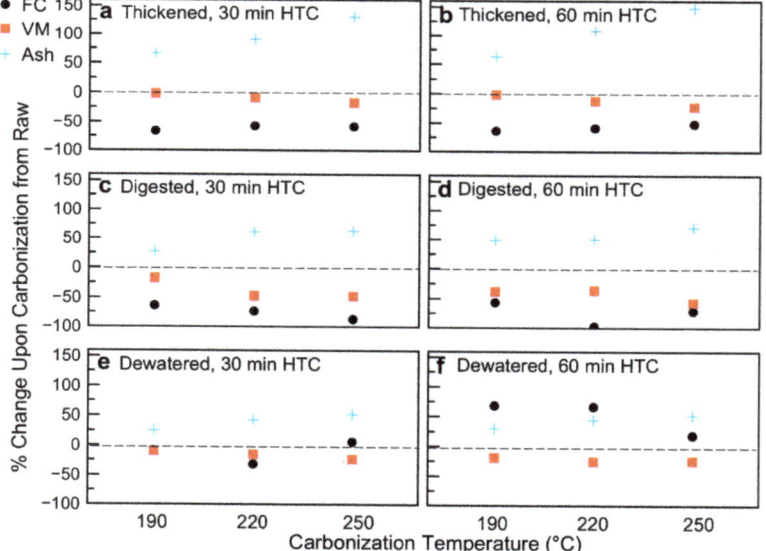

Figure 2. Changes in volatile matter, fixed carbon, and ash composition as a result of hydrothermal carbonization: (**a**) Thickened sludge, 30 min of Hydrothermal Carbonization (HTC); (**b**) Thickened sludge, 60 min HTC; (**c**) Digested sludge, 30 min HTC; (**d**) Digested sludge, 60 min HTC; (**e**) Dewatered sludge, 30 min HTC—note 190 °C Fixed Carbon (FC) and Volatile Matter (VM) data points overlap; (**f**) Dewatered sludge, 60 min HTC.

This behavior is echoed by the ultimate analysis, whereby HTC temperature has a greater impact on organic element composition than time (Figure 3; Figure 4; Table 2; data of Table 2 plotted in Figure S5 in Supplementary Materials). As expected, the biological transformations of the primary sludge lead to overall lower oxygen and carbon contents of the raw digested and dewatered samples [56]. Figure 3 reports the percentage change in content of atomic species C, H, N, and O (all data on a dry basis) due to HTC, in respect to the raw biomasses. As shown in Figure 3, in general the weight percent of the organic elements decrease upon carbonization of the sludge samples, the balance made up by an increasing (relative) inorganic content. The elemental carbon contents for the thickened 190 °C 30 min and 60 min and all dewatered samples remain constant within statistical significance. Interestingly, the relative percent changes of C, H, and O were fairly minimal for the thickened sludge carbonized at 190 °C and 220 °C, suggesting minimal decarboxylation and dehydration, both of which would lower the O:C and H:C ratios, respectively [57]. However, this is not to imply a stagnant system; for the thickened (and indeed all sludge samples) the nitrogen content of the solids decreases considerably upon even mild carbonization, ranging from hydrochars with 40 wt% to those with 80 wt% lower N content than their raw counterparts (Figure 3). This loss of nitrogen supports the hypothesis that there is significant hydrolysis of fat and protein components present in the sludge. The organic nitrogen in the process liquid tends to decrease by 15–85 wt% upon carbonization of all samples. The ammonia nitrogen present in the liquid varies in a certain range, without any clear trend with time or temperature, as shown in Table 1. Given the substantial elemental decrease in N of the hydrochar, and liquid-phase organic nitrogen upon carbonization, closure of a mass balance would suggest either production of nitrogen gases (N_2, NO_x) or formation of nitrate or nitrite in the liquid phase, which could precipitate out as salts onto the hydrochar. As reported by Kruse and co-workers, this concentration would be too low to detect with N content measurements [58].

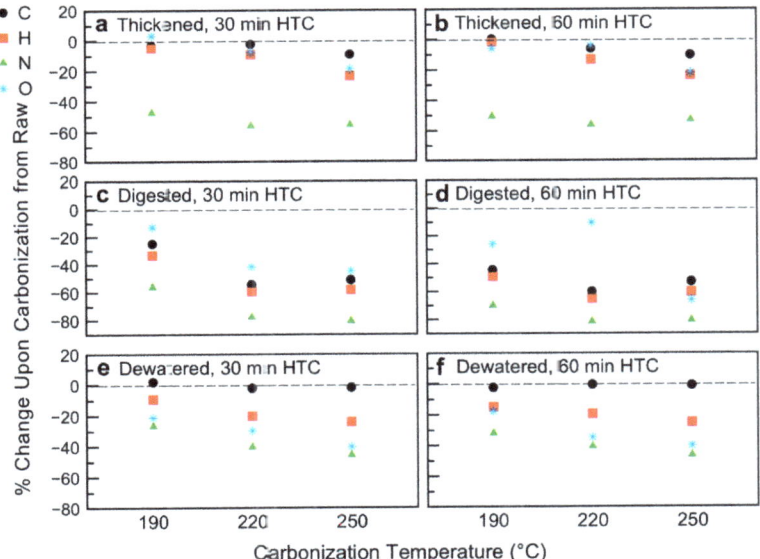

Figure 3. Changes in ultimate analysis as a result of hydrothermal carbonization: (**a**) Thickened sludge, 30 min HTC; (**b**) Thickened sludge, 60 min HTC; (**c**) Digested sludge, 30 min HTC; (**d**) Digested sludge, 60 min HTC; (**e**) Dewatered sludge, 30 min HTC; (**f**) Dewatered sludge, 60 min HTC.

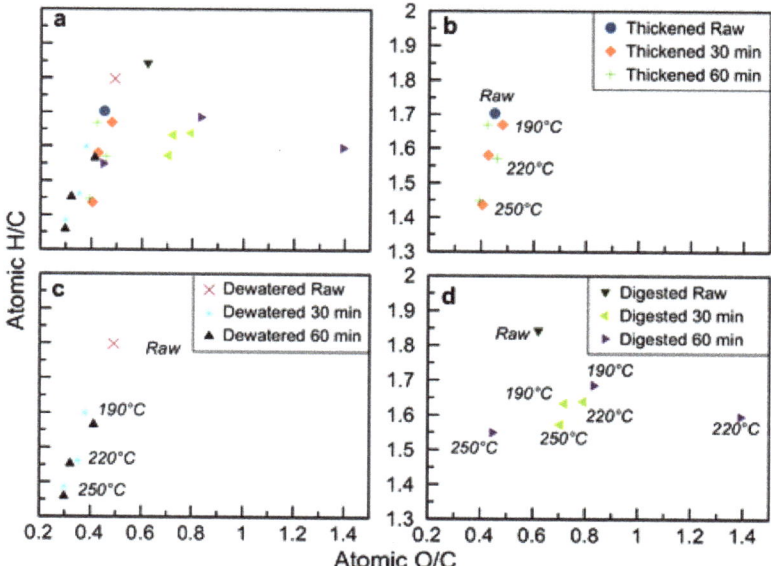

Figure 4. Van Krevelen diagram of raw and hydrothermally carbonized sludge samples with: (**a**) all samples; (**b**) thickened sludge; (**c**) dewatered sludge; (**d**) digested sludge.

The carbon and fixed carbon values reported in Table 2 and the trends of the same variables shown in Figures 2 and 3 testify to the peculiarity of these kinds of substrates which behave differently during HTC in respect to the vast majority of the other biomasses. While HTC applied to agro-waste, lignocellulosic feedstock, organic fraction of municipal solid waste, or compost [20,45,53,59,60] results in an increase in the values of C and FC, this is not the case for the majority of the sludge samples here investigated. The FC content decreases after HTC for thickened and digested sludges (Figure 2). The C content in the hydrochars is equal to or lower than C content in the raw sludges (Figure 3). This apparently strange behavior depends on the very high ash content of the sludge samples and was also previously reported in the literature [31,57,61]: in all these cases, the C and FC contents were expressed on a dry basis. Reverse C and FC trends were reported when the data was expressed on a dry ash free (daf) basis [23]. Here, on a daf basis, thickened and dewatered sludge increase their C content after HTC, and this applies to the digestate treated at the highest HTC temperature of 250 °C, too; FC data, conversely, even on a daf basis, remain still lower in the hydrochars in respect to the parent biomasses. To summarize, the organics in the sludge carbonized as should be expected in an HTC process: the increase in C content is evident when the basis of calculation is represented by the organics themselves (i.e., data on a daf basis), and the HTC was sufficiently severe. On a dry basis, conversely, the relative increase in ash content often prevails over the relative increase in C, as recently also discussed by Ferrentino et al. [62].

3.2. Distribution of Nutrients and Inorganics: Potential for Use as Soil Amendment

As shown in Table 1, the total phosphorous concentrated in the solid hydrochar (all samples/processing conditions) from 5 wt% in the raw samples to up to 10 wt% in the carbonized samples, with the thickened and dewatered 250 °C 30 min and 60 min samples having the highest amounts. The thickened sludge (all times/temperatures) showed lower organic nitrogen concentrations in the liquid phase, suggesting it remains in the solid phase or exits as a gas, as mentioned previously. This, coupled with the higher FC and VM contents of the thickened and dewatered hydrochars, suggests they would make better soil amendments than the digested sludge hydrochars [63–65].

Thermally treated biomasses tend to show enhanced P fertilizer values as a result of various mechanisms, including structural surface changes and improved association of P to inorganics such as Mg, Ca, and Al [66]. In the present work, the type of sludge sample had a larger impact on the retention and concentration of these nutrients than the carbonization conditions. As shown in Figure 5, the dewatered samples had higher overall Al concentrations—from 4 mg_{Al}/g_{sludge} for the raw sample as compared to ~ 2 mg_{Al}/g_{sludge} for the thickened and digested raw samples. Carbonization at 220 and 250 °C for both 30 min and 60 min doubled all of these concentrations, to 8 mg_{Al}/g_{sludge} for the carbonized dewatered sample and ~ 4 mg_{Al}/g_{sludge} for the others. Conversely, the digested sludge had almost twice the concentration of magnesium than the thickened or dewatered samples (which again almost doubled upon carbonization at 220 °C and 250 °C). The calcium concentration was quite similar for all sludge samples at 23, 33, and 28 mg_{Ca}/g_{sludge}, for the thickened, digested, and dewatered raw samples, respectively. Upon carbonization at 250 °C for 30 min, the Ca content increased by over 100% for the thickened sample but only by 39% and 43% for the digested and dewatered raw samples, respectively. While soil incubation studies are beyond the scope of the present work, prior work by Thomsen and co-workers [66] suggests that the more heavily oxidized hydrochar samples (e.g., those that released more CO_2 and thus had higher gas yield) containing more Mg, Ca, and Al would have a higher P availability. This corresponds to the digested 250 °C 60 min and dewatered 220 °C and 250 °C for 30 min and 60 min samples. Future work will investigate the degree to which hydrothermal carbonization plays a role in P plant availability, as well as the impact of HTC processing conditions itself on the volatilization and re-condensation of P species on the hydrochar surfaces.

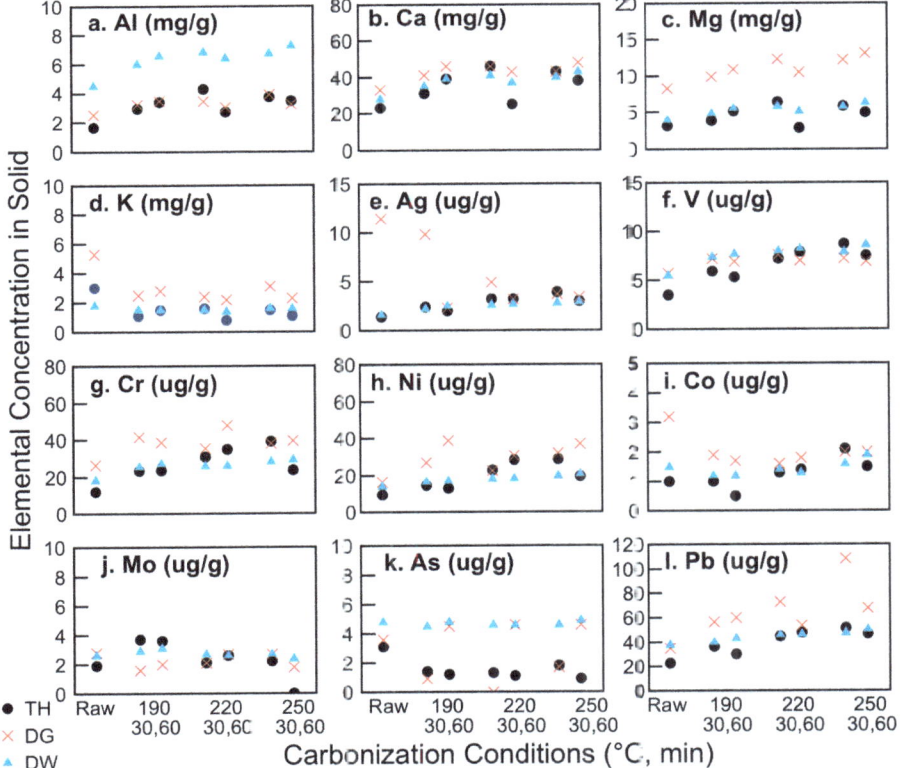

Figure 5. Impact of hydrothermal carbonization on nutrient and heavy metals concentrations for: (**a**) Al; (**b**) Ca; (**c**) Mg; (**d**) K; (**e**) Ag; (**f**) V; (**g**) Cr; (**h**) Ni; (**i**) Co; (**j**) Mo; (**k**) As; (**l**) Pb.

The concentrations of a series of additional inorganic elements in the sludges and hydrochars were measured. Figure 5 shows a representative set of these elements (all data available in Supplementary Materials: Tables S1 and S2, Figures S6 and S7). Silver was included in the analysis, given its increasing prevalence in consumer materials as an incorporated nanomaterial and resulting detection in wastewater treatment systems [67]. Prior work indicates that silver will more likely accumulate in the biosolids than in the WWTP effluent [68]. Here, considerably higher levels of silver in the digested sludge are found as compared to the thickened or dewatered sludge, suggesting that the digestion indeed concentrates the silver in the biosolid, but that the dewatering process shifts the silver to the liquid phase. The silver concentration in the digested hydrochar decreases with increasing carbonization. Both of these indicate that the silver is in an easily mobile state in the digested sludge, equilibrating with its aqueous phase to lower solid concentration.

Arsenic and vanadium were included in this analysis given their high prevalence in Italian drinking water sources; the acceptable limit for arsenic in the Trentino-Alto Adige region increases to 50 µg L^{-1} over the nationwide 10 µg L^{-1} limit due to natural lithology [69]. As seen in Figure 5, the arsenic concentrations in the solid samples decreased with increasing harshness for the thickened and digested samples (with the exception of the 190 °C and 250 °C, 60 min digested samples). All other hydrochars had arsenic concentrations below 2 µg g^{-1}. The International Biochar Initiative (IBI) [70] suggests an acceptable range for As in biochars of between 13 and 100 µg g^{-1} if they are to be applied as a soil amendment.

In Sicily, concentrations of V in drinking water sources are routinely above E.U. and U.S. maximum contaminant levels (MCLs) due to the underlying geology [71–73]. As such, it is important to ensure that concentrations of these metals were below MCLs before considering the potential for hydrochars to be land-applied, which could exacerbate the metal contamination issue. The vanadium concentration was slightly increased upon carbonization, but never exceeded 10 µgg^{-1}; the MCL for drinking water in Italy is 50 µgL^{-1}. V is known to co-precipitate with iron (III) [74]; while the oxidation states are not known here, the concentrations of iron in all solids exceeded 2 mgg^{-1} (therefore higher by several orders of magnitude), such that it is possible that V is present in the hydrochars in an iron (hydr)oxide precipitate.

As shown in Figure 5, the concentrations of chromium, cobalt, nickel, molybdenum, and lead (and a series of additional inorganics, as given in SI) are only modestly affected by carbonization and sludge withdrawal point. The cobalt concentration of the digested sludge was notably higher than the thickened or dewatered, but upon carbonization reached the same levels as the other samples of hydrochars, all at less than 3 µgg^{-1}. The concentrations of all of these metals are below IBI maximum allowable thresholds [70], making them reasonable candidates for use as soil amendments on the basis of heavy metal content.

3.3. Energy Content and Oxidative Reactivity: Potential for Use as Solid Fuel

As Table 2 shows, the solid hydrochars have similar higher heating values to their raw counterparts. In this case, HTC has not substantially improved the "energy density" of the solid fuel. However, HTC does enable a more efficient solid–liquid separation and dewaterability. For the raw dewatered sludge, CST was not complete after 60 min, while it was complete for the 190 °C 60 min hydrochar after only 380 s. With an increase in HTC temperature, CST was 209 and 90 s at 220 °C and 250 °C (60 min), respectively. Images of the CST trials are available in Supplementary Materials Figure S8. This improved dewaterability demonstrates a reduction of volume for transport of the solid waste and reduced moisture content for potential combustion applications.

Figure 6 plots the DTG curves of the hydrochar samples produced at each of the three carbonization temperatures at 30 min, alongside an Illinois No. 6 coal sample. As can be seen, the hydrochar sludge samples are considerably more reactive than the coal sample. Their peak DTG temperatures (highest conversion rates) occur at hundreds of degrees less than the coal sample and at higher conversion rates. The highest peak mass loss rates occur for all three hydrochars produced from the thickened sludge, prior to any anaerobic digestion. Post-digestion, while the shape of the DTG curves changes, the peak rates are quite similar for both the thickened and dewatered sludge. For all three samples, the 190 °C carbonized hydrochars display the highest reactivity compared to the other HTC temperatures. Especially at such mild conditions, HTC does not significantly carbonize the sample—oftentimes, the original materials microstructure is preserved, whereas higher temperatures lead to a more complete destruction of the carbon matrix [53,75].

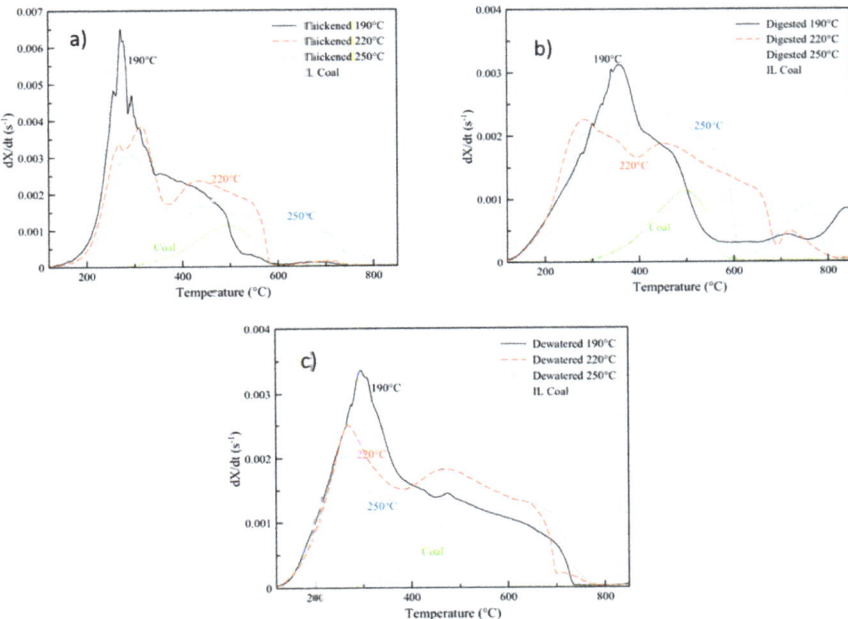

Figure 6. Derivative thermogravimetric curves of oxidation: (a) thickened sludge and hydrochars; (b) digested sludge and hydrochars; (c) dewatered sludge and hydrochars, produced at 30 min alongside Illinois No 6. Coal.

Given the relatively high reactivity of any of the sludge hydrochars, it may be difficult to combust them for electricity generation in current boilers designed for solid fuels such as coal [76]. While the higher heating values of especially thickened sludge (18–21 MJ kg^{-1}) are suitable for such combustion schemes, their lower ignition and peak reactivity temperatures are considerably lower than that of most bituminous coals, and therefore may lead to loss of efficiency in boilers [77]. That, combined with the higher ash content that could result in slagging and fouling, suggests that the sludge hydrochars may perform better in co-combustion scenarios [78].

It was previously shown that blending biofuels with similar characteristics at ratios less than 20 wt% with coal mitigates fuel segregation and efficiency loss issues while increasing the share of renewables in energy generation portfolios [48,49,76]. Recent work in the literature suggests that hydrochars can be co-combusted with a variety of coals in economically, environmentally, and energetically viable schemes in existing infrastructure [79–81] and may even improve the emissions profile at optimized blending ratios [82]. It was recently demonstrated that hydrochars with similar reactivities can be oxidized with Illinois No. 6 coal at ratios of 10 wt% hydrochar, balance coal, without causing significant fuel segregation [83]. In summation, the high reactivity and ash content (though high HHV) of the thickened and dewatered hydrochars temper enthusiasm for their use as a combustible fuel. The low HHV of the digested sludge—below low-rank coals—makes it difficult to envision a combustion scenario where this would be a valuable solid fuel. As such, the potential for using sludge-based hydrochars as a drop-in solid fuel is likely minimal, though this could be accomplished without the need for anaerobic digestion of the sludge if used in a co-fired fuel scenario.

3.4. Further Considerations

This work examined the impact of WWTP withdrawal point and hydrothermal carbonization conditions on resulting hydrochar properties. Table S3 in Supplementary Materials provides a summary of the overall results discussed. While there is considerable potential to use HTC to convert sewage sludge to renewable fuels and/or use the hydrochars for nutrient recovery redistribution, several questions about the feasibility of implementing such a process remain unanswered.

One question is what to do with the process water remaining after treatment, which contains unreacted feedstock and/or chemical intermediates that are potentially hazardous [84]. While this is beyond the scope of the present work, others have previously demonstrated several potential management options. Process liquors remaining after hydrothermal treatments have been shown to be suitable feedstocks for aerobic, anoxic, and anaerobic processes [85–87]. The experimental data of the present investigation testify that RBCOD (readily biodegradable COD) for HTC performed at 190 °C reached values above 85% of total COD, while increasing the temperature led to values below 15 % at 250 °C. While wet air oxidation liquors have shown some inhibitory behaviors for strictly anaerobic treatments [86], this has not been the case for some studies of HTC liquors [87,88], although others suggest methanogenesis inhibition at high inoculum concentrations (>25 $g_{COD}L^{-1}$) [89]. As demonstrated by Qiao and co-workers [90], HTC process water may actually increase the efficiency of the solid hydrolysis step, a rate-limiting step in the digestion process [91,92], and enhance methane production in the digester [93]. The use of nitrates as oxidants can catalyze COD and dissolved organic carbon removal [94], and, as it was recently demonstrated, hydrochars themselves can be used to enhance anaerobic digestion [95].

Second, the fate of some heavy metals during HTC, especially Cu and Cr, may complicate the use of these materials as a soil amendment, as successive WWT and carbonization concentrates them within the hydrochar. While it was not found here that heavy metal concentrations were above recommended IBI limits in the hydrochars, the concentrations of heavy metals in wastewater solids and effluents vary widely across the globe and can pose health risks in certain areas [96]. To address this potential concern, Shi and co-workers [97] demonstrated that Cd in sludge hydrochars can be immobilized by the synergistic nature of apatite P present in the chars and addition of hydroxyapatite. However, as both acid and alkaline leaching have been shown to dissolve phosphate in sludges and sludge ashes [98], the long-term stability of the Cd immobilized by the method of hydroxyapatite addition is not clear. Prior work by Yoshizaki and Tomida [99] demonstrated that such heavy metals could be removed by phosphoric acid and hydrogen peroxide for downstream reuse and recovery. Their method had enhanced recovery and was more environmentally and economically viable than treatment with hydrochloric acid or sulfuric acid, opening a potential pathway for extraction of the metals from hydrochars. Such treatment would likely act as a porogen to increase the surface area of the hydrochars [100], increasing their ability to retain water and slow-release nutrients when used as a solid amendment [101], and may open up possibilities for conversion of the materials to activated carbons for use in water treatment, battery electrodes, and other high-value materials [102–105].

4. Conclusions

The present study probed the impact of hydrothermal processing conditions and sewage sludge withdrawal point on resulting hydrochars. The hydrothermal carbonization (HTC) of sludge proceeds similarly to many other wet biomasses; as harshness (time and temperature) of the process increases, the solid yield decreases, the ash (inorganic) content increases, total COD decreases but soluble COD increases, and solid pH decreases. However, there are distinct differences between the hydrochars produced from primary (thickened) versus secondary (digested and dewatered) sludge. Thickened sludge carbonized at moderate conditions (220 °C, 30 min) produced the most viable solid fuel with the highest HHV, moderate ash content, and high volatile matter content. However, with O:C and H:C ratios higher than typical bituminous coals of similar heating content, the thermal reactivity of the hydrochar was significantly higher than coals typically combusted. This suggests that sludge hydrochars could be co-fired with coal but are not ideal solid fuels. On the other hand, hydrochars produced from secondary sludge are more viable as potential soil amendments. The carbonized digested sludges show relatively neutral pH, low COD, and enhanced phosphorous, along with enhanced Ca, Mg, and Al concentrations to help mobilize P. Their heavy metal composition is well below International Biochar Initiative standards, though the elemental oxygen content and lower volatile matter content warrant future inquiry into this pathway.

Supplementary Materials: The following are available online at http://www.mdpi.com/1996-1073/13/11/2890/s1, Figure S1: Product distribution among solid, gas and liquid (by difference) phases following hydrothermal carbonization of three sludge samples, Figure S2: Impact of hydrothermal carbonization on total COD (top) and soluble COD (bottom) of three sludge samples, Figure S3: pH of raw and hydrothermally carbonized sludge samples, Figure S4: Proximate analysis of hydrothermally carbonized sludge samples with black circles indicating fixed carbon, red squares are volatile matter, and blue plus symbols are ash content (a: Thickened sludge, 30 min HTC; b: Thickened sludge, 60 min HTC; c: Digested sludge, 30 min HTC; d: Digested sludge, 60 min HTC; e: Dewatered sludge, 30 min HTC; f: Dewatered sludge, 60 min HTC), Figure S5: Ultimate analysis of hydrothermally carbonized sludge samples, Figure S6: Inorganic element distribution in raw sludge samples with: a high concentrations and b: low concentrations; error bars indicate 95% confidence interval, FigureS7: Inorganic elements as a function of sludge sample and carbonization conditions, Figure S8: Visual evidence of hydrochar dewaterability, Table S1: ICP-OES analysis of inorganics present in three sludge feedstocks, Table S2: ICP-OES analysis of inorganics present in hydrothermally carbonized sludge samples, Table S3: Summary of observations of sludge hydrochars.

Author Contributions: Conceptualization, J.G., G.A. and L.F.; methodology, F.M., J.G., G.A. and L.F.; validation, F.M., M.V. and T.M.; formal analysis, F.M., J.G., T.M. and M.V.; investigation, F.M.; resources, J.G., G.A. and L.F.; data curation, F.M., J.G. and M.V.; writing—original draft preparation, F.M. and J.G.; writing—review and editing, J.G., M.V. and L.F.; visualization, F.M. and J.G.; supervision, G.A. and L.F.; project administration, G.A. and L.F.; funding acquisition, G.A. and L F. All authors have read and agreed to the published version of the manuscript.

Funding: This work was partially supported by Atzwanger Spa http://www.atzwanger.net/.

Acknowledgments: The authors appreciate the help of Lihui Gao and Giulia Ischia in running the thermogravimetric analysis and of Fabio Valentinuzzi and Stefano Cesco for inductively coupled plasma-optical emission spectroscopy. J. Goldfarb acknowledges support of the U.S.-Italy Fulbright Commission.

Conflicts of Interest: The authors declare no conflict of interest. The funders had no role in the design of the study; in the collection, analyses, or interpretation of data; in the writing of the manuscript, or in the decision to publish the results.

References

1. Steichen, R. Council Directive of 21 May 1991 concerning urban waste water treatment (91/271/EEC). *Off. J. Eur. Communities* **1991**. Available online: https://www.tarimorman.gov.tr/SYGM/Belgeler/ab%20mevzuat%C4%B1/91-271-EEC.pdf (accessed on 4 June 2020).
2. Bianchini, A.; Bonfiglioli, L.; Pellegrini, M.; Saccani, C. Sewage sludge management in Europe: A critical analysis of data quality. *Int. J. Environ. Waste Manag.* **2016**, *18*, 226. [CrossRef]
3. EUROSTAT. *Sewage Sludge Production and Disposal from Urban Wastewater (in Dry Substance (d.s))*. 2017. Available online: https://data.europa.eu/euodp/en/data/dataset/hzWkcfKt5mxEaFijeoA (accessed on 4 June 2020).
4. Zhang, Q.; Hu, J.; Lee, D.; Chang, Y.; Lee, Y. Sludge treatment: Current research trends. *Bioresour. Technol.* **2017**, *243*, 1159–1172. [CrossRef]
5. European Commission. *Eighth Report on the Implementation Status and the Programmes for Implementation (as required by Article 17) of Council Directive 91/271/EEC Concerning Urban Waste Water Treatment*; European Commission: Brussels, Belgium, 2016.
6. Yanagida, T.; Fujimoto, S.; Minowa, T. Application of the severity parameter for predicting viscosity during hydrothermal processing of dewatered sewage sludge for a commercial PFBC plant. *Bioresour. Technol.* **2010**, *101*, 2043–2045. [CrossRef]

7. Campbell, H.W.; Pacific, C.; Technologies, E. Sludge management—Future issues and trends. *Water Sci. Technol.* **2000**, *41*, 1–8. [CrossRef]
8. European Commission. *Disposal and Recycling Routes for Sewage Sludge Part 1—Sludge Use Acceptance Report*; European Commission: Brussels, Belgium, 2001; ISBN 9289417986.
9. Rulkens, W. Sewage Sludge as a Biomass Resource for the Production of Energy: Overview and Assessment of the Various Options. *Energy Fuels* **2008**, *44*, 9–15. [CrossRef]
10. Tyagi, V.K.; Lo, S.L. Sludge: A waste or renewable source for energy and resources recovery? *Renew. Sustain. Energy Rev.* **2013**, *25*, 708–728. [CrossRef]
11. Cao, Y.; Pawłowski, A. Sewage sludge-to-energy approaches based on anaerobic digestion and pyrolysis: Brief overview and energy efficiency assessment. *Renew. Sustain. Energy Rev.* **2012**, *16*, 1657–1665. [CrossRef]
12. Merzari, F.; Langone, M.; Andreottola, G.; Fiori, L. Methane production from process water of sewage sludge hydrothermal carbonization. A review. Valorising sludge through hydrothermal carbonization. *Crit. Rev. Environ. Sci. Technol.* **2019**, *49*, 947–988. [CrossRef]
13. Johansson, K.; Perzon, M.; Fröling, M.; Mossakowska, A.; Svanström, M. Sewage sludge handling with phosphorus utilization—Life cycle assessment of four alternatives. *J. Clean. Prod.* **2008**, *16*, 135–151. [CrossRef]
14. Merzari, F.; Lucian, M.; Volpe, M.; Andreottola, G.; Fiori, L. Hydrothermal carbonization of biomass: Design of a bench-Scale reactor for evaluating the heat of reaction. *Chem. Eng. Trans.* **2018**, *65*, 43–48.
15. Akiya, N.; Savage, P.E. Roles of water for chemical reactions in high-temperature water. *Chem. Rev.* **2002**, *102*, 2725–2750. [CrossRef] [PubMed]
16. Kritzer, P. Corrosion in high-temperature and supercritical water and aqueous solutions: A review. *J. Supercrit. Fluids* **2004**, *29*, 1–29. [CrossRef]
17. Lucian, M.; Volpe, M.; Gao, L.; Piro, G.; Goldfarb, J.L.; Fiori, L. Impact of hydrothermal carbonization conditions on the formation of hydrochars and secondary chars from the organic fraction of municipal solid waste. *Fuel* **2018**, *233*, 257–268. [CrossRef]
18. Lucian, M.; Fiori, L. Hydrothermal carbonization of waste biomass: Process design, modeling, energy efficiency and cost analysis. *Energies* **2017**, *10*, 211. [CrossRef]
19. Volpe, M.; Messineo, A.; Mäkelä, M.; Barr, M.R.; Volpe, R.; Corrado, C.; Fiori, L. Reactivity of cellulose during hydrothermal carbonization of lignocellulosic biomass. *Fuel Process. Technol.* **2020**, *206*, 106456. [CrossRef]
20. Volpe, M.; Fiori, L.; Volpe, R.; Messineo, A. Upgrading of Olive Tree Trimmings Residue as Biofuel by Hydrothermal Carbonization and Torrefaction: A Comparative Study. *Chem. Eng. Trans.* **2016**, *50*, 13–18.
21. Mäkelä, M.; Volpe, M.; Volpe, R.; Fiori, L.; Dahl, O. Spatially resolved spectral determination of polysaccharides in hydrothermally carbonized biomass. *Green Chem.* **2018**, *20*, 1114–1120. [CrossRef]
22. Lucian, M.; Volpe, M.; Fiori, L. Hydrothermal Carbonization Kinetics of Lignocellulosic Agro-Wastes: Experimental Data and Modeling. *Energies* **2019**, *12*, 516. [CrossRef]
23. Zhao, P.; Shen, Y.; Ge, S.; Yoshikawa, K. Energy recycling from sewage sludge by producing solid biofuel with hydrothermal carbonization. *Energy Convers. Manag.* **2014**, *78*, 815–821. [CrossRef]
24. Danso-Boateng, E.; Holdich, R.G.; Wheatley, A.D.; Martin, S.J.; Shama, G. Hydrothermal Carbonization of Primary Sewage Sludge and Synthetic Faeces: Effect of Reaction Temperature and Time on Filterability. *Environ. Prog. Sustain. Energy* **2015**, *34*, 1279–1290. [CrossRef]
25. Khalil, W.A.S.; Shanableh, A.; Rigby, P.; Kokot, S. Selection of hydrothermal pre-treatment conditions of waste sludge destruction using multicriteria decision-making. *J. Environ. Manag.* **2005**, *75*, 53–64. [CrossRef]
26. Catallo, W.J.; Comeaux, J.L. Reductive hydrothermal treatment of sewage sludge. *Waste Manag.* **2008**, *28*, 2213–2219. [CrossRef] [PubMed]
27. Zhao, P.; Chen, H.; Ge, S.; Yoshikawa, K. Effect of the hydrothermal pretreatment for the reduction of NO emission from sewage sludge combustion. *Appl. Energy* **2015**, *111*, 199–205. [CrossRef]
28. Becker, G.C.; Wüst, D.; Köhler, H.; Lautenbach, A.; Kruse, A. Novel approach of phosphate-reclamation as struvite from sewage sludge by utilising hydrothermal carbonization. *J. Environ. Manag.* **2019**, *238*, 119–125. [CrossRef] [PubMed]
29. Shi, Y.; Luo, G.; Rao, Y.; Chen, H.; Zhang, S. Hydrothermal conversion of dewatered sewage sludge: Focusing on the transformation mechanism and recovery of phosphorus. *Chemosphere* **2019**, *228*, 619–628. [CrossRef]
30. Song, E.; Park, S.; Kim, H. Upgrading Hydrothermal Carbonization (HTC) Hydrochar from Sewage Sludge. *Energies* **2019**, *12*, 2383. [CrossRef]

31. Danso-Boateng, E.; Sharma, G.; Wheatley, A.D.; Martin, S.J.; Holdich, R.G. Hydrothermal carbonisation of sewage sludge: Effect of process conditions on product characteristics and methane production. *Bioresour. Technol.* **2015**, *177*, 318–327. [CrossRef]
32. Brookman, H.; Gievers, F.; Zelinski, V.; Ohlert, J.; Loewen, A. Influence of Hydrothermal Carbonization on Composition, Formation and Elimination of Biphenyls, Dioxins and Furans in Sewage Sludge. *Energies* **2018**, *11*, 1582. [CrossRef]
33. Breulmann, M.; Schulz, E.; Van Afferden, M.; Müller, R.A. Hydrochars derived from sewage sludge: Effects of pre-treatment with water on char properties, phytotoxicity and chemical structure. *Arch. Agron. Soil Sci.* **2017**, *64*, 860–872. [CrossRef]
34. Liu, T.; Liu, Z.; Zheng, Q.; Lang, Q.; Xia, Y.; Peng, N. Effect of hydrothermal carbonization on migration and environmental risk of heavy metals in sewage sludge during pyrolysis. *Bioresour. Technol.* **2018**, *247*, 282–290. [CrossRef]
35. Chen, C.; Liu, G.; An, Q.; Lin, L.; Shang, Y.; Wan, C. From wasted sludge to valuable biochar by low temperature hydrothermal carbonization treatment: Insight into the surface characteristics. *J. Clean. Prod.* **2020**, *263*, 1–9. [CrossRef]
36. Zhai, Y.; Peng, C.; Xu, B.; Wang, T.; Li, C. Hydrothermal carbonisation of sewage sludge for char production with different waste biomass: Effects of reaction temperature and energy recycling. *Energy* **2017**, *127*, 167–174. [CrossRef]
37. Wang, R.; Wang, C.; Zhao, Z.; Jia, J.; Jin, Q. Energy recovery from high-ash municipal sewage sludge by hydrothermal carbonization: Fuel characteristics of biosolid products. *Energy* **2019**, *186*, 115848. [CrossRef]
38. Zheng, X.; Jiang, Z.; Ying, Z.; Song, J.; Chen, W.; Wang, B. Role of feedstock properties and hydrothermal carbonization conditions on fuel properties of sewage sludge-derived hydrochar using multiple linear regression technique. *Fuel* **2020**, *271*, 1–11. [CrossRef]
39. Xu, Z.-X.; Song, H.; Li, P.-J.; He, Z.-X.; Wang, Q.; Wang, K.; Duan, P.-G. Hydrothermal carbonization of sewage sludge: Effect of aqueous phase recycling. *Chem. Eng. J.* **2020**, *387*, 1–12. [CrossRef]
40. Vardon, D.R.; Sharma, B.K.; Scott, J.; Yu, G.; Wang, Z.; Schideman, L.; Zhang, Y.; Strathmann, T.J. Chemical properties of biocrude oil from the hydrothermal liquefaction of Spirulina algae, swine manure, and digested anaerobic sludge. *Bioresour. Technol.* **2011**, *102*, 8295–8303. [CrossRef]
41. Hao, X.D.; Li, J.; van Loosdrecht, M.C.M.; Li, T.Y. A sustainability-based evaluation of membrane bioreactors over conventional activated sludge processes. *J. Environ. Chem. Eng.* **2018**, *6*, 2597–2605. [CrossRef]
42. Manser, R.; Gujer, W.; Siegrist, H. Membrane bioreactor versus conventional activated sludge system: Population dynamics of nitrifiers. *Water Sci. Technol.* **2005**, *52*, 417–425. [CrossRef]
43. APHA. *Standard Methods for the Examination of Water and Wastewater*, 22th ed.; American Public Health Association: Washington, DC, USA, 2012.
44. Fiori, L.; Basso, D.; Castello, D.; Baratieri, M. Hydrothermal carbonization of biomass: Design of a batch reactor and preliminary experimental results. *Chem. Eng. Trans.* **2014**, *37*, 55–60.
45. Basso, D.; Weiss-Hortala, E.; Patuzzi, F.; Castello, D.; Baratieri, M.; Fiori, L. Hydrothermal carbonization of off-specification compost: A byproduct of the organic municipal solid waste treatment. *Bioresour. Technol.* **2015**, *182*, 217–224. [CrossRef]
46. Basso, D.; Patuzzi, F.; Castello, D.; Baratieri, M.; Rada, C.E.; Weiss-Hortala, E.; Fiori, L. Agro-industrial waste to solid biofuel through hydrothermal carbonization. *Waste Manag.* **2016**, *47*, 114–121. [CrossRef] [PubMed]
47. Hunt, J.E. *Argonne Premium Coal Sample Program*; Argonne: Lemont, IL, USA, 2007. Available online: https://publications.anl.gov/anlpubs/2007/04/58856.pdf (accessed on 4 June 2020).
48. Celaya, A.A.M.; Lade, A.T.A.; Goldfarb, J.J.L. Co-combustion of brewer's spent grains and Illinois No. 6 coal: Impact of blend ratio on pyrolysis and oxidation behavior. *Fuel Process. Technol.* **2015**, *129*, 39–51. [CrossRef]
49. Xue, J.; Chellappa, T.; Ceylan, S.; Goldfarb, J.L. Enhancing biomass + coal Co-firing scenarios via biomass torrefaction and carbonization: Case study of avocado pit biomass and Illinois No. 6 coal. *Renew. Energy* **2018**, *122*, 152–162. [CrossRef]
50. Andreottola, G.; Foladori, P.; Ferrai, M.; Ziglio, G. *Respirometria Applicata Alla Depurazione Delle Acque; Principi e metodi*; Department of Civil, Environmental and Mechanical Engineering, University of Trento: Trento, Italy, 2002.

51. Ren, W.; Zhou, Z.; Jiang, L.M.; Hu, D.; Qiu, Z.; Wei, H.; Wang, L. A cost-effective method for the treatment of reject water from sludge dewatering process using supernatant from sludge lime stabilization. *Sep. Purif. Technol.* **2015**, *142*, 123–128. [CrossRef]
52. Al Momani, F.A.; Örmeci, B. Optimization of polymer dose based on residual polymer concentration in dewatering supernatant. *Water Air Soil Pollut.* **2014**, *225*. [CrossRef]
53. Volpe, M.; Goldfarb, J.L.; Fiori, L. Hydrothermal carbonization of Opuntia ficus indica cladodes: Role of process parameters on hydrochar properties. *Bioresour. Technol.* **2018**, *247*, 310–318. [CrossRef]
54. Jomaa, S.; Shanableh, A.; Khalil, W.; Trebilco, B. Hydrothermal decomposition and oxidation of the organic component of municipal and industrial waste products. *Adv. Environ. Res.* **2003**, *7*, 647–653. [CrossRef]
55. Kocsisová, T.; Juhasz, J.; Cvengroš, J. Hydrolysis of fatty acid esters in subcritical water. *Eur. J. Lipid Sci. Technol.* **2006**, *108*, 652–658. [CrossRef]
56. Díaz, E.; Pintado, L.; Faba, L.; Ordóñez, S.; González-LaFuente, J.M. Effect of sewage sludge composition on the susceptibility to spontaneous combustion. *J. Hazard. Mater.* **2019**, *361*, 267–272. [CrossRef]
57. Berge, N.D.; Ro, K.S.; Mao, J.; Flora, J.R.V.; Chappell, M.A.; Bae, S. Hydrothermal Carbonization of Municipal Waste Streams. *Environ. Sci. Technol.* **2011**, *45*, 5696–5703. [CrossRef]
58. Kruse, A.; Koch, F.; Stelzl, K.; Wüst, D.; Zeller, M. Fate of Nitrogen during Hydrothermal Carbonization. *Energy Fuels* **2016**, *30*, 8037–8042. [CrossRef]
59. Volpe, M.; Fiori, L. From olive waste to solid biofuel through hydrothermal carbonisation: The role of temperature and solid load on secondary char formation and hydrochar energy properties. *J. Anal. Appl. Pyrolysis* **2017**, *124*, 63–72. [CrossRef]
60. Volpe, M.; Wüst, D.; Merzari, F.; Lucian, M.; Andreottola, G.; Kruse, A.; Fiori, L. One stage olive mill waste streams valorisation via hydrothermal carbonisation. *Waste Manag.* **2018**, *80*, 224–234. [CrossRef] [PubMed]
61. Ekpo, U.; Ross, A.B.; Camargo-Valero, M.A.; Williams, P.T. A comparison of product yields and inorganic content in process streams following thermal hydrolysis and hydrothermal processing of microalgae, manure and digestate. *Bioresour. Technol.* **2016**, *200*, 951–960. [CrossRef] [PubMed]
62. Ferrentino, R.; Ceccato, R.; Marchetti, V.; Andreottola, G.; Fiori, L. Sewage Sludge Hydrochar: An Option for Removal of Methylene Blue from Wastewater. *Appl. Sci.* **2020**, *10*, 3445. [CrossRef]
63. Zornoza, R.; Moreno-Barriga, F.; Acosta, J.A.; Muñoz, M.A.; Faz, A. Stability, nutrient availability and hydrophobicity of biochars derived from manure, crop residues, and municipal solid waste for their use as soil amendments. *Chemosphere* **2016**, *144*, 122–130. [CrossRef]
64. Wu, H.; Lai, C.; Zeng, G.; Liang, J.; Chen, J.; Xu, J.; Dai, J.; Li, X.; Liu, J.; Chen, M.; et al. The interactions of composting and biochar and their implications for soil amendment and pollution remediation: A review. *Crit. Rev. Biotechnol.* **2017**, *37*, 754–764. [CrossRef]
65. Melo, T.M.; Bottlinger, M.; Schulz, E.; Leandro, W.M.; Filho, A.M.D.A.; Wang, H.; Ok, Y.S.; Rinklebe, J. Plant and soil responses to hydrothermally converted sewage sludge (sewchar). *Chemosphere* **2018**, *206*, 338–348. [CrossRef] [PubMed]
66. Thomsen, P.T.; Zsuzsa, S.; Ahrenfeldt, J.; Henriksen, U.B.; Frandsen, F.J.; Müller-st, D.S. Changes imposed by pyrolysis, thermal gasification and incineration on composition and phosphorus fertilizer quality of municipal sewage sludge. *J. Environ. Manag.* **2017**, *198*, 308–318. [CrossRef]
67. Dwivedi, A.D.; Dubey, S.P.; Sillanp, M.; Kwon, Y.N.; Lee, C.; Varma, R.S. Fate of engineered nanoparticles: Implications in the environment. *Coord. Chem. Rev.* **2015**, *287*, 64–78. [CrossRef]
68. Wang, Y.; Westerhoff, P.; Hristovski, K.D. Fate and biological effects of silver, titanium dioxide, and C60(fullerene) nanomaterials during simulated wastewater treatment processes. *J. Hazard. Mater.* **2012**, *201–202*, 16–22. [CrossRef] [PubMed]
69. Dinelli, E.; Lima, A.; Albanese, S.; Birke, M.; Cicchella, D.; Giaccio, L.; Valera, P.; De Vivo, B. Major and trace elements in tap water from Italy. *J. Geochem. Explor.* **2012**, *112*, 54–75. [CrossRef]
70. International Biochar Initiative. *Standardized Product Definition and Product Testing Guidelines for Biochar That Is Used in Soil.* 2015. Available online: https://www.biochar-international.org/wp-content/uploads/2018/04/IBI_Biochar_Standards_V2.1_Final.pdf (accessed on 4 June 2020).
71. Giammanco, S.; Valenza, M.; Pignato, S.; Giammanco, G. Mg, Mn, Fe, and V Concentration in the Ground Waters of Mount Etna (Sicily). *Water Res.* **1996**, *30*, 378–386. [CrossRef]
72. Roccaro, P.; Barone, C.; Mancini, G.; Vagliasindi, F.G.A. Removal of manganese from water supplies intended for human consumption: A case study. *Desalination* **2007**, *210*, 205–214. [CrossRef]

73. Varrica, D.; Tamburo, E.; Dongarrà, G. Sicilian bottled natural waters: Major and trace inorganic components. *Appl. Geochem.* **2013**, *34*, 102–113. [CrossRef]
74. Roccaro, P.; Vagliasindi, F.G.A. Coprecipitation of vanadium with iron(III) in drinking water: A pilot-scale study. *Desalin. Water Treat.* **2015**, *55*, 799–809. [CrossRef]
75. Kruse, A.; Funke, A.; Titirici, M.-M. Hydrothermal conversion of biomass to fuels and energetic materials. *Curr. Opin. Chem. Biol.* **2013**, *17*, 515–521. [CrossRef]
76. Goldfarb, J.L.; Liu, C. Impact of blend ratio on the co-firing of a commercial torrefied biomass and coal via analysis of oxidation kinetics. *Bioresour. Technol.* **2013**, *149*, 208–215. [CrossRef]
77. Khan, A.A.; de Jong, W.; Jansens, P.J.; Spliethoff, H. Biomass combustion in fluidized bed boilers: Potential problems and remedies. *Fuel Process. Technol.* **2009**, *90*, 21–50. [CrossRef]
78. Haykiri-Acma, H.; Yaman, S.; Kucukbayrak, S. Does carbonization avoid segregation of biomass and lignite during co-firing? Thermal analysis study. *Fuel Process. Technol.* **2015**, *137*, 312–319. [CrossRef]
79. Cartmell, E.; Gostelow, P.; Riddell-black, D.; Simms, N.; Oakey, J.; Morris, J.O.E.; Jeffrey, P.; Howsam, P. Biosolids-A Fuel or a Waste? An Integrated Appraisal of Five Co-combustion Scenarios with Policy Analysis. *Environ. Sci. Technol.* **2006**, *40*, 649–658. [CrossRef]
80. Xiao, H.; Ma, X.; Liu, K. Co-combustion kinetics of sewage sludge with coal and coal gangue under different atmospheres. *Energy Convers. Manag.* **2010**, *51*, 1976–1980. [CrossRef]
81. Otero, M.; Calvo, L.F.; Gil, M.V.; Garcia, A.I.; Moran, A. Co-combustion of different sewage sludge and coal: A non-isothermal thermogravimetric kinetic analysis. *Bioresour. Technol.* **2008**, *99*, 6311–6319. [CrossRef]
82. Parshetti, G.K.; Liu, Z.; Jain, A.; Srinivasan, M.P.; Balasubramanian, R. Hydrothermal carbonization of sewage sludge for energy production with coal. *Fuel* **2013**, *111*, 201–210. [CrossRef]
83. Gao, L.; Volpe, M.; Lucian, M.; Fiori, L.; Goldfarb, J.L. Does hydrothermal carbonization as a biomass pretreatment reduce fuel segregation of coal-biomass blends during oxidation? *Energy Convers. Manag.* **2019**, *181*, 93–104. [CrossRef]
84. Fregolente, L.G.; Miguel, T.B.A.R.; de Castro Miguel, E.; de Almeida Melo, C.; Moreira, A.B.; Ferreira, O.P.; Bisinoti, M.C. Toxicity evaluation of process water from hydrothermal carbonization of sugarcane industry by-products. *Environ. Sci. Pollut. Res.* **2019**, *26*, 27579–27589. [CrossRef] [PubMed]
85. Barlindhaug, J.; Ødegaard, H. Thermal hydrolysis for the production of carbon source for denitrification. *Water Sci. Technol.* **1996**, *34*, 371–378. [CrossRef]
86. Friedman, A.A.; Smith, J.E.; DeSantis, J.; Ptak, T.; Ganley, R.C. Characteristics of Residues from Wet Air Oxidation of Anaerobic Sludges. *J. (Water Pollut. Control Fed.)* **1988**, *60*, 1971–1978.
87. Wirth, B.; Reza, T.; Mumme, J. Influence of digestion temperature and organic loading rate on the continuous anaerobic treatment of process liquor from hydrothermal carbonization of sewage sludge. *Bioresour. Technol.* **2015**, *198*, 215–222. [CrossRef]
88. Aragón-briceño, C.; Ross, A.B.; Camargo-valero, M.A. Evaluation and comparison of product yields and bio-methane potential in sewage digestate following hydrothermal treatment. *Appl. Energy* **2017**, *208*, 1357–1369. [CrossRef]
89. Villamil, J.A.; Mohedano, A.F.; Rodriguez, J.J.; de la Rubia, M.A. Valorisation of the liquid fraction from hydrothermal carbonisation of sewage sludge by anaerobic digestion. *J. Chem. Technol. Biotechnol.* **2018**, *93*, 450–456. [CrossRef]
90. Qiao, W.; Peng, C.; Wang, W.; Zhang, Z. Biogas production from supernatant of hydrothermally treated municipal sludge by upflow anaerobic sludge blanket reactor. *Bioresour. Technol.* **2011**, *21*, 9904–9911. [CrossRef]
91. Appels, L.; Dewil, R.; Baeyens, J.; Degreve, J. Ultrasonically enhanced anaerobic digestion of waste activated sludge. *Int. J. Sustain. Eng.* **2008**, *1*, 94–104. [CrossRef]
92. Vavilin, V.A.; Rytov, S.V.; Lokshina, L.Y. A Description of Hydrolysis Kinetics in Aanaerobic Degradation of Particulate Organic Matter. *Bioresour. Technol.* **1996**, *56*, 229–237. [CrossRef]
93. Nuchdang, S.; Frigon, J.; Roy, C.; Pilon, G.; Phalakornkule, C.; Guiot, S.R. Hydrothermal post-treatment of digestate to maximize the methane yield from the anaerobic digestion of microalgae. *Waste Manag.* **2018**, *71*, 683–688. [CrossRef] [PubMed]
94. Stutzenstein, P.; Weiner, B.; Köhler, R.; Pfeifer, C.; Kopinke, F.D. Wet oxidation of process water from hydrothermal carbonization of biomass with nitrate as oxidant. *Chem. Eng. J.* **2018**, *339*, 1–6. [CrossRef]

95. Luz, F.C.; Volpe, M.; Fiori, L.; Manni, A.; Cordiner, S.; Mulone, V.; Rocco, V. Spent Coffee Enhanced Biomethane Potential via an Integrated Hydrothermal Carbonization-Anaerobic Digestion Process. *Bioresour. Technol.* **2018**, *256*, 102–109.
96. Singh, A.; Sharma, R.K.; Agrawal, M.; Marshall, F.M. Risk assessment of heavy metal toxicity through contaminated vegetables from waste water irrigated area of Varanasi, India. *Trop. Ecol.* **2010**, *51*, 375–387.
97. Shi, L.; Zhang, G.; Wei, D.; Yan, T.; Xue, X.; Shi, S.; Wei, Q. Preparation and utilization of anaerobic granular sludge-based biochar for the adsorption of methylene blue from aqueous solutions. *J. Mol. Liq.* **2014**, *198*, 334–340. [CrossRef]
98. Stark, K.; Plaza, E.; Hultman, B. Phosphorus release from ash, dried sludge and sludge residue from supercritical water oxidation by acid or base. *Chemosphere* **2006**, *62*, 827–832. [CrossRef]
99. Yoshizaki, S.; Tomida, T. Principle and Process of Heavy Metal Removal from Sewage Sludge. *Environ. Sci. Technol.* **2000**, *34*, 1572–1574. [CrossRef]
100. Hotova, G.; Slovak, V.; Soares, O.S.G.P.; Figueiredo, J.L.; Pereira, M.F.R. Oxygen surface groups analysis of carbonaceous samples pyrolysed at low temperature. *Carbon* **2018**, *134*, 255–263. [CrossRef]
101. Liang, B.; Lehmann, J.; Solomon, D.; Kinyangi, J.; Grossman, J.; Skjemstad, J.O.; Thies, J.; Luiza, F.J.; Petersen, J.; Neves, E.G. Black Carbon Increases Cation Exchange Capacity in Soils. *Soil Sci. Soc. Am. J.* **2006**, 1719–1730. [CrossRef]
102. Goldfarb, J.L.; Dou, G.; Salari, M.; Grinstaff, M.W. Biomass-Based Fuels and Activated Carbon Electrode Materials: An Integrated Approach to Green Energy Systems. *ACS Sustain. Chem. Eng.* **2017**, *5*, 3046–3054. [CrossRef]
103. Isitan, S.; Ceylan, S.; Topcu, Y.; Hintz, C.; Tefft, J.; Chellappa, T.; Guo, J.; Goldfarb, J.L. Product quality optimization in an integrated biorefinery: Conversion of pistachio nutshell biomass to biofuels and activated biochars via pyrolysis. *Energy Convers. Manag.* **2016**, *127*, 576–588. [CrossRef]
104. Goldfarb, J.L.; Buessing, L.; Gunn, E.; Lever, M.; Billias, A.; Casoliba, E.; Schievano, A.; Adani, F. Novel Integrated Biorefinery for Olive Mill Waste Management: Utilization of Secondary Waste for Water Treatment. *ACS Sustain. Chem. Eng.* **2017**, *5*, 876–884. [CrossRef]
105. Gopu, C.; Gao, L.; Volpe, M.; Fiori, L.; Goldfarb, J.L. Valorizing municipal solid waste: Waste to energy and activated carbons for water treatment via pyrolysis. *J. Anal. Appl. Pyrolysis* **2018**, *133*, 48–58. [CrossRef]

© 2020 by the authors. Licensee MDPI, Basel, Switzerland. This article is an open access article distributed under the terms and conditions of the Creative Commons Attribution (CC BY) license (http://creativecommons.org/licenses/by/4.0/).

Article

Experimental and Computational Evaluation of Heavy Metal Cation Adsorption for Molecular Design of Hydrothermal Char

Louise Delahaye [1,†], **John Thomas Hobson** [1,†], **Matthew Peter Rando** [1], **Brenna Sweeney** [1], **Avery Bernard Brown** [1], **Geoffrey Allen Tompsett** [1], **Ayten Ates** [2], **N. Aaron Deskins** [1] and **Michael Thomas Timko** [1,*]

1. Department of Chemical Engineering, Worcester Polytechnic Institute, Worcester, MA 01609, USA; lcdelahaye@wpi.edu (L.D.); jthobson@wpi.edu (J.T.H.); mprando@wpi.edu (M.P.R.); bes117@pitt.edu (B.S.); abbrown@wpi.edu (A.B.B.); gtompsett@wpi.edu (G.A.T.); nadeskins@wpi.edu (N.A.D.)
2. Department of Chemical Engineering, Engineering Faculty, Sivas Cumhuriyet University, 58140 Sivas, Turkey; aytates@gmail.com
* Correspondence: mttimko@wpi.edu
† These authors contributed equally to this work.

Received: 16 July 2020; Accepted: 10 August 2020; Published: 14 August 2020

Abstract: A model hydrochar was synthesized from glucose at 180 °C and its Cu(II) sorption capacity was studied experimentally and computationally as an example of molecular-level adsorbent design. The sorption capacity of the glucose hydrochar was less than detection limits (3 mg g^{-1}) and increased significantly with simple alkali treatments with hydroxide and carbonate salts of K and Na. Sorption capacity depended on the salt used for alkali treatment, with hydroxides leading to greater improvement than carbonates and K$^+$ more than Na$^+$. Subsequent zeta potential and infrared spectroscopy analysis implicated the importance of electrostatic interactions in Cu(II) sorption to the hydrochar surface. Computational modeling using Density Functional Theory (DFT) rationalized the binding as electrostatic interactions with carboxylate groups; similarly, DFT calculations were consistent with the finding that K$^+$ was more effective than Na$^+$ at activating the hydrochar. Based on this finding, custom-synthesized hydrochars were synthesized from glucose-acrylic acid and glucose-vinyl sulfonic acid precursors, with subsequent improvements in Cu(II) adsorption capacity. The performance of these hydrochars was compared with ion exchange resins, with the finding that Cu(II)-binding site stoichiometry is superior in the hydrochars compared with the resins, offering potential for future improvements in hydrochar design.

Keywords: hydrochar; alkali treatment; copper ions; adsorption; computational

---

1. Introduction

According to the World Health Organization (WHO), approximately 785 million people lack access to clean drinking water, mainly in poor countries [1], but also in some rural and even highly urbanized areas [2] in the developed world. Heavy metal contamination of the water supply is a persistent problem that dates back to antiquity [3], and some researchers speculate that lead contamination of the water supply may have played a role in the downfall of the Roman empire [3]. More recently, Fernández-Luqueño et al. [4] summarized the health effects of heavy metals, listing their contributions to disease ranging from cancer to lung failure. Despite widespread acknowledgement of these negative human health outcomes, providing uniform access to drinking water free of heavy metal contamination has proven remarkably difficult to achieve, as evidenced by recent widely reported examples [2].

Different technologies have been developed to remove heavy metals from water [5] including precipitation [5–7], sedimentation [8], flotation [8,9], membrane processes [8,10–12], electrochemical processes [13–15], adsorption [16,17], and ion exchange [18,19]. Most of these technologies are either costly, wasteful, reliant on non-renewable resources, energy inefficient, or unable to achieve sufficiently low levels of metal concentration on their own. Of the available options, adsorption and the kindred technique of ion exchange are energy efficient and capable of achieving suitably low metal concentrations [20]. However, commercial adsorbents and ion exchange resins are derived from non-renewable resources, either petrochemicals or coal, meaning that their use has associated negative environmental impacts, creating a tradeoff between clean drinking water and mitigating climate change [21]. Accordingly, development of renewable, cost-effective, and high-capacity metal adsorbates has potential to greatly expand access to clean drinking water while minimizing other negative environmental impacts [22,23].

Recently, pyrolysis biochar has emerged as a renewable sorbent for heavy metal removal [24,25]. Unlike most activated carbons, which are produced from coal, pyrolysis biochar is produced from biomass or agricultural wastes [26]. In some cases, heavy metal capacity on biochar can reach 40 mg g^{-1}, which is comparable with activated carbons (20–80 mg g^{-1}) [25] or ion exchange resin capacity (20–30 mg g^{-1}) [25]. Surface precipitation and electrostatic interactions are thought to be the key adsorption mechanisms onto biochar, and abundant oxygenated functional groups (OFG) are associated with effective cation adsorption [27,28]. Unfortunately, pyrolysis at typical conditions (>400 °C) tends to be ineffective for OFG formation, meaning that biochar must usually be activated to increase its adsorption capacity [29]. Various activation procedures can increase the sorption capacity of biochar, but only by adding cost, energy use, or waste generation [30]. Lastly, pyrolysis is performed in the vapor phase and requires a dry feedstock, negatively impacting the process energy balance for utilization of abundant wet wastes [31,32].

Hydrothermal carbonization (HTC) [33–35] of carbon-rich feeds, including carbohydrates, biomass, and food waste, at moderate temperature (130–250 °C) and autogenous pressures is a versatile, low-energy, and renewable way to produce carbon-rich materials with abundant OFGs that are known as hydrothermal chars (hydrochars) [36,37]. Possibly due to the abundance of OFGs present in hydrochars, they have greater metal sorption capacity than most pyrolysis biochars, making them especially attractive for drinking water purification [5,38,39]. For example, Regmi et al. [33] reported a hydrochar with greater Cu(II) sorption capacity than conventional activated carbon (4.8 compared with 1.8 mg g^{-1}) [39]. Similarly, HTC compatibility with wet feeds eliminates the need for drying, benefiting process-level energy balance especially for agricultural and food waste streams [40,41].

Although HTC is a promising technology, process costs and uncertainties must be reduced to de-risk further investment [42]. Similarly, HTC can benefit by maximizing its value, which in the case of sorbent manufacture, can be accomplished by maximizing hydrochar sorption capacity [16]. Accordingly, a persistent mystery in the field of hydrochar sorption is the high capacity that the material has for metal cations despite its relatively low (<10 m^2 g^{-1}) measurable surface area. For this reason, many reports describe methods to increase hydrochar surface area [29,37,43]. Unfortunately, hydrochar activation to increase surface area again produces wastes and requires energy; moreover, the resulting capacity of the activated material often decreases on a per area basis; for example, Jain et al. [44] reported that the phenol capacity of activated carbon produced from hydrochar decreased with increasing surface area, from 0.16 to 0.13 mg g^{-1} [16].

While pyrolytic treatment and various chemicals can increase hydrochar sorption capacity, immersion in an alkali solution at room temperature reportedly increases OFG abundance and heavy metal adsorption capacity with minimal energy requirements while generating minimal amounts of waste [28,33,37]. The mechanism of the alkali-treatment promotion effect is not clear since mild alkali treatment does not increase hydrochar surface area [45], and the conditions are not sufficient to make or break covalent bonds. Moreover, alkali treatment is sometimes reported as a necessary step for hydrochar to exhibit any heavy metal sorption capacity [29,37], while for others, alkali treatment is not reported [33]. Understanding alkali treatment is, therefore, one goal of this work.

Complicating analysis further, hydrochar OFG abundance, and sorption capacity vary depending on the properties of the feed and reaction conditions. Guo et al. [46,47] have shown that, starting with a lignocellulosic raw material, a hydrochar can be produced with greater OFG abundance than produced from carbohydrate precursors. For example, HTC from wood (initially consisting of 52.7 wt% C) produces a hydrochar with 77 wt% C [48], whereas the carbon content of hydrochar produced from glucose typically contains 53–62 wt% C [49]. Hydrochar OFG abundance tends to decrease with increasing reaction conditions such as reaction time, temperature, and water-to-biomass ratio [46–48]. For many feeds [50–54], HTC can be performed over the temperature range from 160–180 °C and at reaction times <12 h to promote formation of OFGs. More aggressive conditions (≥200 °C) are preferred for other applications [55].

The relationship between feed properties, reaction conditions, and sorption capacities provides an opportunity to synthesize a hydrochar tailor-made for a specific application, such as heavy metal adsorption. For example, Demir-Cakan et al. [56] reported that co-HTC of glucose and acrylic acid produced a hydrochar with exceptionally high OFG abundance; the resulting materials exhibited sorption capacities up to 350 mg g^{-1} for Pb(II) and 90 mg g^{-1} for Cd(II). In comparison, Xue et al. [16] reported peanut-hull based hydrochar sorption capacity for Cd(II) as 12.38 mg g^{-1}.

Ideally, hydrochars could be designed for a specific application at the molecular level. Computational methods have proven effective at understanding sorption mechanisms and thereby enabling the molecular level design of metal-organic frameworks for sorption of perfluoroalkyl substances [57], nanopores for CO$_2$ adsorption [58], and ion exchange resins and activated carbon for heavy metal sorption [59,60]. Density Functional Theory (DFT) is an especially valuable tool for studying the geometry and energetics of sorbate binding to the active site, provided that molecular structures are known that can be used as targets of rational design. Unfortunately, the structural models of hydrochar have only recently converged [61–63], making attempts at molecular level design or computational modeling of sorbate-sorbent binding difficult until now.

Recent work by Brown et al. [61] reconciled several disparate models proposed for the structure of hydrochar synthesized from glucose. Previous models inferred from infrared spectroscopy and Raman microscopy [62,64] indicated that hydrochar structures resembled activated carbon and consisted of fused aromatic cores, comprised of many aromatic rings and with OFGs present primarily as side chains. In contrast, solid-state Nuclear Magnetic Resonance (NMR) [65] and Near-Edge X-ray Absorption Fine Structure (NEXAFS) [63] indicate a structure consisting primarily of individual furan and arene groups, polymerized via short alkyl chains, and decorated with OFGs. Brown et al. [61] recognized that previously reported Raman spectra of hydrochar contained artifacts due to laser-induced pyrolysis of the hydrochar material, causing it to collapse into a condensed aromatic structure. DFT simulation of hydrochar Raman vibrations [61] then indicated that artifact-free Raman spectra were indeed consistent with the furan-arene polymer previously inferred from NMR and NEXAFS [62–64,66]. This paves the way for a molecular-level study of metal binding to hydrochar as a furan-rich polymer, thereby enabling rational design of hydrochar.

The objective of this work was molecular-level design of a hydrochar adsorbent using both experiments and simulation. To focus on generalizable mechanisms, we studied a model hydrochar synthesized from glucose for sorption of a model heavy metal, Cu(II) cations. Sorption capacity was studied before and after alkali treatment and compared with capacities measured for several activated carbon materials. Similarly, the hydrochar was characterized for OFG type and density using Fourier transforms-infrared spectroscopy (FT-IR), solid-state titration, and zeta potential measurement. Metal-hydrochar binding interactions and geometries were evaluated using DFT simulations. Chars with different types and/or densities of OFGs and other metal binding groups were custom synthesized for comparison with glucose hydrochar. The experimental and simulation results described here establish a new method for the rational design of hydrochar sorbents at the molecular level.

2. Materials and Methods

2.1. Materials

All reagents were analytical grade, including: D-(+)-glucose (≥99.5%-Sigma Aldrich), acrylic acid monomer (≥99.0%-TCI chemicals), vinyl sulfonic acid (≥99.0%-TCI Chemicals), hydrochloric acid (0.1 M, ≥99.0%-Acros), sodium hydroxide (0.1 M, ≥95.0%), and the anhydrous salts sodium carbonate (≥99.5%-Sigma Aldrich), sodium hydroxide (≥97.0%-EM Science), potassium carbonate (≥99.0%-Alfar Aesar), and potassium hydroxide (≥85%-Sigma Aldrich). The carbon materials were donations from Norit (since acquired by Cabot) and MeadWestvaco. Specific activated carbon samples included wood-based carbons, Norit Darco® KB-G, Norit Darco® KB-WJ and Nuchar® (MeadWestvaco), and a peat-based carbon, Norit® SX-1. Amberlyst® 15 was used in its hydrogen form and AG® 50 W-X4 in its hydrogen form was purchased from Bio-Rad. $Cu(NO_3)_2 \cdot 2.5\ H_2O$ and elemental standards, 5% HNO_3, 10 µg/mL were acquired from PerkinElmer. Deionized (DI) water was purified to a minimum resistivity of 17.9 MΩ cm prior to use.

2.2. Hydrochar Synthesis

Hydrochar was prepared from a precursor solution formed by dissolving 28.152 g of D-(+)-glucose in 100 mL of DI water. The solution was loaded into a 160 cm^3 PTFE-lined, stainless-steel autoclave, which was then placed in a room-temperature oven that was heated to 180 °C at a heating rating of approximately 10 °C min^{-1}, held at 180 °C for 8 h, before allowing to cool for 12 h. The reaction protocol was selected to replicate those that favor OFG generation, as reported previously by Brown et al. [61] and others [46]. After reaction, the resulting slurry was mixed first with a solution of 100 mL of ethanol and 100 mL of water, and then filtered to remove soluble organic materials from the solid hydrochar. The solid hydrochar was recovered by filtration and rinsed again with ethanol and water. The washing and filtering steps were repeated twice. The hydrochar was then placed in a crucible, dried in an oven at 65 °C for 24 h, and stored in airtight vials before further analysis or use.

Functionalized hydrochars were custom synthesized by preparing a precursor solution consisting of glucose and either acrylic acid or vinyl sulfonic acid and subjecting it to a modified HTC treatment. For the synthesis of acrylic acid-hydrochars (AA-hydrochar), the precursor solution consisted of 10 g of glucose, 10 g of acrylic acid, and 80 g of DI water. The HTC reaction time was extended to 16 h for AA-hydrochar synthesis (at 190 °C), as 8 h reaction time yielded a material that could not be recovered by filtration [56]. For synthesis of vinyl sulfonic acid-hydrochars (VSA-hydrochar), the precursor solution consisted of 36.1 g of glucose, 7.22 g of vinyl sulfonic acid, and 150 mL of water. After some preliminary trials to evaluate the effects of reaction temperature and time on hydrochar yield, the synthesis of VSA-hydrochar was performed at 190 °C for 24 h to yield a solid that could be recovered by filtration. As described previously for hydrochar, both AA-hydrochar and VSA-hydrochar were recovered by filtration, with ethanol and water washing, and oven drying at 65 °C.

Hydrochars were activated by mixing 2.0 g of the synthesized material with 500 mL of alkali solution (2 N). The effects of alkali solutions of Na_2CO_3, K_2CO_3, NaOH, KOH were evaluated. After several hours at room temperature, the material was recovered by filtration, and placed in DI water where the pH was neutralized by dropwise addition of HCl (1 N) and NaOH (1 N) until the pH stabilized. The final product was washed 3 times with DI water, dried in a 100 °C oven, ground, and stored in airtight glass vials.

2.3. Hydrochar Characterization

The surface areas of the samples were determined by N_2 physisorption at 77 K on a Micromeritics ASAP 2000 apparatus, using N_2 as adsorbate. N_2 physisorption on similar instruments has been reported in more detail elsewhere [67]. Surface areas were determined using the Brunauer–Emmett–Teller (BET) model [68]. Prior to adsorption-desorption experiments, all the samples were degassed at 120 °C for 12 h.

Diffuse Reflectance Infrared Fourier Transform Spectroscopy (DRIFTS) was performed on powder samples using a Thermo-Fisher FT-IR 6700 with DRIFTS accessory, described previously in the literature [69]. The spectral resolution was 2 cm^{-1} and all samples were purged with N_2 gas for 2 min before analysis to exclude atmospheric CO_2 and H_2O from the sample space. A background spectrum was obtained prior to each measurement and results were obtained by scanning 1024 times and taking their average. Spectra were analyzed using MagicPlot software and plotted by normalization with the baseline.

Zeta potential measurements were determined using a zeta meter (Malvern Zetasizer-Nano-Z) that has been previously described elsewhere [70]. For each test, 0.005 g of the solid sample was suspended in 100 cm^3 of de-ionized water containing 0.1 N NaCl followed by homogenization for 2 h in an ultrasonic bath. After ultrasonication, the aqueous suspension was equilibrated at different pH values for 30 min. Zeta potential results are reported as the average and standard deviation of three measurements.

The combined densities of strong and weak acid groups were determined using the Boehm titration method, described previously in the literature [29,71,72]. In brief, a carbon sample (0.5 g) was placed in $NaHCO_3$ solution (20 mL, 0.1 N), agitated for 48 h, and the carbon was removed by filtration. The resulting filtrate was degassed for at least 30 min using N_2 to remove CO_2 and was then titrated to determine the acid site density of carbon-rich materials [73].

2.4. Hydrochar Adsorption Tests

For single-point Cu(II) cation adsorption tests, 0.2 g of sorbent was suspended in 10 mL of an aqueous solution of Cu(II) (0.08 M) and placed within an high density polyethylene sample vial (Celltreat). Initially, the supernatant liquid was clear and pale blue, a consequence of the Cu(II) present in the mixture. The resulting slurry was agitated at room temperature using a wrist-action shaker (Burrell) for 24 h. Preliminary tests indicated that adsorption equilibrium was reached after 12 h. After shaking and centrifugation, the supernatant liquid remained clear and the pale blue coloration was visibly fainter. The hydrochar was discarded and Cu(II) concentrations in the supernatant liquid were measured using a Perkin Elmer, Nexior 350 × Inductively Coupled Plasma (ICP) spectrometer. The ICP response was calibrated using standard solutions. Samples were diluted prior to each measurement to ensure that the concentration fell within the pre-determined calibration range. Adsorption capacity was then estimated using the initial (mass$_{copper,i}$) and final (mass$_{copper,f}$) Cu(II) concentrations and the mass of sorbate (mass$_{sorbent}$):

$$\text{Adsorption Capacity } (mg\ g^{-1}) = \frac{\text{mass}_{copper,i} - \text{mass}_{copper,f}}{\text{mass}_{sorbent}} * 1000$$

Each adsorption measurement was performed at least in duplicate and ICP concentration measurements were performed in triplicate. Average values are reported here. Control runs were performed in the absence of sorbent and the loss to the vial was equivalent to <1 mg g^{-1} of sorbent.

2.5. Computational Modeling

DFT simulations were performed to study cation-hydrochar binding energies and geometries. The proposed hydrochar structures were modeled using DFT with the Gaussian 09 program [74] run via WebMO [75]. All the geometries were optimized using the Becke, 3-parameter, Lee–Yang–Parr (B3LYP) [76,77] hybrid functional, with the 6-311+G(d,p) basis set. Because the physical experiments included a water solvent, the polarizable continuum model (PCM) [78] was chosen to implicitly include the solvent in these calculations, assuming its dielectric constant was equal to 78. Model adsorption reactions were modeled and compared with one another to determine their energies and relative favorability. For each adsorption reaction, the products were hydrochar with a bound copper atom, and the respective ion, either hydrogen, sodium, or potassium, in solution. Multiple initial configurations were studied to confirm that the final geometry captured a global minimum, rather than a local one.

3. Results and Discussion

The objective of this study development of rational methods to design hydrochar for heavy metal adsorption at the molecular level [79]. The study consists of several components: (1) measurements of sorption capacity of glucose hydrochar and comparison with activated carbon; (2) characterization and DFT modeling of glucose hydrochar; (3) custom-synthesis of hydrochars with tailored heavy metal capacity.

3.1. Glucose Hydrochar Sorption Capacity and Characterization

As a starting point, hydrochar was synthesized from a glucose precursor solution. The base structure of glucose hydrochar was recently reconciled between several models [61–63], with the finding that it consists of a furan-rich polymer decorated with OFG groups. Accordingly, and following previous reports, we hypothesized that the OFG groups would encourage heavy metal sorption. Unfortunately, glucose hydrochar exhibited negligible Cu(II) sorption capacity, <3 mg g^{-1}, as shown in Table 1. The negligible Cu(II) capacity measured for glucose hydrochar seemingly contradicted previous literature descriptions of hydrochar absorption capacity [80–82], thus, motivating further examination.

Table 1. Adsorption Capacity and Surface Area of Hydrochar and Activated Carbon. BET = Brunauer–Emmett–Teller model.

Material		Post Treatment	Adsorption Capacity (mg Cu g^{-1})		BET Surface Area (m^2 g^{-1})
Glucose Hydrochar		None	<3		4 ± 2
		Na$_2$CO$_3$	20.0	± 0.4	
		K$_2$CO$_3$	29	±5	
		NaOH	35	±4	
		KOH	40	±4	
Activated Carbons	Norit® SX1	None	19	±4	800 ± 100
		KOH	21	±6	
	Nuchar®	None	29	±3	1700 ± 200
		KOH	11	±7	
	Darco® KB-G	None	9	±0.1	1500 ± 100
		KOH	5	±0.8	
	Darco® KB-WJ	None	6	±0.2	1600 ± 100
		KOH	21	±4	

Sun et al. [37] reported that alkali treatment increased the sorption capacity of hydrochar by 2–3 times, motivating the study of alkali treatment in the current study. As shown in Table 1, alkali treatment greatly increased the Cu(II) capacity, by at least an order of magnitude compared with the original glucose hydrochar. Several different bases were evaluated, with the finding that strong bases (hydroxides) outperformed weak ones (carbonates) and that bases featuring the potassium cation outperformed ones possessing sodium.

We compared the capacity of alkali-activated glucose hydrochar with several different activated carbons (Table 1), selected to cover a range of properties [83]. Interestingly, the activated carbons exhibited much greater sorption capacity than glucose hydrochar without activation, but less capacity than their alkali-activated forms. Alkali treatment was evaluated for two of the activated carbons, and it was found that the treatment either had no effect (Norit® SX1) or even negative effect (Nuchar®) on sorption capacity. The different response to alkali treatment observed for activated carbon and hydrochar clearly points to differences in the mechanism that must be understood for molecular-level hydrochar design.

To understand the adsorption results presented in Table 1, surface areas were measured using N$_2$ sorption and the BET isotherm fitting method. Consistent with previous reports [29,37,43], the measured BET surface area of hydrochar was <10 m^2 g^{-1}. Alkali treatment had no effect on the measured hydrochar surface area, allowing us to reject the hypothesis that the effect of the treatment was to open up the hydrochar pore structure [82,84]. Similarly, consistent with previous reports [83], the BET surface areas of the activated carbons were >800 m^2 g^{-1} and not affected by the dilute alkali treatment. Accordingly, while the capacity of glucose hydrochar is comparable to activated carbon on a mass basis, on a surface area basis the capacity is orders of magnitudes greater. This observation clearly points to a specific hydrochar-sorbate interaction that can be engineered to maximize adsorption.

The strongest common sorbate-hydrochar interaction is electrostatic [85], which can be understood as the interaction between the positively charged metal cation and negatively charged functional groups on the hydrochar surface. Accordingly, as a way to understand and quantify hydrochar surface charge, we measured hydrochar zeta potential before and after alkali activation and over a wide range of pH, from 2–12. Figure 1 presents the results, showing that zeta potential of alkali activated hydrochar was much more negative in the pH range of interest (pH < 7) than the parent hydrochar. Under strongly alkali pH, the zeta potential of the parent and the alkali treated material are the same to within the limits of experimental uncertainty, which is consistent with expectations given that the alkali treatment is simply immersion in an alkali solution with pH > 9.

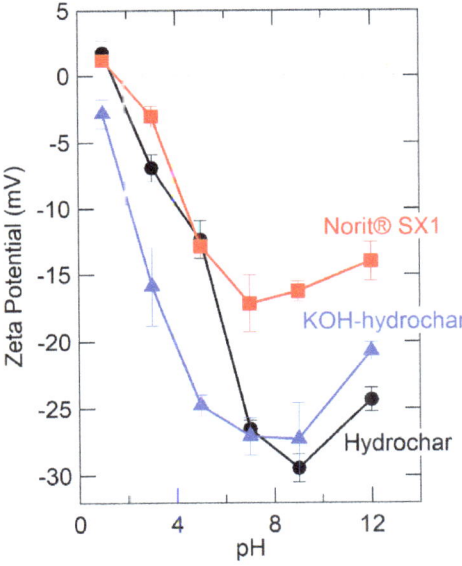

Figure 1. Comparison of zeta potential of different sorbent materials as a function of pH.

Zeta potential measurements support electrostatic interaction as the primary basis of cation sorption to the hydrochar, providing a valuable clue for rational design. For further comparison, we measured the zeta potential of one of the aforementioned activated carbons (Norit® SX-1) and include these data in Figure 1. The zeta potential of the activated carbon was much less negative than glucose hydrochar, even before alkali treatment. Again, this points to a qualitatively different sorption mechanism for activated carbon compared to hydrochar, with cation sorption to activated carbon likely occurring due to cation-π interactions, which appear to be less important than electrostatic interactions for cation binding to hydrochar [86].

Rational sorbent design requires understanding the molecular binding sites. Figure 1 clearly implicates the importance of groups that ionize on alkali treatment, which naturally suggests carboxylic acids, acid anhydrides, and strongly acidic aromatic alcohols, such as phenol [16,24,73]. The fact that strong bases were more activating than weak ones (Table 1) seems consistent with de-protonation of weak acids but does not provide sufficient molecular detail for sorbent design. Accordingly, glucose hydrochar was analyzed using FT-IR for identification of ionizable OFGs. Figure 2 provides FT-IR spectra divided into the fingerprint region, 1000–2000 cm^{-1} (Figure 2a) and the C–H and O–H stretching region, 2400–4000 cm^{-1} (Figure 2b). Prior to alkali treatment, glucose hydrochar exhibits bands attributable to carbonyl (1720 cm^{-1}) and hydroxyl (3200–3400 cm^{-1}) groups associated with carboxylic acids. Other features at 2900, 1600, and 1200 cm^{-1} are attributable to C–H stretches, arenes/furans breathing modes, and C–O stretches, respectively [65] These spectroscopic attributions are broadly consistent with the structural models previously inferred from NMR [65], Raman [61], and NEXAFS [62–64,66]. Interestingly, the C–H and O–H stretches are sufficiently differentiated to suggest that carboxylic acid exists in its free, rather than dimerized, state [87].

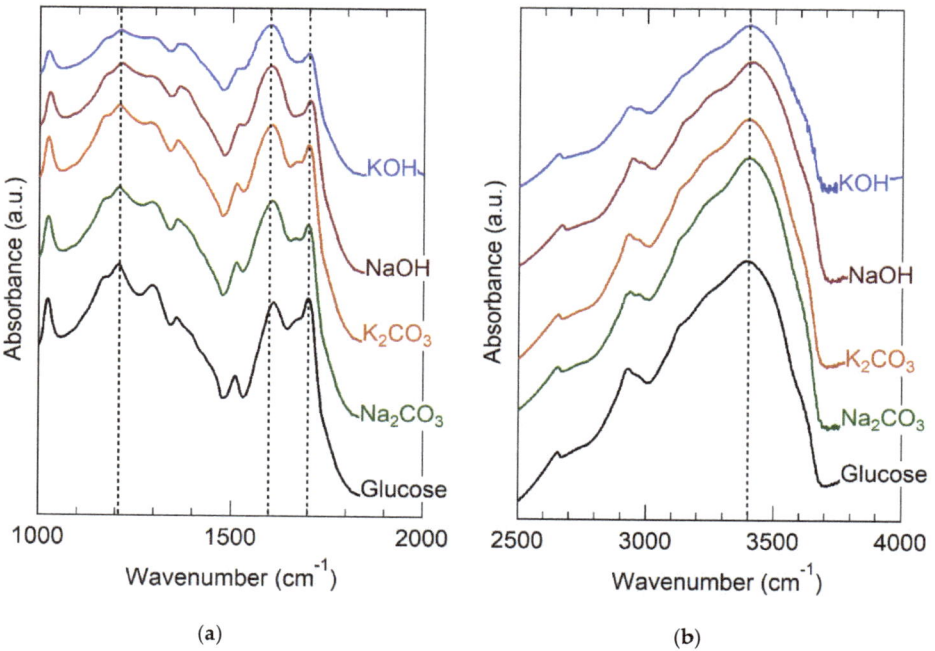

Figure 2. FT-IR spectra of glucose hydrochar as synthesized and after activation by various bases. (a) the fingerpint region (1000–2000 cm^{-1}) (b) the O–H and C–H stretch region (2500–4000 cm^{-1}).

After alkali treatment, the intensities of the hydroxyl band at 3200–3400 cm^{-1} and carbonyl band at 1720 cm^{-1} become much less intense. Simultaneously with these changes, the intensity of the band at 1600 cm^{-1} increases and the feature broadens noticeably. The C–O stretch present at approximately 1200 cm^{-1} becomes less intense and broader after alkali treatment. The effects are more noticeable for treatment with the hydroxides than the carbonates, consistent with their relative basicities and with the observed effects on sorption capacity, noted in Table 1.

All of the aforementioned changes observed in the FT-IR spectra of glucose hydrochar after alkali treatment are attributable to deprotonation of carboxylic acid groups to form carboxylates [37,88]. Specifically, deprotonation involves a shift of the main carbonyl band from approximately 1700 to about 1600 cm^{-1}; [84] a reduction of intensity of the C–O stretch at 1200 cm^{-1}; and a reduction of the intensity of the O–H stretch at 3200–3400 cm^{-1}. The last of these is consistent with partial removal of the H atoms involved with O–H stretches, as expected for de-protonation. The carboxylate feature at 1600 cm^{-1} overlaps with the furan/arene breathing mode that is characteristic of hydrochar [61,63,89].

The fact that alkali treated hydrochar still exhibits an O−H stretching band is consistent either with incomplete de-protonation of acid groups or with the presence of multiple forms of O−H in the structure (i.e., alcohol groups that are not sufficiently acidic to be deprotonated).

Since the carboxylic acid groups present in glucose hydrochar appear to be primarily in their protonated forms (Figure 1) and since the pH of the HTC reaction mixture is about 3 [90], the pKa of these acid groups must be greater than approximately 3–otherwise, they would be present in hydrochar in their deprotonated forms. Alkali treatment then deprotonates these groups, resulting in formation of the alkali carboxylate. Because hydrochar is a complex material and because localized induction and steric effects can influence pKa [91], carboxylic acids present in hydrochar likely possess a range of pKa's. In fact, this assertion is supported from the broad zeta potential curve observed for glucose hydrochar and shown in Figure 1. Treatment with carbonates may therefore deprotonate only the strongest carboxylic acids present in hydrochar, while treatment with hydroxides deprotonates both strong and weak carboxylic acids.

We considered the possibility of alternative ionizable groups, aside from carboxylic acid. Treatment with hydroxide would partially deprotonate any strongly acidic alcohol groups (e.g., phenols) present in the hydrochar structure; however, the FT-IR spectra show no direct evidence to support the formation of phenolate ions, nor do reported structural models suggest the presence of phenol in hydrochar [24,61,63]. Accordingly, metal-carboxylate binding appears to be the primary cation adsorption mechanism underlying glucose hydrochar sorption, providing a clear target for molecular simulation.

Hydrochar is thought to be composed of furan/arene polymers connected by alkyl spacers. Mild alkali treatment is insufficient to break or form covalent bonds present in this structure [82], which is consistent with the negligible change in surface area associated with alkali treatment (see Table 1). That stated, Mihajlovic et al. [92] proposed that hydrolytic degradation of OFGs can sometimes occur during alkali treatment, and re-arrangement of the hydrochar structure from its hypothesized form would complicate attempts to model the binding site. Accordingly, we searched for evidence of bond breaking in the spectra shown in Figure 2. Inspection of the FT-IR spectra suggests that important hydrochar features [28,29] at 1020 cm^{-1} (C−OH alcohol and/alkyl-substituted ethers groups), 1600 cm^{-1} (furan/arene breathing modes), and 2900–3000 cm^{-1} (C−H stretch) are not affected by alkali treatment, consistent with the main effect of alkali treatment being confined to deprotonation rather than making and/or breaking of covalent bonds. This observation permits use of published hydrochar structural models to recreate the local environment of the metal-carboxylate binding site for DFT simulation.

3.2. DFT Simulations of the Metal-Carboxylate Interactions

Zeta potential measurements and FT-IR analysis clearly implicate metal-carboxylate binding. DFT simulations of the metal-carboxylate complex were performed to examine this hypothesis in more detail. In particular, we sought to answer three questions: (1) since a metal cation should be able to displace protons from carboxylic acid, why is alkali treatment required to activate the hydrochar? (2) Why do potassium salts outperform sodium salts? (3) What is the geometry of the binding site?

Simulating cation-carboxylate binding requires recreating a plausible local environment. The hydrochar molecule, pictured in Figure 3a, was created as a composite structure based on previous literature. Titirici et al. [93] demonstrated via NMR that the majority of the aromatic functionality of hydrochars synthesized from glucose at temperatures less than 200 °C can be attributed to furan groups. Latham et al. [63] supported this via NEXAFS while showing that carbonyl groups are also important. The previously mentioned IR spectra also indicate the presence of carbonyl groups and are in agreement with the model proposed by Latham et al. [63]. Accordingly, we recreated the local adsorption environment as a furanic dimer configuration to be consistent with published hydrochar structures [61]. The carboxylic acid/carboxylate group resides as a side chain on the alkyl linker between adjacent furan rings, consistent with the observation using FT-IR and the importance of carboxylate groups inferred from sorption capacity measurements presented here. The local environment experienced by a metal cation during adsorption also includes water solvent molecules. Here, we recreated the water solvation effect using an implicit cavity model of the appropriate dielectric constant (taken as 78). Future work can improve the accuracy of our calculations by including explicit water molecules in the simulation.

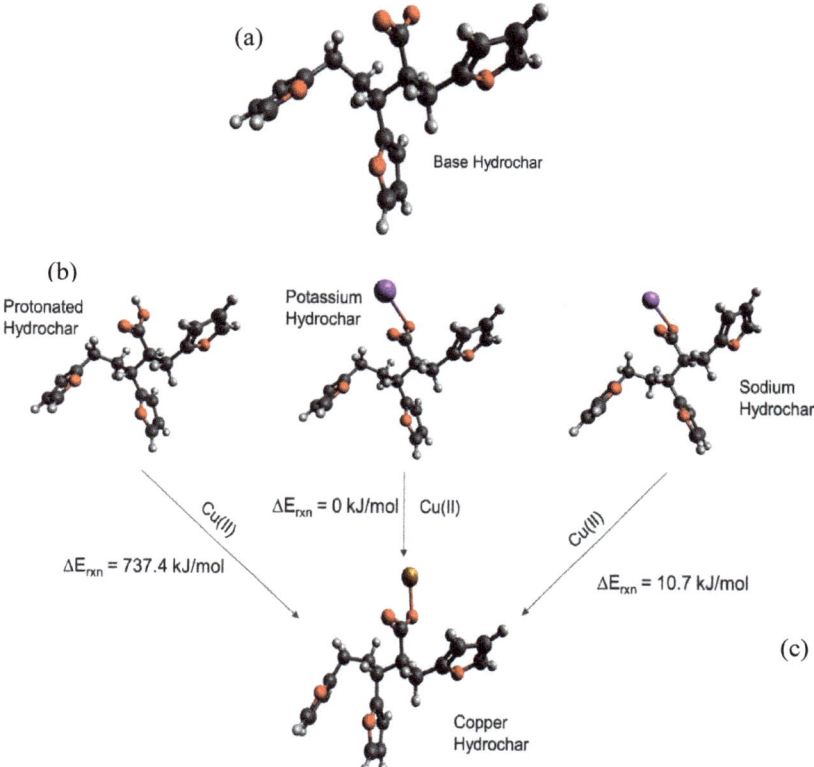

Figure 3. Carboxylate-containing hydrochar structures optimized using Density Functional Theory (DFT). (**a**) Model of the base hydrochar molecule. The different structures in (**b**) involve carboxylate binding with hydrogen, potassium, and sodium, respectively, with the carboxylate, which make up the reactants of the modeled adsorption reactions. (**c**) Illustrates the interaction between Cu(II) and the carboxylate group. Estimated adsorption energies are provided as shown. Legend: ● carbon; ○ hydrogen; ● oxygen; ● potassium; ● sodium; ● copper.

We then simulated a series of possible cation-carboxylate structures, starting with the hydrochar model shown in Figure 3a. The focus of these calculations was to answer the aforementioned questions that focus on elucidation of trends, rather than quantitative energy estimates. We then simulated cation binding, shown stoichiometrically in Figure 3b, by replacing either H^+, K^+, or Na^+ with the Cu(II) cation to form the final structure shown in Figure 3c.

Consistent with experimental observations (Table 1), we find that replacing H^+ with Cu(II) is energetically unfavorable, whereas replacing K^+ and Na^+ is energetically favorable. The simulated energies are consistent with the observation that glucose hydrochar requires alkali treatment prior to activation. Moreover, DFT simulations predict that replacing K^+ is energetically more favorable than replacing Na^+ by 10.72 kJ mol^{-1}, which is consistent with the observation that KOH is a more effective activating salt than NaOH and K$_2$CO$_3$ is more effective than Na$_2$CO$_3$. That stated, the calculated energy difference between K$^+$ and Na$^+$ substitution is relatively modest, which is again consistent with experimental observation. Note that for these reactions the cations in solution may not be properly modeled by implicit solvation, which is why for instance the replacement of a hydrogen by Cu(II) is so endothermic. Nonetheless the trends in cation exchange are captured by the DFT calculations.

Figure 3c shows the optimized geometry of a Cu-hydrochar structure. Here, the Cu-carboxylate bond length is approximately 1.85 Å, slightly longer than that associated with the distance between the proton and carboxylate group in carboxylic acid. The longer bond is consistent with the size of the Cu(II) ion compared with the proton [37].

The DFT simulations summarized in Figure 3 explain that alkali treatment removes the proton to activate the sorption capacity of glucose hydrochar. Physically, the proton is more tightly bonded to the carboxylate group than the metal cations, owing to the differences in ionic radii and the strong effect of ion-ion distance on the strength of electrostatic interactions [94]. Similarly, the differences observed between potassium and sodium can be ascribed to their relative ionic radii.

Interestingly, alkali treatment is not always reported as a necessary step for observation of hydrochar sorption capacity. This may be due to differences in the reaction mixture pH for different precursors and/or the presence of alkali salts in many hydrochar starting materials [95,96]. Accordingly, subtle differences in the reaction mixture and the composition of the precursor may decide whether or not alkali treatment is required to activate a given hydrochar for metal adsorption. Alternatively, the alkali step may not be uniformly reported, even when it is required. We recommend more consistent reporting of alkali treatment and reaction mixture pH in future work in this area.

3.3. Custom-Synthesis of Hydrochar for Heavy Metal Adsorption

Experiments and DFT simulations clearly implicate the importance of metal-carboxylate interactions in hydrochar adsorption of Cu(II). Accordingly, our next step was custom synthesis of a hydrochar for heavy metal adsorption. Following the work of Demir-Cakan et al. [56], we elected to synthesize a hydrochar by co-processing glucose and acrylic acid. Acrylic acid possesses a polymerizable double bond, which can form covalent linkages with the alkyl linker groups in the hydrochar structure, thereby increasing the density of carboxylate groups in the resulting material. We term the resulting material acrylic acid-hydrochar, or simply AA-hydrochar. Demir-Cakan et al. [56] reported synthesis of a series of AA-hydrochars, starting with different amounts of acrylic acid in the precursor mixture. Here, we selected a precursor mixture with composition similar to the optimum reported by Demir-Cakan et al. [56] as a proof of concept.

Table 2 provides the sorption capacity and surface area measurements for AA-hydrochar. As expected from DFT simulations, without activation the Cu(II) sorption capacity of AA-hydrochar is negligible. Interestingly, Demir-Cakan et al. [56] did not report alkali activation of their materials, which might be attributable to their study of Pb(IV) and Cd(II) whereas we studied Cu(II) or the aforementioned impact of subtle differences in reaction mixture pH on hydrochar protonation and subsequent sorption capacity. Regardless, after alkali activation, sorption capacity increases by at least an order of magnitude for the AA-hydrochar, and strong bases are again more effective than weak bases. AA-hydrochar capacity for Cu(II) sorption is greater than that observed for standard glucose hydrochar (50 ± 4 compared with 40 ± 4 mg g^{-1}). Again, the effect is not as pronounced as reported by Demir-Cakan et al. [56], but it is consistent with the design concept.

Table 2. Adsorption capacity and surface area of custom-synthesized hydrochar and ion exchange resins.

	Material	Post Treatment	Adsorption Capacity (mg Cu g^{-1})		BET Surface Area (m^2 g^{-1})
Hydrochars	Acrylic acid-hydrochar (AA-hydrochar)	None	<3		5–10
		Na$_2$CO$_3$	30	±7	
		K$_2$CO$_3$	26	±6	
		NaOH	50	±4	
		KOH	45	±7	
	Vinyl sulfonic acid (VSA-hydrochar)	None	<2		<1
		NaOH	34	±5	
		KOH	51	±3	
Resins	Amberlyst®	None	128	±4	53 [97]
	AG® 50W-X4	None	109	±7	<1

The effect of AA and glucose co-processing to produce hydrochar was consistent with our expectations, but consistency does not imply confirmation and we considered alternative hypotheses. Table 2 shows that the surface area of AA-hydrochar was similar to the glucose hydrochar, eliminating surface area changes as a major difference between these materials. To understand further, we studied the OFGs of AA-hydrochar using FT-IR. Figure 4 provides the FT-IR spectra obtained for AA-hydrochar before and after KOH treatment. The FT-IR of glucose hydrochar is included in Figure 4 for direct comparison to show that the AA-hydrochar spectrum exhibits much more intense bands associated with carboxylic acids at 1720 (carbonyl) and 1200 cm^{-1} (C-O stretch) than glucose hydrochar. In fact, the carboxylic acid bands dominate the AA-hydrochar spectrum and appear as the most prominent features. The band at 1600 cm^{-1}, which is characteristic of furans and arenes, appears only as a minor, though distinct, feature in the AA-hydrochar spectrum. In comparison, the furan/arene band is one of the most prominent features in the glucose hydrochar spectrum. Similarly, after alkali treatment, the carbonyl band shifts to approximately 1550 cm^{-1} and becomes the most prominent feature in the AA-hydrochar spectrum. Correspondingly, the C-O stretch feature shifts and broadens. Taken together, these observations clearly indicate that AA-hydrochar has abundant carboxylic acid groups that deprotonate after alkali treatment.

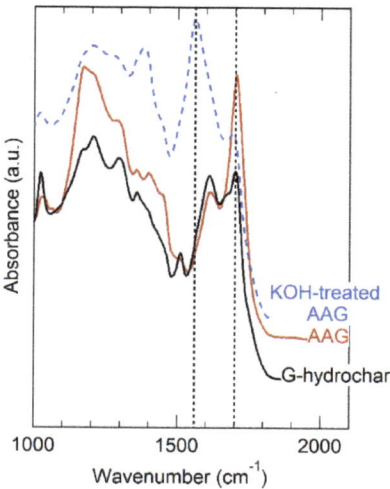

Figure 4. FT-IR spectra of AA-hydrochar before and after KOH treatment. G-hydrochar is synthesized entirely of glucose precursor, shown before alkali treatment. AA-hydrochar synthesized from co-feed of acrylic acid and glucose, shown before and after alkali treatment using KOH.

The performance of hydrochar sorbents is often compared with activated carbon [25,39]. However, our findings indicate that hydrochar adsorption is mediated by metal-carboxylate binding interactions that are more similar to what occurs on an ion exchange resin, rather than activated carbon. Accordingly, we measured Cu(II) sorption capacity of two commercial ion exchange resins, Amberlyst®-15 and AG® 50W-X4. Capacity results for these resins are provided in Table 2. Interestingly, these resins far outperform activated carbon (Table 1) and outperform by about a factor of two the AA-hydrochar. Moreover, the ion exchange resins did not require alkali activation, unlike the hydrochars. Since the resins outperform hydrochar, even AA-hydrochar, we sought to understand the differences between the resins and the hydrochar as part of our rational design approach.

An obvious potential difference between the resins and the hydrochar is surface area. Table 2 provides N_2 sorption-based BET surface areas for Amberlyst®-15 and AG® 50W-X4. Interestingly, Amberlyst®-15 exhibits much greater surface area than any of the hydrochars, which could explain its superior performance. However, the surface area measured for AG® 50W-X4 was less than any of the other materials (<1 $m^2 g^{-1}$, the instrument detection limit), meaning that surface area considerations alone cannot explain the performance of the resins–at least not the surface area measured by N_2 sorption and estimated by BET analysis of the isotherm. In fact, the swelling behavior of ion exchange resins has been studied carefully in water and other solvents [98–100]; swelling in the presence of water likely opens the pore structure of AG® 50W-X4 (and possibly the other sorbents), accounting for its sorption capacity despite negligible N_2 sorption surface area. Understanding the effects of hydrochar swelling on surface area available for cation sorption is an area that should be studied in the future.

The binding site in both Amberlyst®-15 and AG® 50W-X4 is a sulfonate group [101], whereas the findings presented here indicate that carboxylate groups are mainly responsible for binding in glucose hydrochar and especially AA-hydrochar. The sulfonic acid group is at least 1000× stronger than the carboxylic acid group, meaning that this difference could explain sorption behavior and the need for alkali activation. Accordingly, we modified the acrylic acid synthesis procedure for incorporation of a sulfonate group into the hydrochar structure by co-processing glucose and vinyl sulfonic acid. Like acrylic acid, vinyl sulfonic acid possesses a polymerizable double bond that can be incorporated in the hydrochar-alkylated backbone. Unlike acrylic acid, though, vinyl sulfonic acid can introduce a sulfonate group into the hydrochar instead of the carboxylic acid introduced by acrylic acid. Accordingly, we term this new char vinyl sulfonic acid-hydrochar, or VSA-hydrochar.

Table 2 provides the Cu(II) cation sorption capacity of VSA-hydrochar. Interestingly, despite the strength of the vinyl sulfonic acid precursor (pK_a < 1 compared with 4.35 for acrylic acid) [98,102], we observed negligible Cu(II) sorption capacity for VSA-hydrochar before treatment with alkali. After treatment with KOH, the Cu(II) capacity of VSA-hydrochar increased substantially to 51 ± 3 mg g^{-1}. Interestingly, NaOH was much less effective at increasing Cu(II) sorption capacity than KOH, consistent with the aforementioned trend observed and simulated for carboxylate binding.

As before, the measured BET surface area of VSA-hydrochar was in the same range as the other hydrochars and <10 $m^2 g^{-1}$. Similarly, Figure 5 shows the FT-IR spectrum of VSA-hydrochar. Unlike carboxylic acid and carboxylate groups that have intense and well-differentiated vibrational bands, sulfonic acid and sulfonate give rise to weak and broad bands that are not easily differentiated from other features [103]. That stated, the FT-IR VSA-hydrochar spectrum contains bands in the range expected for sulfonic acid (1100–1300 cm^{-1}). The carboxylate/carboxylic acid bands are less intense in VSA-hydrochar than glucose hydrochar, indicating substitution of the weak acid in AA-hydrochar for the strong acid in VSA-hydrochar.

Cu(II)-sulfonate structures were simulated using DFT methods, similar to those previously presented for carboxylate binding. Figure 6a shows the sulfonic acid- hydrochar geometry, which consisted of two furan groups bonded to a sulfonic acid group. As before, binding was simulated as an exchange of Cu(II) for H^+, K^+, and Na^+. Despite the strength of the sulfonic acid, DFT calculations found that replacing H^+ with Cu(II) was thermodynamically unfavorable, consistent with the need to activated VSA-hydrochar with alkali. Figure 6b summarizes this result. Similarly, the distance of the Cu–O bonded to sulfonate (shown in Figure 6c) is 1.95 Å, somewhat greater than the Cu–O bond in carboxylate hydrochar (1.85 Å). As before, the cations in solution may not be properly modeled by implicit solvation, which is why some energies may be so large, despite DFT identifying the trends in cation exchange.

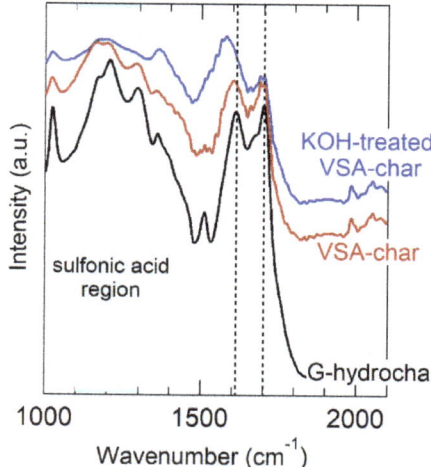

Figure 5. FT-IR spectrum of VSA-hydrochar, before and after KOH treatment. Vertical lines mark the carboxylic acid and carboxylate vibration bands. The region where sulfonic acid vibrations appear is indicated. The spectrum of glucose hydrochar is reproduced from Figure 4 as a point of reference.

Surface area, FT-IR, and DFT simulations provide further evidence of cation-sulfonate binding in the VSA-hydrochar, but do not explain why the performance of neither VSA-hydrochar nor AA-hydrochar can match the commercial ion exchange resins. As a final hypothesis, we quantified the density of surface acids present on the various sorbents, with the expectation that differences in the density of surface acids might explain observed differences in sorption capacity. For these experiments, hydrochars were first treated with strong acid (HCl) to protonate fully all available acid groups. Then, the acid group density was measured of the protonated sorbent using Boehm titration methods [29,71,72].

Table 3 summarizes the carboxylic acid site density measurements. As expected, the density of acid functional groups on the glucose hydrochar is much greater than on the activated carbons considered here, consistent with the different adsorption mechanisms for the two materials (primarily electrostatic vs. primarily π-cation). The ion exchange resins have much greater acid concentrations than any of the other sorbents, consistent with their superior performance and indicating that the AA- and VSA-hydrochars function as designed, albeit with fewer acid binding groups than are available on the ion exchange resins tested here. Nonetheless, the Cu(II) adsorption performance of the designer hydrochars is comparable to the ion exchange resins (to within a factor of two) and superior to activated carbon, meaning that strategies to increase acid functional group density can be effective for synthesis of task-specific hydrochar sorbents.

Table 3 provides qualitative evidence of the importance of acid group density on sorption performance and permits analysis of a critical parameter: the binding stoichiometry of the metal-acid complex formed during adsorption. Binding stoichiometry is important for quantifying sorbent performance since the ideal absorbent will possess high density of binding sites and utilize them as efficiently as possible. Simultaneously achieving high binding site density and binding site utilization may not be possible, since densely spaced binding sites may promote bidentate binding instead of monodentate binding, which is less efficient binding site utilization. We analyzed the sorption and acid site density data to evaluate these effects in hydrochar, activated carbon, and ion exchange resins.

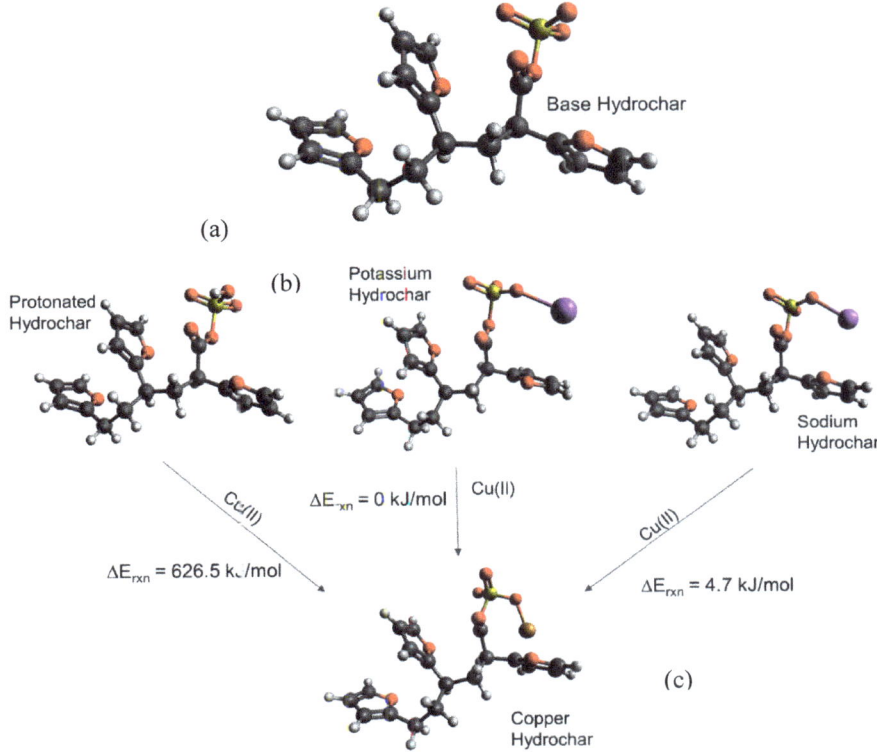

Figure 6. Sulfonate-containing hydrochar structures optimized using DFT. The structure in (**a**) is the sulfonated hydrochar molecule. (**b**) Depicts the reactants which are the interactions between the sulfonate group and either hydrogen, potassium, or sodium, respectively. Structure (**c**) shows binding of Cu(II) to the sulfonated hydrochar molecule. Adsorption energies are provided as shown. Legend: ● carbon; ○ hydrogen; ● oxygen; ● sulfur; ● potassium; ● sodium; ● copper.

Table 3. Titration results of sorption materials.

Material		Carboxylic Acid Density (mmol g^{-1})		Carboxylic Acid Density [b] (mmol m^{-2})	
Hydrochars	glucose-hydrochar	1	±0.07	0.15	±0.02
	VSA-hydrochar	0.9	±0.1	0.12	±0.01
	AA-hydrochar	2.8	±0.3	0.4	±0.1
Activated Carbons	Norit® SX1	<0.1 [a]			
	Norit® SX Ultra	<0.1 [a]		<0.0001	
	Nuchar®	<0.1 [a]			
Resins	Amberlyst-15®	5	±0.01	0.1	±0.0002
	AG® 50W-X4	1.71	±0.1	2.6	±0.1

[a] the estimated acid site detection limit. [b] based on estimated BET surface area.

To use Table 3 data to understand stoichiometry, we plotted Cu(II) sorption capacity as a function of measured acid group density, converting both to molar quantities, as shown in Figure 7. For comparison, lines of constant ion-binding site stoichiometry (two Cu ions per acid, 1:1, and 1:2) are shown. Data for the activated carbons cluster around the origin and fall entirely off the stoichiometric trend lines, as expected given that the sorption mechanism to activated carbons is likely cation-π interactions and is, therefore, independent of acid site density. In contrast, sorption for the hydrochars falls between the 1:1 and 1:2 stoichiometry lines, indicating that—on average—each acid group binds approximately 0.75 Cu ions. This again is further quantitative evidence of the importance of electrostatic interactions for binding to hydrochar.

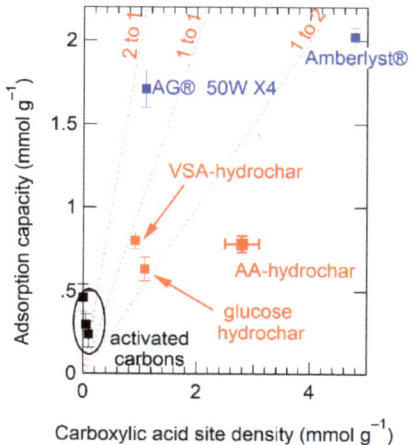

Figure 7. Measured Cu(II) adsorption capacity plotted as a function of measured carboxylate group density. Lines of constant cation-anion stoichiometry are shown as dotted lines at 2 Cu(II) ions per binding site (2:1), 1 Cu(II) ions per binding site, and 1 Cu(II) ion for every two binding sites. Legend: ■ depicts activated carbon, ■ ion exchange resins, ■ hydrochar.

The stoichiometry inferred from Figure 7 shows that the designer hydrochars outperform Amberlyst®-15 on a per acid site basis. This is an important finding since increasing binding site utilization efficiency is an effective means of increasing sorption capacity, along with increasing the density of binding sites themselves. Interestingly, AG® 50W-X4 far exceeds all other sorbents on effectiveness per acid site, with nearly two Cu ions associated with every acid site (Figure 6). The difference between AG® 50W-X4 and Amberlyst®-15 is noteworthy as both sorbents are described in the literature as polymerized styrene backbones with periodic sulfonic acid group substitution [101]. The difference in their performance must be due either to (1) the ability of the sorbent to hold charge, which could be saturated for Amberlyst®-15 limiting its sorption capacity, (2) differences in acid site accessibility in the swollen resins and hydrochars, or (3) differences in the spatial proximity of the acid binding groups in the different sorbents. The performance of AG® 50W-X4 suggests further engineering of the hydrochar structure to optimize sorption capacity.

To understand the origins of stoichiometry between Cu(II) ions and carboxylate or sulfonate groups, we performed simulations to compare monodentate with bidentate binding of Cu(II) to carboxylate and sulfonate groups. To make the calculation accessible using DFT, we simplified the structure previously used in Figures 3 and 6 to remove the furan groups. Figure 8a shows the optimized geometry for the monodentate binding structures, and Figure 8b shows the optimized geometry for the bidentate binding structures. As expected, bidentate binding is much more energetically favorable than binding to a single acid functional group. For the sulfonate site, bidentate binding is more stable by 201 kJ mol^{-1}, and for the carboxylate site, bidentate binding is more stable by 126 kJ mol^{-1}. These values indicate a clear thermodynamic preference for bidentate binding. As measured by the Cu–O distance, the Cu(II) ion is roughly equidistant between the two sulfonate groups; as a result, the Cu–O distance in bidentate binding complex is greater than found in the geometry optimized for single Cu-acid stoichiometry (shown previously in Figure 6). This clearly shows that Cu(II) (and presumably other double charged cations) will prefer bidentate binding, when such an option is available. Since bidentate binding is a less efficient use of sites than monodentate binding, rational design of hydrochars should attempt to achieve uniform acid spacing to minimize acid-acid interaction and the ability of cations to bind simultaneously to multiple acid sites.

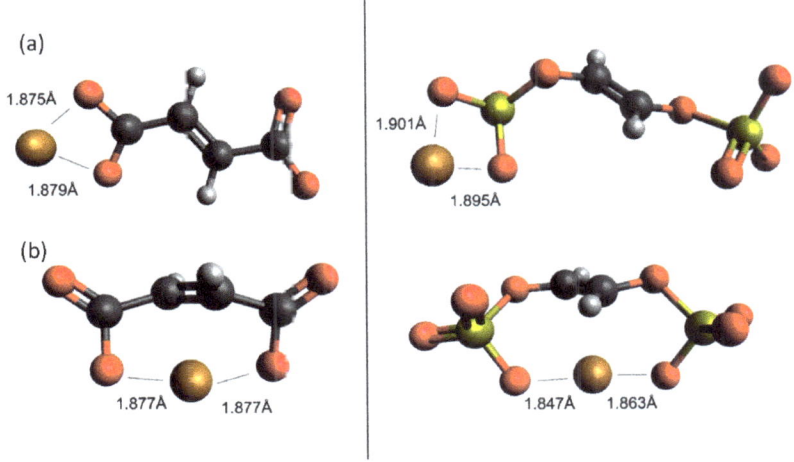

Figure 8. Optimized geometries of monodentate (**a**) and bidentate (**b**) binding of Cu(II) to hydrochar carboxylate and sulfonate groups simulated using DFT. Oxygen-Cu(II) distances are shown for reference. Legend: ● carbon; ● hydrogen; ● oxygen; ● sulfur; ● copper.

When functional group precursors with polymerizable double bonds are co-fed to the HTC reactor with glucose, the functional group bearing molecules will polymerize primarily with each other, rather than with groups present in the hydrochar. Because they are formed by co-feeding glucose and vinyl groups, VSA-hydrochar is not engineered to achieve the desired spacing, which may explain why it falls short of the desired 1:1 Cu-binding site stoichiometry. More uniform spacing of the binding groups has potential to improve binding site utilization by forcing binding to occur via the preferred monodentate arrangement rather than via the thermodynamically preferred bidentate geometry. Furthermore, utilization of vinyl sulfonic acid as a source of binding groups detracts from the renewable and green characteristics of hydrochar. Accordingly, future work in this area should seek to utilize feeds that are naturally abundant in anionic binding sites and/or functional groups that are converted into anionic binding sites during HTC. Questions of binding site access and cooperative effects should be addressed for hydrochars synthesized from renewable or waste resources, using similar methods as shown here for rational sorbent design.

As a final analysis, we evaluated cation-π binding to the furan backbone itself in the absence of acid groups, as a comparison with the arene backbone present in commercial exchange resins. By providing a secondary stabilizing interaction, optimizing the cation-π binding interaction can potentially improve the utilization efficiency of the anionic binding sites—a desired goal as explained previously. In particular, we were interested to understand the effect of locating the cation between nearby rings as compared with interacting with a single ring individually—in the absence of anionic binding groups, such as sulfonate or carboxylate. Figure 9 provides the results of these calculations; Figure 9a,b depict arene binding and Figure 9c,d depict furan binding, respectively. In both cases, locating the Cu(II) between two nearby aromatic rings (either furan or arene) is more stable than interaction with a single aromatic ring. Interestingly, the energy difference is greater for furan-cation interactions (Figure 9c,d) than arene-cation interactions (Figure 9a,b), by approximately 23 kJ mol^{-1}. Accordingly, a final design consideration for custom-synthesis of hydrochar sorbents is inclusion of geometries, which permit formation of furan "pockets" for optimized cation-π interaction. When combined with electrostatic interactions, cation-π interactions can provide a secondary stabilizing force to optimize hydrochar sorption capacity.

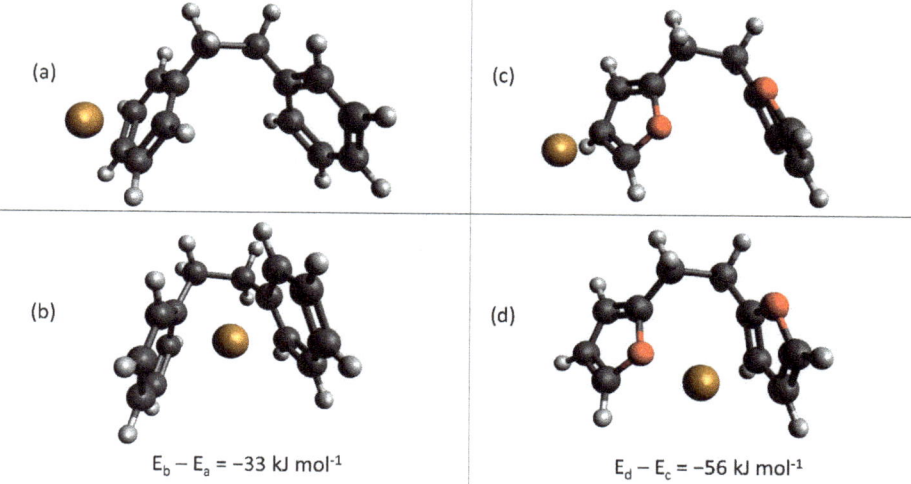

Figure 9. Optimized geometries calculated for arene, (**a,b**), and furan, (**c,d**), cation-π binding, in the absence of anionic binding groups. Legend: ● carbon; ○ hydrogen; ● oxygen; ● sulfur; ● copper. Differences in energy are shown directly in the Figure.

Figures 3 and 6–9 describe a combined experimental and simulation approach for rational design of hydrochar sorbents to exploit electrostatic interactions between anionic functional groups and metal cations. Maximizing the effectiveness of each functional group can be achieved by spacing them uniformly throughout the material, thus emphasizing monodentate binding over bidentate binding. Presumably, highly effective hydrochar sorbents, such as those reported by Demir-Cakan et al. [56], who first demonstrated the acrylic acid co-HTC approach, and Xue et al. [16], who activated peanut hull hydrochar using hydrogen peroxide, already exploit these principles. Likewise, the accuracy of the computational approach in particular will benefit as hydrochar structure is further resolved, especially for materials produced from precursors other than glucose. The result of the rational design approach will be hydrochars with maximized value; thus, making them as competitive as possible with sorbents obtained from non-renewable resources. Although not within the scope of this work, computational modeling should be appropriate for guiding selection of conditions for hydrochar regeneration, for example by using alkali solutions to remove the heavy metal adsorbates. Applying our adsorption capacity results directly to metals other than Cu(II) is not recommended; however, the combined experimental and computational approach should be amenable to any metal cation of interest. Similar analysis can be applied in the future to understand the adsorption of organic substances to hydrochar, as organic pollutants will exhibit different hydrochar interactions than metal cations [104].

4. Conclusions

Glucose hydrochar was studied as a model renewable sorbent for heavy metals, using Cu(II) as a test case for custom-designing a hydrochar sorbent. Glucose hydrochar required alkali activation to exhibit Cu(II) sorption capacity; a strong base (hydroxide) was more effective than a weak base (carbonate) and K^+ counter ions were more effective than Na^+ for activation. In comparison, activated carbon sorption was less than observed for activated glucose hydrochar, despite significant differences in their measured BET surface areas (<10 m^2 g^{-1} compared with >800 m^2 g^{-1}). Similarly, activated carbon did not require alkali treatment to promote cation sorption, consistent with entirely different sorption mechanisms for these two common sorbents. Zeta potential measurements indicated that Cu(II) sorption to hydrochar was due to electrostatic interactions and FT-IR analysis implicated a key role for carboxylate groups. DFT simulations provided further information on Cu-hydrochar binding to the carboxylate site, suggesting beneficial synergy with nearby furan groups.

These results were used as the basis for molecular level design of two hydrochars bearing either carboxylate or sulfonate groups. The designer hydrochars exhibited approximately 50 m^2 g^{-1} Cu(II) sorption capacity, consistent with the role of carboxylate and sulfonate groups in cation binding. Nonetheless, the capacity of the hydrochar sorbents failed to yield the expected increase in sorption capacity. In accordance with the electrostatic binding mechanism, sorption capacity of two ion exchange resins, Amberlyst®-15 and AG® 50W-X4, was studied for comparison with hydrochar. The ion exchange resins outperformed hydrochar on a per mass basis. Interestingly, the hydrochars bound approximately 0.75 Cu ions per acid site, whereas Amberlyst®-15 bound only 0.5 Cu ions per acid site. The superior performance of Amberlyst®-15 compared with hydrochar was therefore attributable entirely to differences in acid site density. AG® 50W-X4, on the other hand, bound nearly 2 Cu ions per acid site, which accounted for its superior performance compared with hydrochar. DFT simulations confirmed that bidentate binding is preferred whenever possible, meaning, uniform spacing of the binding groups will maximize their binding efficiency on a per site basis. Similarly, cooperative effects from nearby aromatic groups (either furan or arene) can promote cation-π interactions that improve binding and acid site utilization. The combination of experimental investigation and computer simulation provide a clear starting point for molecular-level design of hydrochar for use as sorbents, that can be used in future work.

Author Contributions: Conceptualization, M.T.T. and N.A.D.; Methodology, A.B.B., and G.A.T.; Validation, L.D. and B.S.; Formal Analysis L.D. and M.P.R.; Investigation J.T.H., L.D., A.A., and M.P.R.; Data Curation, L.D.; Writing (original draft), Writing (review and editing), N.A.D., M.T.T., A.B.B., A.A, J.T.H. and G.A.T.; Visualization, B.S., L.D., M.T.T.; Supervision, M.T.T., N.A.D.; Project Administration, M.T.T., A.B.B., N.A.D.; Funding Acquisition, M.T.T. All authors have read and agreed to the published version of the manuscript.

Funding: The work was funded by the U.S. National Science Foundation (ENG/CBET 1605916).

Acknowledgments: Rediet Tegegne and Maksim Tyufekchiev supported experimental efforts by synthesizing and titrating several of the hydrochars.

Conflicts of Interest: The authors declare no conflict of interest.

References

1. Christian-Smith, J.; Gleick, P.H.; Cooley, H. Twenty-first century US water policy. In *21st Century US Water Policy*; Oxford University Press: Oxford, UK, 2012.
2. Bellinger, D.C. Lead contamination in Flint—An abject failure to protect public health. *N. Engl. J. Med.* **2016**, *374*, 1101–1103. [CrossRef]
3. Retief, F.P.; Cilliers, L. Lead poisoning in ancient Rome. *Acta Theol.* **2006**, *26*, 147–164. [CrossRef]
4. Fernandez-Luqueno, F.; López-Valdez, F.; Gamero-Melo, P.; Luna-Suárez, S.; Aguilera-González, E.N.; Martínez, A.I.; García-Guillermo, M.; Hernández-Martínez, G.; Herrera-Mendoza, R.; Álvarez-Garza, M.A. Heavy metal pollution in drinking water-a global risk for human health: A review. *Afr. J. Environ. Sci. Technol.* **2013**, *7*, 567–584.

5. Fu, F.; Wang, Q. Removal of heavy metal ions from wastewaters: A review. *J. Environ. Manag.* **2011**, *92*, 407–418. [CrossRef] [PubMed]
6. Wen, T.; Zhao, Y.; Zhang, T.; Xiong, B.; Hu, H.; Zhang, Q.; Song, S. Effect of anions species on copper removal from wastewater by using mechanically activated calcium carbonate. *Chemosphere* **2019**, *230*, 127–135. [CrossRef] [PubMed]
7. Charerntanyarak, L. Heavy metals removal by chemical coagulation and precipitation. *Water Sci. Technol.* **1999**, *39*, 135–138. [CrossRef]
8. Blöcher, C.; Dorda, J.; Mavrov, V.; Chmiel, H.; Lazaridis, N.; Matis, K. Hybrid flotation—Membrane filtration process for the removal of heavy metal ions from wastewater. *Water Res.* **2003**, *37*, 4018–4026. [CrossRef]
9. Aldrich, C.; Feng, D. Removal of heavy metals from wastewater effluents by biosorptive flotation. *Miner. Eng.* **2000**, *13*, 1129–1138. [CrossRef]
10. Al-Rashdi, B.; Johnson, D.; Hilal, N. Removal of heavy metal ions by nanofiltration. *Desalination* **2013**, *315*, 2–17. [CrossRef]
11. Mukherjee, R.; Bhunia, P.; De, S. Impact of graphene oxide on removal of heavy metals using mixed matrix membrane. *Chem. Eng. J.* **2016**, *292*, 284–297. [CrossRef]
12. Oehmen, A.; Viegas, R.; Velizarov, S.; Reis, M.A.; Crespo, J.G. Removal of heavy metals from drinking water supplies through the ion exchange membrane bioreactor. *Desalination* **2006**, *199*, 405–407. [CrossRef]
13. Wang, A.; Zhou, K.; Zhang, X.; Zhou, D.; Peng, C.; Chen, W. Arsenic removal from highly-acidic wastewater with high arsenic content by copper-chloride synergistic reduction. *Chemosphere* **2020**, *238*, 124675. [CrossRef] [PubMed]
14. Tran, T.-K.; Chiu, K.-F.; Lin, C.-Y.; Leu, H.-J. Electrochemical treatment of wastewater: Selectivity of the heavy metals removal process. *Int. J. Hydrogen Energy* **2017**, *42*, 27741–27748. [CrossRef]
15. Hunsom, M.; Pruksathorn, K.; Damronglerd, S.; Vergnes, H.; Duverneuil, P. Electrochemical treatment of heavy metals (Cu^{2+}, Cr^{6+}, Ni^{2+}) from industrial effluent and modeling of copper reduction. *Water Res.* **2005**, *39*, 610–616. [CrossRef] [PubMed]
16. Xue, Y.; Gao, B.; Yao, Y.; Inyang, M.; Zhang, M.; Zimmerman, A.R.; Ro, K.S. Hydrogen peroxide modification enhances the ability of biochar (hydrochar) produced from hydrothermal carbonization of peanut hull to remove aqueous heavy metals: Batch and column tests. *Chem. Eng. J.* **2012**, *200*, 673–680. [CrossRef]
17. Ihsanullah, D.; Abbas, A.; Al-Amer, A.M.; Laoui, T.; Al-Marri, M.J.; Nasser, M.S.; Khraisheh, M.; Atieh, M.A. Heavy metal removal from aqueous solution by advanced carbon nanotubes: Critical review of adsorption applications. *Sep. Purif. Technol.* **2016**, *157*, 141. [CrossRef]
18. Bakkaloglu, I.; Butter, T.; Evison, L.; Holland, F.; Hancockt, I. Screening of various types biomass for removal and recovery of heavy metals (Zn, Cu, Ni) by biosorption, sedimentation and desorption. *Water Sci. Technol.* **1998**, *38*, 269–277. [CrossRef]
19. Rengaraj, S.; Yeon, J.-W.; Kim, Y.; Jung, Y.; Ha, Y.-K.; Kim, W.-H. Adsorption characteristics of Cu (II) onto ion exchange resins 252H and 1500H: Kinetics, isotherms and error analysis. *J. Hazard. Mater* **2007**, *143*, 469–477. [CrossRef]
20. Lin, S.H.; Lai, S.L.; Leu, H.G. Removal of heavy metals from aqueous solution by chelating resin in a multistage adsorption process. *J. Hazard. Mater* **2000**, *76*, 139–153. [CrossRef]
21. Pourhashem, G.; Adler, P.R.; Spatari, S. Time effects of climate change mitigation strategies for second generation biofuels and co-products with temporary carbon storage. *J. Clean. Prod.* **2016**, *112 Pt 4*, 2642. [CrossRef]
22. Bulut, Y. Removal of heavy metals from aqueous solution by sawdust adsorption. *J. Environ. Sci.* **2007**, *19*, 160–166. [CrossRef]
23. Javaid, A.; Bajwa, R.; Shafique, U.; Anwar, J. Removal of heavy metals by adsorption on Pleurotus ostreatus. *Biomass Bioenergy* **2011**, *35*, 1675–1682. [CrossRef]
24. Kılıç, M.; Kırbıyık, C.; Çepelioğullar, Ö.; Pütün, A.E. Adsorption of heavy metal ions from aqueous solutions by bio-char, a by-product of pyrolysis. *Appl. Surf. Sci.* **2013**, *283*, 856–862. [CrossRef]
25. Inyang, M.; Gao, B.; Yao, Y.; Xue, Y.; Zimmerman, A.R.; Pullammanappallil, P.; Cao, X. Removal of heavy metals from aqueous solution by biochars derived from anaerobically digested biomass. *Bioresour. Technol.* **2012**, *110*, 50–56. [CrossRef]
26. Tripathi, M.; Sahu, J.N.; Ganesan, P. Effect of process parameters on production of biochar from biomass waste through pyrolysis: A review. *Renew. Sustain. Energy Rev.* **2016**, *55*, 467–481. [CrossRef]

27. Wiedner, K.; Rumpel, C.; Steiner, C.; Pozzi, A.; Maas, R.; Glaser, B. Chemical evaluation of chars produced by thermochemical conversion (gasification, pyrolysis and hydrothermal carbonization) of agro-industrial biomass on a commercial scale. *Biomass Bioenergy* **2013**, *59*, 264–278. [CrossRef]
28. Sun, Y.; Gao, B.; Yao, Y.; Fang, J.; Zhang, M.; Zhou, Y.; Chen, H.; Yang, L. Effects of feedstock type, production method, and pyrolysis temperature on biochar and hydrochar properties. *Chem. Eng. J.* **2014**, *240*, 574–578. [CrossRef]
29. Sevilla, M.; Ferrero, G.A.; Fuertes, A.B. Beyond KOH activation for the synthesis of superactivated carbons from hydrochar. *Carbon* **2017**, *114*, 50–58. [CrossRef]
30. Wahi, R.; Zuhaidi, N.; Yusof, Y.; Jamel, J.; Kanakaraju, D.; Ngaini, Z. Chemically treated microwave-derived biochar: An overview. *Biomass Bioenergy* **2017**, *107*, 411–421. [CrossRef]
31. Verma, D.; Fortunati, E.; Jain, S.; Zhang, X. *Biomass, Biopolymer-Based Materials, and Bioenergy: Construction, Biomedical, and Other Industrial Applications*; Woodhead Publishing: Cambridge, UK, 2019.
32. Zhang, J.; Zhang, X. 15-The thermochemical conversion of biomass into biofuels. In *Biomass, Biopolymer-Based Materials, and Bioenergy*; Verma, D., Fortunati, E., Jain, S., Zhang, X., Eds.; Woodhead Publishing: Cambridge, UK, 2019; pp. 327–368.
33. Regmi, P.; Moscoso, J.L.G.; Kumar, S.; Cao, X.; Mao, J.; Schafran, G. Removal of copper and cadmium from aqueous solution using switchgrass biochar produced via hydrothermal carbonization process. *J. Environ. Manag.* **2012**, *109*, 61–69. [CrossRef]
34. Marfíl, A.P.; Ocampo-Pérez, R.; Collins-Martínez, V.H.; Flores-Vélez, L.M.; Gonzalez-Garcia, R.; Medellín-Castillo, N.A.; Labrada-Delgado, G.J. Synthesis and characterization of hydrochar from industrial Capsicum annuum seeds and its application for the adsorptive removal of methylene blue from water. *Environ. Res.* **2020**, *184*, 109334. [CrossRef] [PubMed]
35. Funke, A.; Ziegler, F. Hydrothermal carbonization of biomass: A summary and discussion of chemical mechanisms for process engineering. *Biofuels Bioprod. Biorefining* **2010**, *4*, 160–177. [CrossRef]
36. Titirici, M.-M. *Sustainable Carbon Materials from Hydrothermal Processes*; John Wiley & Sons: Hoboken, NJ, USA, 2013.
37. Sun, K.; Tang, J.; Gong, Y.; Zhang, H. Characterization of potassium hydroxide (KOH) modified hydrochars from different feedstocks for enhanced removal of heavy metals from water. *Environ. Sci. Pollut. Res.* **2015**, *22*, 16640–16651. [CrossRef] [PubMed]
38. Poo, K.-M.; Son, E.-B.; Chang, J.-S.; Ren, X.; Choi, Y.-J.; Chae, K.-J. Biochars derived from wasted marine macro-algae (Saccharina japonica and Sargassum fusiforme) and their potential for heavy metal removal in aqueous solution. *J. Environ. Manag.* **2018**, *206*, 364–372. [CrossRef] [PubMed]
39. Karnib, M.; Kabbani, A.; Holail, H.; Olama, Z. *Heavy Metals Removal Using Activated Carbon, Silica and Silica Activated Carbon Composite*; Elsevier Ltd.: Amsterdam, The Netherlands, 2014; Volume 50, pp. 113–120.
40. Oliveira, I.; Blöhse, D.; Ramke, H.-G. Hydrothermal carbonization of agricultural residues. *Bioresour. Technol.* **2013**, *142*, 138–146. [CrossRef]
41. Kruse, A.; Funke, A.; Titirici, M.-M. Hydrothermal conversion of biomass to fuels and energetic materials. *Curr. Opin. Chem. Biol.* **2013**, *17*, 515–521. [CrossRef]
42. Saqib, N.U.; Sharma, H.B.; Baroutian, S.; Dubey, B.; Sarmah, A.K. Valorisation of food waste via hydrothermal carbonisation and techno-economic feasibility assessment. *Sci. Total Environ.* **2019**, *690*, 261–276. [CrossRef]
43. Yu, X.; Liu, S.; Lin, G.; Yang, Y.; Zhang, S.; Zhao, H.; Zheng, C.; Gao, X. Promotion effect of KOH surface etching on sucrose-based hydrochar for acetone adsorption. *Appl. Surf. Sci.* **2019**, *496*, 143617. [CrossRef]
44. Jain, A.; Balasubramanian, R.; Srinivasan, M.P. Tuning hydrochar properties for enhanced mesopore development in activated carbon by hydrothermal carbonization. *Microporous Mesoporous Mater.* **2015**, *203*, 178–185. [CrossRef]
45. Sevilla, M.; Fuertes, A.B. A green approach to high-performance supercapacitor electrodes: The chemical activation of hydrochar with potassium bicarbonate. *Chem. Sus. Chem.* **2016**, *9*, 1880–1888. [CrossRef]
46. Guo, S.; Dong, X.; Liu, K.; Yu, H.; Zhu, C. Chemical, energetic, and structural characteristics of hydrothermal carbonization solid products for lawn grass. *Bio. Resour.* **2015**, *10*, 4613–4625. [CrossRef]
47. Guo, S.; Dong, X.; Wu, T.; Shi, F.; Zhu, C. Characteristic evolution of hydrochar from hydrothermal carbonization of corn stalk. *J. Anal. Appl. Pyrolysis* **2015**, *116*, 1–9. [CrossRef]
48. Guo, S.; Dong, X.; Wu, T.; Zhu, C. Influence of reaction conditions and feedstock on hydrochar properties. *Energy Convers. Manag.* **2016**, *123*, 95–103. [CrossRef]

49. Baccile, N.; Falco, C.; Titirici, M.-M. Characterization of biomass and its derived char using 13 C-solid state nuclear magnetic resonance. *Green Chem.* **2014**, *16*, 4839–4869. [CrossRef]
50. Aragón-Briceño, C.; Ross, A.B.; Camargo-Valero, M.A. Evaluation and comparison of product yields and bio-methane potential in sewage digestate following hydrothermal treatment. *Appl. Energy* **2017**, *208*, 1357–1369. [CrossRef]
51. Aragón-Briceño, C.I.; Grasham, O.; Ross, A.B.; Dupont, V.; Camargo-Valero, M.A. Hydrothermal carbonization of sewage digestate at wastewater treatment works: Influence of solid loading on characteristics of hydrochar, process water and plant energetics. *Renew. Energy* **2020**, *157*, 959–973. [CrossRef]
52. Jackowski, M.; Niedzwiecki, L.; Lech, M.; Wnukowski, M.; Arora, A.; Tkaczuk-Serafin, M.; Baranowski, M.; Krochmalny, K.; Veetil, V.K.; Seruga, P. HTC of Wet Residues of the Brewing Process: Comprehensive Characterization of Produced Beer, Spent Grain and Valorized Residues. *Energies* **2020**, *13*, 2058. [CrossRef]
53. Urbanowska, A.; Kabsch-Korbutowicz, M.; Wnukowski, M.; Seruga, P.; Baranowski, M.; Pawlak-Kruczek, H.; Serafin-Tkaczuk, M.; Krochmalny, K.; Niedzwiecki, L. Treatment of Liquid By-Products of Hydrothermal Carbonization (HTC) of Agricultural Digestate Using Membrane Separation. *Energies* **2020**, *13*, 262. [CrossRef]
54. Wilk, M.; Magdziarz, A.; Jayaraman, K.; Szymańska-Chargot, M.; Gökalp, I. Hydrothermal carbonization characteristics of sewage sludge and lignocellulosic biomass. A comparative study. *Biomass Bioenergy* **2019**, *120*, 166–175. [CrossRef]
55. Liu, Z.; Quek, A.; Kent Hoekman, S.; Srinivasan, M.P.; Balasubramanian, R. Thermogravimetric investigation of hydrochar-lignite co-combustion. *Bioresour. Technol.* **2012**, *123*, 646–652. [CrossRef]
56. Demir-Cakan, R.; Baccile, N.; Antonietti, M.; Titirici, M.M. Carboxylate-rich carbonaceous materials via one-step hydrothermal carbonization of glucose in the presence of acrylic acid. *Chem. Mater.* **2009**, *21*, 484–490. [CrossRef]
57. Liu, K.; Zhang, S.; Hu, X.; Zhang, K.; Roy, A.; Yu, G. Understanding the adsorption of PFOA on MIL-101 (Cr)-based anionic-exchange metal–organic frameworks: Comparing DFT calculations with aqueous sorption experiments. *Environ. Sci. Technol.* **2015**, *49*, 8657–8665. [CrossRef] [PubMed]
58. Vishnyakov, A.; Ravikovitch, P.I.; Neimark, A.V. Molecular Level Models for CO_2 Sorption in Nanopores. *Langmuir* **1999**, *15*, 8736–8742. [CrossRef]
59. Gao, J.; Liu, F.; Ling, P.; Lei, J.; Li, L.; Li, C.; Li, A. High efficient removal of Cu (II) by a chelating resin from strong acidic solutions: Complex formation and DFT certification. *Chem. Eng. J.* **2013**, *222*, 240–247. [CrossRef]
60. Gupta, N.K.; Sengupta, A.; Boda, A.; Adya, V.C.; Ali, S.M. Oxidation state selective sorption behavior of plutonium using N, N -dialkylamide functionalized carbon nanotubes: Experimental study and DFT calculation. *Rsc. Adv.* **2016**, *6*, 78692–78701. [CrossRef]
61. Brown, A.B.; McKeogh, B.J.; Tompsett, G.A.; Lewis, R.; Deskins, N.A.; Timko, M.T. Structural analysis of hydrothermal char and its models by density functional theory simulation of vibrational spectroscopy. *Carbon* **2017**, *125*, 614–629. [CrossRef]
62. Chuntanapum, A.; Matsumura, Y. Formation of tarry material from 5-HMF in subcritical and supercritical water. *Ind. Eng. Chem. Res.* **2009**, *48*, 9837–9846. [CrossRef]
63. Latham, K.G.; Simone, M.I.; Dose, W.M.; Allen, J.A.; Donne, S.W. Synchrotron based NEXAFS study on nitrogen doped hydrothermal carbon: Insights into surface functionalities and formation mechanisms. *Carbon* **2017**, *114*, 566–578. [CrossRef]
64. Sevilla, M.; Fuertes, A.B. Chemical and Structural Properties of Carbonaceous Products Obtained by Hydrothermal Carbonization of Saccharides. *Chem. A Eur. J.* **2009**, *15*, 4195–4203. [CrossRef]
65. Falco, C.; Perez Caballero, F.; Babonneau, F.; Gervais, C.; Laurent, G.; Titirici, M.-M.; Baccile, N. Hydrothermal Carbon from Biomass: Structural Differences between Hydrothermal and Pyrolyzed Carbons via 13C Solid State NMR. *Langmuir* **2011**, *27*, 14460–14471. [CrossRef]
66. Baccile, N.; Laurent, G.; Coelho, C.; Babonneau, F.; Zhao, L.; Titirici, M.-M. Structural Insights on Nitrogen-Containing Hydrothermal Carbon Using Solid-State Magic Angle Spinning 13C and 15N Nuclear Magnetic Resonance. *J. Phys. Chem. C* **2011**, *115*, 8976–8982. [CrossRef]
67. Dieguez-Alonso, A.; Funke, A.; Anca-Couce, A.; Rombolà, A.G.; Ojeda, G.; Bachmann, J.; Behrendt, F. Towards biochar and hydrochar engineering—Influence of process conditions on surface physical and chemical properties, thermal stability, nutrient availability, toxicity and wettability. *Energies* **2018**, *11*, 496. [CrossRef]

68. Naderi, M. Chapter Fourteen-Surface Area: Brunauer–Emmett–Teller (BET). In *Progress in Filtration and Separation*; Tarleton, S., Ed.; Academic Press: Oxford, UK, 2015; pp. 585–608.
69. Sheng, K.; Zhang, S.; Liu, J.; Shuang, E.; Jin, C.; Xu, Z.; Zhang, X. Hydrothermal carbonization of cellulose and xylan into hydrochars and application on glucose isomerization. *J. Clean Prod.* **2019**, *237*, 117831. [CrossRef]
70. Clogston, J.D.; Patri, A.K. Zeta Potential Measurement. In *Characterization of Nanoparticles Intended for Drug Delivery*; McNeil, S.E., Ed.; Humana Press: Totowa, NJ, USA, 2011; pp. 63–70.
71. Yu, G.X.; Jin, M.; Sun, J.; Zhou, X.L.; Chen, L.F.; Wang, J.A. Oxidative modifications of rice hull-based carbons for dibenzothiophene adsorptive removal. *Catal. Today* **2013**, *212*, 31–37. [CrossRef]
72. Seredych, M.; Lison, J.; Jans, U.; Bandosz, T.J. Textural and chemical factors affecting adsorption capacity of activated carbon in highly efficient desulfurization of diesel fuel. *Carbon* **2009**, *47*, 2491–2500. [CrossRef]
73. Boehm, H.P. Surface oxides on carbon and their analysis: A critical assessment. *Carbon* **2002**, *40*, 145–149. [CrossRef]
74. Frisch, M.J.; Trucks, G.W.; Schlegel, H.B.; Scuseria, G.E.; Robb, M.A.; Cheeseman, J.R.; Scalmani, G.; Barone, V.; Petersson, G.A.; Nakatsuji, H.; et al. Gaussian 09. 2009. Available online: https://gaussian.com/g09citation/ (accessed on 1 May 2020).
75. Schmidt, J.R.; Polik, W.F. *WebMO Enterprise, Version 20.0*; WebMO LLC: Holland, MI, USA. Available online: https://www.webmo.net (accessed on 20 May 2020).
76. Becke, A.D. Density-functional exchange-energy approximation with correct asymptotic behavior. *Phys. Rev. A* **1988**, *38*, 3098. [CrossRef]
77. Lee, C.; Yang, W.; Parr, R.G. Development of the Colle-Salvetti correlation-energy formula into a functional of the electron density. *Phys. Rev. B* **1988**, *37*, 785. [CrossRef]
78. Tomasi, J.; Mennucci, B.; Cammi, R. Quantum mechanical continuum solvation models. *Chem. Rev.* **2005**, *105*, 2999–3094. [CrossRef]
79. Kunin, R.; Meitzner, E.F.; Oline, J.A.; Fisher, S.A.; Frisch, N. Characterization of amberlyst 15 macroreticular sulfonic acid cation exchange resin. *Ind. Eng. Chem. Prod. Res. Dev.* **1962**, *1*, 140–144. [CrossRef]
80. Xia, Y.; Liu, H.; Guo, Y.; Liu, Z.; Jiao, W. Immobilization of heavy metals in contaminated soils by modified hydrochar: Efficiency, risk assessment and potential mechanisms. *Sci. Total Environ.* **2019**, *685*, 1201–1208. [CrossRef] [PubMed]
81. Han, L.; Sun, H.; Ro, K.S.; Sun, K.; Libra, J.A.; Xing, B. Removal of antimony (III) and cadmium (II) from aqueous solution using animal manure-derived hydrochars and pyrochars. *Bioresour. Technol.* **2017**, *234*, 77–85. [CrossRef] [PubMed]
82. Shi, Y.; Zhang, T.; Ren, H.; Kruse, A.; Cui, R. Polyethylene imine modified hydrochar adsorption for chromium (VI) and nickel (II) removal from aqueous solution. *Bioresour. Technol.* **2018**, *247*, 370–379. [CrossRef] [PubMed]
83. Timko, M.T.; Wang, J.A.; Burgess, J.; Kracke, P.; Gonzalez, L.; Jaye, C.; Fischer, D.A. Roles of surface chemistry and structural defects of activated carbons in the oxidative desulfurization of benzothiophenes. *Fuel* **2016**, *163*, 223–231. [CrossRef]
84. Petrović, J.T.; Stojanović, M.D.; Milojković, J.V.; Petrović, M.S.; Šoštarić, T.D.; Laušević, M.D.; Mihajlović, M.L. Alkali modified hydrochar of grape pomace as a perspective adsorbent of Pb2+ from aqueous solution. *J. Environ. Manag.* **2016**, *182*, 292–300. [CrossRef]
85. Israelachvili, J.N. *Intermolecular and Surface Forces*; Academic press: Cambridge, MA, USA, 2011.
86. Tran, H.; You, S.-J.; Chao, H.-P. Insight into adsorption mechanism of cationic dye onto agricultural residues-derived hydrochars: Negligible role of π-π interaction. *Korean J. Chem. Eng.* **2017**, *34*, 1708–1720. [CrossRef]
87. Doan, V.; Köppe, R.; Kasai, P.H. Dimerization of Carboxylic Acids and Salts: An IR Study in Perfluoropolyether Media. *J. Am. Chem. Soc.* **1997**, *119*, 9810–9815. [CrossRef]
88. Semerciöz, A.S.; Göğüş, F.; Celekli, A.; Bozkurt, H. Development of carbonaceous material from grapefruit peel with microwave implemented-low temperature hydrothermal carbonization technique for the adsorption of Cu (II). *J. Clean. Prod.* **2017**, *165*, 599–610. [CrossRef]
89. Elaigwu, S.E.; Greenway, G.M. Microwave-assisted and conventional hydrothermal carbonization of lignocellulosic waste material: Comparison of the chemical and structural properties of the hydrochars. *J. Anal. Appl. Pyrolysis* **2016**, *118*, 1–8. [CrossRef]

90. Liang, J.; Liu, Y.; Zhang, J. Effect of Solution pH on the Carbon Microsphere Synthesized by Hydrothermal Carbonization. *Procedia Environ. Sci.* **2011**, *11*, 1322–1327. [CrossRef]
91. Feng, Z.; Odelius, K.; Rajarao, G.K.; Hakkarainen, M. Microwave carbonized cellulose for trace pharmaceutical adsorption. *Chem. Eng. J.* **2018**, *346*, 557–566. [CrossRef]
92. Mihajlović, M.; Petrović, J.; Kragović, M.; Stojanović, M.; Milojković, J.; Lopičić, Z.; Koprivica, M. Effect of KOH activation on hydrochar surface: FT-IR analysis. *RAD Assoc. J.* **2017**, *2*, 65–67.
93. Libra, J.A.; Ro, K.S.; Kammann, C.; Funke, A.; Berge, N.D.; Neubauer, Y.; Titirici, M.-M.; Führer, C.; Bens, O.; Kern, J. Hydrothermal carbonization of biomass residuals: A comparative review of the chemistry, processes and applications of wet and dry pyrolysis. *Biofuels* **2011**, *2*, 71–106. [CrossRef]
94. Xia, Y.; Yang, T.; Zhu, N.; Li, D.; Chen, Z.; Lang, Q.; Liu, Z.; Jiao, W. Enhanced adsorption of Pb(II) onto modified hydrochar: Modeling and mechanism analysis. *Bioresour. Technol.* **2019**, *288*, 121593. [CrossRef]
95. Fornes, F.; Belda, R.M.; Lidón, A. Analysis of two biochars and one hydrochar from different feedstock: Focus set on environmental, nutritional and horticultural considerations. *J. Clean. Prod.* **2015**, *86*, 40–48. [CrossRef]
96. Fang, J.; Gao, B.; Chen, J.; Zimmerman, A.R. Hydrochars derived from plant biomass under various conditions: Characterization and potential applications and impacts. *Chem. Eng. J.* **2015**, *267*, 253–259. [CrossRef]
97. Dupont, AMBERLYST™ 15DRY Polymeric Catalyst, product Data Sheet. 2019. Available online: https://www.dupont.com/content/dam/dupont/amer/us/en/water-solutions/public/documents/en/45-D00927-en.pdf (accessed on 1 December 2019).
98. Nandan, D.; Gupta, A. Solvent sorption isotherms, swelling pressures, and free energies of swelling of polystyrenesulfonic acid type cation exchanger in water and methanol. *J. Phys. Chem.* **1977**, *81*, 1174–1179. [CrossRef]
99. Davies, C.; Yeoman, G. Swelling equilibria with some cation exchange resins. *Trans. Faraday Soc.* **1953**, *49*, 968–974. [CrossRef]
100. Marcus, Y.; Naveh, J. Anion exchange of metal complexes. XVII. Selective swelling of the exchanger in mixed aqueous-organic solvents. *J. Phys. Chem.* **1969**, *73*, 591–596. [CrossRef]
101. Howery, D.G.; Shore, L.; Kohn, B.H. Proton magnetic resonance studies of the structure of water in Dowex 50 W. *J. Phys. Chem.* **1972**, *76*, 578–581. [CrossRef]
102. Mao, Y.; Fung, B.M. A Study of the Adsorption of Acrylic Acid and Maleic Acid from Aqueous Solutions onto Alumina. *J. Colloid Interface Sci.* **1997**, *191*, 216–221. [CrossRef] [PubMed]
103. Sun, B.; Zhao, Y.; Wu, J.-G.; Yang, Q.-C.; Guang-Xian, X. Crystal structure and FT-IR study of cesium 4-methylbenzenesulfonate. *J. Mol. Struct.* **1998**, *471*, 63–66. [CrossRef]
104. Zhang, X.; Gao, B.; Fang, J.; Zou, W.; Dong, L.; Cao, C.; Zhang, J.; Li, Y.; Wang, H. Chemically activated hydrochar as an effective adsorbent for volatile organic compounds (VOCs). *Chemosphere* **2019**, *218*, 680–686. [CrossRef] [PubMed]

© 2020 by the authors. Licensee MDPI, Basel, Switzerland. This article is an open access article distributed under the terms and conditions of the Creative Commons Attribution (CC BY) license (http://creativecommons.org/licenses/by/4.0/).

Article

Sewage Sludge Valorization via Hydrothermal Carbonization: Optimizing Dewaterability and Phosphorus Release

Taina Lühmann and Benjamin Wirth *

Biorefineries Department, Deutsches Biomasseforschungszentrum Gemeinnützige GmbH, Torgauer Straße 116, 04347 Leipzig, Germany; taina.luehmann@ru.nl
* Correspondence: benjamin.wirth@dbfz.de; Tel.: +49-341-2434-449

Received: 25 July 2020; Accepted: 26 August 2020; Published: 26 August 2020

Abstract: As the use of sewage sludge as a fertilizer in agriculture is increasingly restricted in the European Union, other ways to utilize this waste stream need to be developed. Sewage sludge is an ideal input material for the process of hydrothermal carbonization, as it can convert wet biomass into a solid energy carrier with increased mechanical dewaterability. Digested sewage sludge was hydrothermally carbonized at 160–200 °C for 30–60 min with initial pH levels of 1.93–8.08 to determine optimal reaction conditions for enhanced dewaterability and phosphorus release into the liquid phase. Design of experiments was used to develop response surface models, which can be applied to optimize the process conditions. For optimal dewaterability and phosphorus release, low initial pH values (pH 1.93) and mild temperatures around 170 °C are favorable. Because holding time had no statistically relevant effect, a dependency of reaction time was investigated. Though it did not yield substantially different results, it could be included in investigations of short reaction times prospectively. Low reaction temperatures and short holding times are desirable considering economic reasons for scale-up, while the high acid consumption necessary to achieve these results is unfavorable.

Keywords: dry matter; moisture; dewatering; sewage sludge; hydrothermal treatment; phosphorus; RSM; DoE; optimization; regression model

---

1. Introduction

Sewage sludge is widely used in agriculture because high contents of phosphorus, nitrogen, and potassium, as well as organic carbon make it a favorable organic fertilizer [1,2]. Because of concerning substances, such as heavy metals, organic residues, microplastics, and various pathogens, stronger regulations regarding the use as fertilizer have been passed. This has been combined with a requirement to recover phosphorus from sewage sludge in countries such as Germany [3,4], so that new ways for valorization are sought after. After landfilling became restricted in the European Union (EU) in 1999, incineration has become the preferred solution in the EU-15 countries [5]. During incineration, sewage sludge can be mono-combusted in specifically designed furnaces, co-combusted in existing waste incineration plants or added in small quantities in cement kilns [6]. Thermal dewatering, an energy intensive process, is the major drawback of incineration [6].

Hydrothermal carbonization (HTC) is a process that can convert sewage sludge into a more easily mechanically dewaterable substance by breaking up the cell structure, thereby releasing bound water [7–10]. Essentially, it is a process that converts biomass in hot pressurized water yielding an upgraded solid biofuel [11,12]. Typical reaction conditions reported in literature are 180 to 250 °C for 1 to 12 h holding time [11] at an elevated pressure due to water saturation pressure and additional pressure by gaseous compounds formed during the reaction. To measure the dewaterability of sewage sludge, different approaches such as filtration and mechanical dewatering were examined. Danso-Boateng et al. [8] used a filtration cell and calculated the specific cake resistance as a measure for dewaterability. They found that the reaction temperature was the defining factor for dewaterability, especially when compared to holding time. When Escala et al. [9] mechanically dewatered hydrochar (HC) from stabilized sewage sludge that was acidified, they were able to reach dry matter contents of 52 ± 5.5%, compared to the dry matter content of sewage sludge of 30%. Besides improved dewaterability, another advantage of applying HTC to sewage sludge is the opportunity for phosphorus recovery, the most valuable component of wastewater. Distribution of phosphorus in liquid and solid phase as well as chemical composition has been increasingly studied [13–16]. Shi et al. [14] investigated the effect of initial pH and reaction temperature on the distribution of phosphorus. They found that higher temperatures caused phosphorus to be more present in the solid phase while adding large amounts of acid could shift phosphorus into the liquid phase.

The aim of this study is to investigate the dewaterability and phosphorus release of hydrothermally treated sewage sludge as minimum required reaction conditions are yet to be determined. A method to investigate the optimal reaction conditions in a defined reaction space is facilitated by a Design of Experiments (DoE)/Response Surface Model (RSM) approach. For HTC, this has been already utilized frequently [13,17–20] and for sewage sludge dewatering after HTC by Danso-Boateng et al. [8]. While they tested the filterability as an indicator for dewatering, the objective of this study was to take a more practical approach by investigating mechanical dewatering using a screw press at lab-scale. This experimental set-up was implemented to reflect a possible industrial application. Sewage sludge was used as received (without any pre-treatment) from a wastewater treatment plant (WWTP) to identify if mild HTC conditions are suitable for valorization.

Although the process parameters reaction temperature, holding time and initial pH all play a role on phosphorus release and dewaterability, they have only been investigated individually so far. To also consider interaction effects, a DoE approach was taken. This allowed for the development of regression models to predict the above-mentioned parameters as well as typical HTC characteristics. Reaction conditions that are close to industrial applicability, for which low reaction temperatures and short holding times are of interest, were investigated.

2. Materials and Methods

2.1. Digested Sewage Sludge

Digested sewage sludge (DSS) was collected from a municipal WWTP in Leipzig, Germany, where the wastewater undergoes mechanical (screening, grit removal, primary sedimentation), chemical (phosphate precipitation) and biological treatment (aeration reactor, secondary sedimentation). Sludge stabilization includes sludge thickening, digestion, and mechanical dewatering, reaching a dry matter content of 24.3%. The composition of the sample used in this study can be found in Table 1. A sample of approximately 30 L was homogenized and separated into 27 subsamples, which were stored at 4 °C. Before being used in the experiments, they were left to warm up at room temperature.

Table 1. Properties of the tested samples with standard deviations given in parentheses.

C [‡]	H [‡]	N [‡]	S [‡]	O [‡]	Ash [†]	HHV$_{meas}$ [†]	HHV$_{calc}$
(%$_{db}$)						(MJ kg$_{db}^{-1}$)	
32.5 (±0.24)	4.7 (±0.05)	5 (±0.04)	1.7 (±0.01)	15.2 (±0.33)	40.9 (±0.11)	14.3 (±0.10)	13.4

‡ = triplicates; † = duplicates; HHV = higher heating value; meas = measured; calc = calculated; db = dry basis.

The calculated higher heating value (HHV) is slightly lower than the actual measured one as high ash contents of a sample can severely impact the results of various calculation formulas [21]. However, the correlation of ultimate analysis data and HHV used here is reported to fit measured HHVs of HCs very well and the value for DSS was added as a comparison.

2.2. Experimental Method

The DoE included the variation of reaction temperature, holding time, and initial pH. A face-centered central composite design (FCCD) was chosen. Corner points were carried out in duplicate and the center point in triplicate, resulting in 27 experiments. The reaction conditions were (i) reaction temperature: 160, 180, 200 °C, (ii) holding time (starting when the reaction temperature was reached): 30, 60, 90 min, and (iii) initial pH: 2, 4.5, 7. Though holding time only accounts for part of the timespan during which reactions take place, it is the common measure for comparison of HTC studies. For sewage sludge, Escala et al. [9] used the timespan above 180 °C for each experiment. Wang et al. [22] concluded that the conversion of sewage sludge starts at around 130–150 °C, observed by a rapid rise of the organic load found in process water. However, they did not observe decomposition of these organic compounds below 180 °C. In this study, it was investigated if using the timespan above 140 °C instead of holding time for analysis would yield substantially different regression models, in accordance to the values provided by Wang et al. [7].

A graphical illustration of the design space can be found in Figure 1 and the randomized experimental design in Table A1 in Appendix A.

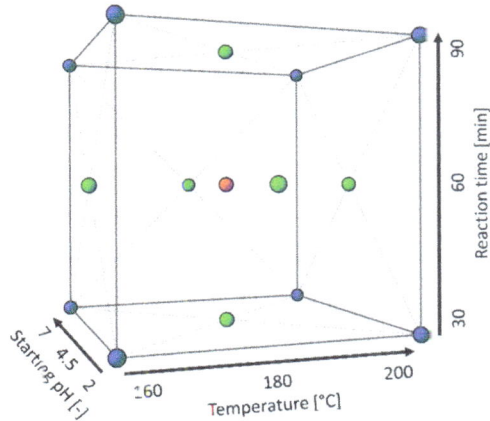

Figure 1. Design space of the face-centered central composite design (FCCD), red dots indicate triplicates, green dots duplicates, and blue dots single repetitions.

In preparation for the pH adjustment, the amount of acid needed for the targeted pH levels was determined according to DIN EN 15933 [23]. After homogenizing a mixture of 10 g of sludge and 100 mL of deionized water on a vibrating plate for 10 min at 130 rpm, concentrated sulfuric acid (18 M) was added whilst stirring until the desired pH levels were reached and held constant for 30 min. The amount of acid needed for the actual amount of sludge could then be extrapolated from these results.

For each experiment, around 350 g of DSS were weighed out and the calculated amount of sulfuric acid (18 M) was added. Of that mixture, 290 g were used for the HTC experiment and 10 g for pH measurement of the acidified sample. The measured pH was later used for regression modeling instead of the targeted pH. Except for the addition of acid, the DSS was not further diluted with water but utilized as received from the WWTP. The experiments were conducted in a 500 mL high-pressure stirred-tank reactor made of stainless steel (BR-500, Berghof Products + Instruments GmbH, Eningen unter Achalm, Germany). The stirrer was set to 100 rpm and the reactor was heated up electrically with a heating rate of 2 K min^{-1}. After reaching the desired temperature, the reactor temperature was kept constant over the holding time. Then, the heater of the reactor was switched off and the reactor was allowed to cool down to room temperature by the environment.

After the process, the HC-water-slurry was separated with a screw press with an integrated scale. The screw press was tightened until a pressure of 300 kg (which corresponds to 2.2 bar) was applied and retightened when the weight dropped by 10 kg. Pressing was stopped when there was no liquid flow for more than one minute. After the pressing process was completed, the HC was divided into two homogeneous samples according to DIN EN ISO 14780 [24] by coning and quartering. The HC was dried and the process water was stored at 4 °C until further use.

2.3. Analytics

In all solid samples (DSS and HC), dry matter content, elemental composition (C, H, N, S) and ash content was analyzed. Additionally, HHV of DSS was measured and the concentrations of phosphorus analyzed in DSS as well as process water.

Dry matter content was determined by placing the samples in a drying oven at 105 °C overnight referring to DIN EN ISO 18134-3 [25]. Elemental carbon, hydrogen, nitrogen, and sulfur (C, H, N, S) were measured according to DIN EN 15104 [26] with an elemental analyzer (vario MACRO CUBE, Elementar Analysensysteme GmbH, Langenselbold, Germany). Ash content was determined by incineration at 550 °C according to DIN EN 14775 [27] and the oxygen content was calculated by difference. HHV was determined by combustion with oxygen in a bomb calorimeter (Parr Instrument GmbH, Moline, IL, USA), according to DIN EN ISO 18125 [28]. Phosphorus was determined according to DIN EN ISO 11885 [29] with a spectrometer (ICP-OES Icap 6300, Thermo Fisher Scientific, Waltham, MA, USA). Every sample was analyzed three times.

2.4. Calculations

Solid yields based on dry matter (DM) were calculated according to Equation (1) with respect to the mass of DSS put into the reactor and the mass of HC after the reaction:

$$\text{Solid yield } (\%_{db}) = \frac{m_{char} \times DM_{char}}{m_{input} \times DM_{input}} \times 100\% \tag{1}$$

For the calculation of carbon yields, Equation (1) was supplemented by the carbon contents of DSS and HC and calculated as follows:

$$\text{Carbon yield } (\%_{db}) = \frac{m_{char} \times DM_{char} \times C_{char}}{m_{input} \times DM_{input} \times C_{input}} \times 100\% \tag{2}$$

Higher heating value (HHV) was calculated from the elemental analysis according to Channiwala and Parikh [30]:

$$\text{HHV } (MJ\ kg_{db}^{-1}) = \\ 0.3491 \times C + 1.1783 \times H - 0.1005 \times S - 0.1034 \times O - 0.0051 \times N - 0.0211 \times Ash \tag{3}$$

The calculated HHV from Equation (3) can be then used to determine the energy yield:

$$\text{Energy yield } (\%_{db}) = \frac{m_{char} \times DM_{char} \times HHV_{char}}{m_{input} \times DM_{input} \times HHV_{input}} \times 100\% \tag{4}$$

To determine the share of phosphorus in the liquid phase (P_{liquid}), the concentrations of phosphorus (c_P) in the DSS and HC are put into relation as follows:

$$P_{liquid}(\%) = \frac{m_{liquid} \times c_{P,liquid}}{m_{input} \times DM_{input} \times c_{P,input}} \times 100\% \tag{5}$$

2.5. Regression Modeling

Because the relationship between certain output variables from HTC reaction conditions was to be determined, a DoE/RSM approach was chosen to obtain the most information out of a limited number of experiments. Regression modelling was performed as described by Montgomery [31]. The FCCD was chosen to aim at fitting a second order model through the following equation:

$$y = \beta_0 + \beta_T x_T - \beta_t x_t + \beta_{pH} x_{pH} + \beta_{Tt} x_T x_t + \beta_{TpH} x_T x_{pH} + \beta_{tpH} x_t x_{pH} + \beta_{T^2} x_T^2 + \beta_{t^2} x_t^2 + \beta_{pH^2} x_{pH}^2 + \varepsilon \quad (6)$$

in which T denotes the reaction temperature, t the holding time after the reaction temperature was reached and pH the initial pH.

Equation (6) can also be expressed as:

$$y = X\beta + \varepsilon \quad (7)$$

where y is a vector of the measured responses (e.g., dry matter content), X is a matrix containing information about the levels of the variables at which the responses were obtained, β is the vector of regression coefficients, and ε is a vector of random errors. With the aim to find the values of the β's that minimize the sum of squares of ε, the least squares estimates are calculated by:

$$\hat{\beta} = (X'X)^{-1} X'y \quad (8)$$

The regression model can now be estimated as:

$$\hat{y} = X \hat{\beta} \quad (9)$$

and the difference between measured responses (y) and fitted values ($\hat{y}$) is given in the vector of residuals:

$$e = y - \hat{y} \quad (10)$$

Analysis of variance was used to refine the regression model by testing the significance of each of the terms in Equation (6) by conducting an F-test. The F-value of a term (e.g., temperature) was calculated by comparing the mean squares of the evaluated term (MS_{Term}) and the remaining residuals ($MS_{Residual}$):

$$F_0 = \frac{MS_{Term}}{MS_{Residual}} = \frac{SS_{Term}/df_{Term}}{SS_{Residual}/df_{Residual}} \quad (11)$$

with SS denoting the sum of squares and df the degrees of freedom. F_0 was tested against $F_{Stat}(\alpha, df_{Term}, df_{Residual})$ on a significance level of $\alpha = 0.1$ and if $F_0 > F_{stat}$, it was concluded that there are significant effects caused by the evaluated term. After the model was refined to include only statistically relevant terms, it was further improved by testing to exclude outliers (results from one or more experiments) from the model. For each term, a t-value was calculated for each run, according to Weisberg [32]:

$$t_i = \frac{\hat{e}_i}{MS_{Residual,i} \sqrt{1 + x_i'(X_i'X_i)^{-1} x_i}} \quad (12)$$

The designation i denotes that data from the ith run was excluded during the calculation and x is essentially a vector containing the first row of X. The t-value was compared with $t_{Stat}(\alpha/2, n - p - 1)$, where n denotes the total number of runs and p the number of parameters in the model. If $|t_i| < t_{Stat}$, the run was excluded from regression modelling.

To give the reader an impression of the regression model quality, the metrics predictive and adjusted R^2 are provided. R^2, the coefficient of determination, expresses how much variation around the mean is explained by the model and is calculated as follows:

$$R^2 = 1 - \frac{SS_{Residual}}{SS_{Model} + SS_{Residual}} \tag{13}$$

where SS_{Model} denotes the sum of squares of the model, which is calculated by adding up all SS_{Term} of the terms included in the model. If the model captures all variation around the mean, R^2 equals one, and if the model cannot account for any variation, R^2 equals zero. R^2 is closest to one when all terms are still in the model and can thus mislead to include terms in the model that do not contribute a statistically significant effect. An adjusted R^2 (R^2_{adj}) accounts for this and, therefore, decreases as the number of model terms increases if the additional terms do not improve the model. Additionally, predictive R^2 (R^2_{pred}) expresses the predictive ability of the model. It takes into account the variation that arises when excluding one run from the model and using it to predict this value [31].

Regression modeling and data plotting were performed with the software packages Design Expert 12 (Stat-Ease, Inc., Minneapolis, MN, USA) and Origin 2020 (OriginLab Corp., Northampton, MA, USA).

3. Results and Discussion

3.1. Regression Modeling

For the parameters of interest, dewaterability, and phosphorus release, as well as general HC characteristics energy, carbon, and solid yield, regression models were developed and are presented in Table 2. The calculation of the F-test (Equation (11)) indicated which model coefficients are statistically relevant, which means that they are adding more information than noise to the model. Factors that were not included in any model ($x_T \times x_t$ and $x_t \times x_{pH}$) were excluded from the table.

Table 2. Coded regression model coefficients for hydrochar (HC) and process water properties.

Target Value	Coded Regression Model Coefficients							Metrics	
	x_T	x_t	x_{pH}	$x_T \times x_{pH}$	x_T^2	x_t^2	x_{pH}^2	R^2_{adj}	R^2_{pred}
Hydrochar Characteristics									
Dry matter (%)	2.70		−3.90	2.22	−1.87			0.89	0.84
Carbon content (%$_{daf}$)	2.12	0.61	2.52	1.31		−0.90		0.92	0.90
HHV (MJ kg$_{db}^{-1}$)	−0.23		−0.16		0.31			0.63	0.53
Solid yield (%$_{db}$)	−3.36		5.19	−4.09				0.72	0.69
Carbon yield (%$_{db}$)	−4.39		4.62	−3.40				0.72	0.69
Energy yield (%$_{db}$)	−4.95		4.04	−4.01				0.78	0.74
Process Water Properties									
P content (mg L^{-1})	−159.55		−2260.21	163.81			1720.72	0.99	0.99
P share (%)	−2.56		−31.46	3.64			24.97	0.99	0.98
Final pH (-)	0.10	0.11	2.17		0.32		−0.87	0.99	0.99

daf = dry and ash-free basis; db = dry basis; P = phosphorus.

The displayed models were further refined by the exclusion of outliers by conducting a *t*-test (Equation (12)). Because of the experimental design (FCCD) and the repetitions of corner and central points, outliers could be excluded without having to limit the order of the model, which led to regression models with good predictive and adjusted R^2 in most cases. The coded regression model coefficients indicate by how much the response is expected to change when the input parameter is changed by one unit. By default, the coded units range from −1 for the lowest level and +1 for the highest level. In this study, the change by one unit relates to a temperature change of 20 °C, holding time change of 30 min or a pH change of 3.1. The comparison of the coefficient's magnitude within a response allows for an estimation of the relative impact that each input parameter has.

Table 2 shows that reaction temperature and initial pH have a statistically significant (at least $p < 0.05$) influence on the output parameters in all models while holding time is only included in two models. As the absolute values of the coefficients of reaction temperature and initial pH are of similar magnitude for HC characteristics, it can be assumed that the effect of changing the temperature by 20 °C is similar to a pH change of 3.1. Even though holding time is included for carbon content, its value is the smallest, which relates to a small influence on the actual model. These trends are different for the investigated process water properties where pH has 12–21 times higher effects compared to temperature. In addition, the final pH is equally influenced by reaction temperature as by holding time. In comparison with other RSM publications, Álvarez-Murillo et al. [19] observed a similar effect of temperature on the solid yield of olive stone ($-5.56 \times x_T$ whereby a temperature change of one unit corresponds to 30 °C). Holding time was included in the model, because their tested maximum duration of 10 h yielded a statistically relevant effect. Mäkelä et al. [17] encountered a much stronger influence on the solid yield (daf) treating industrial mixed sludge from a pulp and paper mill ($-27 \times x_T$ whereas a temperature change of one unit corresponds to 40 °C). This can be attributed to the higher temperatures that were investigated (180–260 °C) and the exclusion of the observed high ash contents for the calculation of the solid yields. Hence, greater differences in yield can be observed. A more in depth analysis of the investigated output parameters is included in the following sections.

3.2. Holding Time vs. Reaction Time

Because holding time does not cover the whole timespan during which reactions take place, a more inclusive measure was used to investigate if this would cause time to become statistically relevant. The time during which temperature in the reactor exceeded 140 °C was chosen as Wang et al. [7] implicated that reactions start taking place between 130 and 150 °C. Reaction time will now be referred to by t_{140} in this study. It should be noted that this value only applies to sewage sludge and should be carefully chosen depending on the input material. Holding time and reaction time for each experiment can be found in Table A1 in Appendix A.

Regression modeling was carried out analogous to the procedure described in the previous sub-chapter. The output parameters for which reaction time is statistically relevant and, thus, included in the regression model are shown in Table 3. Carbon content is still mainly dependent on reaction temperature and initial pH, but the influence of time increased by using reaction time instead of holding time in the model. A comparison of contour plots is provided in Figure A1 in Appendix A and the use of reaction time shows the expected trend that with increasing reaction time and temperature, the carbon content increases This becomes not that clear when holding time is used for the model. While the model for HHV does not include holding time, it does so with reaction time. The general trends that were described in the previous subsection remain as can be observed in the comparison in Figure A2 in Appendix A. For the final pH, reaction temperature is now excluded from the model, but the influence of time is still marginal, compared to initial pH as would be expected. Not much changes when comparing the contour plots in Figure A3 in Appendix A.

Table 3. Coded regression model coefficients for hydrochar (HC) and process water properties with t_{140} reaction time instead of holding time.

Target Value	Coded Regression Model Coefficients						Metrics	
	x_T	$x_{t,140}$	x_{pH}	$x_T{}^*x_{pH}$	$x_{t140}{}^*x_{pH}$	$x_T{}^2$	$x_{pH}{}^2$	$R^2{}_{adj}$ $R^2{}_{pred}$
Hydrochar characteristics								
Carbon content (%$_{daf}$)	1.72	1.04	2.56	1.34		−0.93		0.91 0.89
HHV (MJ kg$_{db}{}^{-1}$)	−0.17	−0.15	−0.17		0.16		0.35	0.73 0.61
Process water properties								
Final pH (-)			0.24	2.17			−0.59	0.99 0.99

daf = dry and ash-free basis; db = dry basis.

The refined data analysis with reaction time instead of holding time did not yield significantly different results. This shows that for batch experiments, variation of holding time does not have a significant influence on most output parameters. It could be considered to be kept constant in future DoEs, unless parameters such as carbon content, HHV, and final pH are of major interest and holding or reaction time will be varied in a much wider range. However, holding time is clearly important when developing continuous processes as it will alter the size of the equipment for a certain targeted mass flow. This has to be addressed with corresponding equipment in lab-scale to allow for a smaller difference between holding time and time above a certain reaction temperature.

3.3. Hydrochar Properties

To verify the results and to form a basis for comparison, a van-Krevelen-diagram was created (Figure 2) and common HC properties such as solid, carbon, and energy yield were calculated (Table 4). For reasons of clarity and comprehensibility, only a selection of HCs as well as DSS is shown in these illustrations. The HCs produced at the mildest (160 °C, 30 min, pH 7) and harshest conditions (200 °C, 90 min, pH 2) are shown to illustrate the range of reaction severity. To estimate the individual effect of the different reaction parameters, each parameter was varied from the mildest to the harshest condition while the other two remained at the mildest.

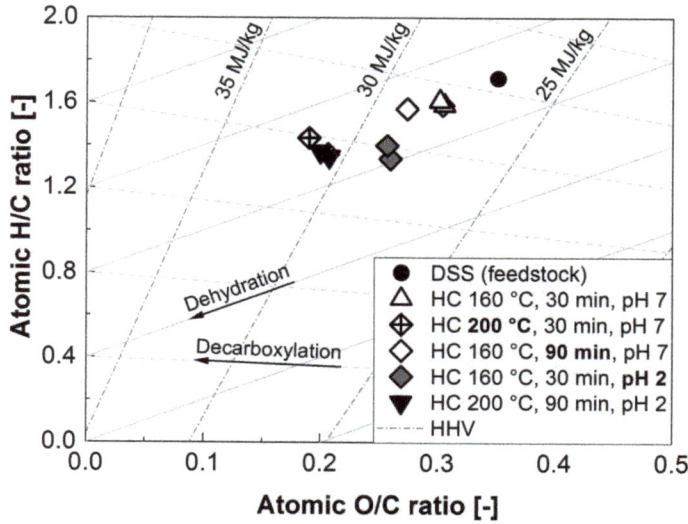

Figure 2. van-Krevelen-diagram of selected hydrochars (HC) and digested sewage sludge (DSS) at different process conditions.

Table 4. Hydrochar (HC) properties in relation to reaction conditions.

T (°C)	t (min)	pH (-)	C Content (%$_{daf}$)	N Content (%$_{daf}$)	HHV (MJ kg$_{db}^{-1}$)	Solid Yield (%$_{db}$)	C Yield (%$_{db}$)	Energy Yield (%$_{db}$)
160	30	7.5 ± 0.23	57.3 ± 0.04	4.6 ± 0.16	13.7 ± 0.27	90.0 ± 0.88	86.4 ± 2.02	92.1 ± 2.71
200	30	7.9 ± 0.23	63.8 ± 0.63	3.7 ± 0.19	13.0 ± 0.52	76.9 ± 0.69	71.9 ± 2.32	74.9 ± 3.68
160	90	7.7 ± 0.07	58.3 ± 1.02	4.4 ± 0.04	13.6 ± 0.00	85.8 ± 7.44	82.2 ± 6.94	87.5 ± 7.58
160	30	2.2 ± 0.14	55.5 ± 0.13	3.6 ± 0.09	14.5 ± 0.62	62.7 ± 18.60	62.4 ± 15.93	67.4 ± 17.24
200	90	2.1 ± 0.05	57.9 ± 0.45	3.1 ± 0.08	13.3 ± 0.07	74.2 ± 1.77	68.1 ± 1.46	73.5 ± 1.37

T = reaction temperature; t = holding time; C = carbon; N = nitrogen; HHV = higher heating value.

By visualizing the atomic H/C versus O/C ratios of DSS and HCs in Figure 2, insights into the governing reactions can be gained. Generally, carbonization decreases both H/C and O/C ratios and, depending on how the ratios change in relation, an assumption about the major governing reaction can be made. In this case, the atomic ratios of the HCs primarily follow dehydration, which confirms the finding of Wang and Li [33] that reactions at temperatures below 180 °C are governed by dehydration. When comparing the HCs that were both produced at 160 °C and pH 7, but at differing holding times of 30 and 90 min, it becomes apparent that varying the holding time between these time intervals does not have a noticeable effect. Contrarily, the variation of pH or temperature leads to a more severe conversion of the feedstock. The pH variation resulted in HCs with lower H/C ratios than caused by dehydration reactions only.

Considering the lower nitrogen contents of pH-adjusted HCs (Table 4), it seems like the pH adjustment favored deamination, which is besides decarboxylation the major conversion mechanism for proteins [22]. This causes the additional removal of hydrogen, resulting in a lower H/C ratio. When analyzing the influence of temperature variation, HC produced at 200 °C possesses almost the same atomic ratios as the HC produced at the harshest combined conditions (200 °C, 90 min, pH 2).

It can be seen from the data that the initial calculated HHV based on dry matter of DSS is slightly higher (13.4 MJ kg$_{db}^{-1}$, Table 1) than the HHV of some of the produced HCs (12.7–15.0 MJ kg$_{db}^{-1}$). This is due to the accumulation of ash in HC (increasing ash content from 40.9%$_{db}$ in DSS to 39.2–53.4%$_{db}$ in HC). Calculating HHV on a dry and ash free basis (daf) results in the expected increase when comparing the resulting chars (26.0–30.9 MJ kg$_{daf}^{-1}$) with the initial DSS (22.6 MJ kg$_{daf}^{-1}$). The HC characteristics solid, carbon, and energy yield were calculated according to the described approach (Equations (1), (2), and (4), respectively). They follow the expected trajectory that more severe reactions result in lower yields. Conducting an analysis of variance for the respective yields showed for reaction temperature ($p < 0.01$), initial pH ($p < 0.001$), and their interaction ($p < 0.05$) to have statistically significant effects on the yields. The regression models in terms of actual factors (T in °C, t in min and pH) are as follows:

$$\text{Solid yield (\%}_{db}) = 40.14 + 0.16x_T + 13.64x_{pH} - 0.066x_T \times x_{pH} \quad (14)$$

$$\text{Carbon yield (\%}_{db}) = 56.28 + 0.057x_T + 11.45x_{pH} - 0.055x_T \times x_{pH} \quad (15)$$

$$\text{Energy yield (\%}_{db}) = 59.31 + 0.079x_T + 13.04x_{pH} - 0.065x_T \times x_{pH} \quad (16)$$

In contrast to the majority of studies, holding time was found not to be statistically relevant in these experiments. This is because others investigated a much longer time span [18,19]. Therefore, it can be concluded that for solid, carbon, and energy yield, a variation of holding time between 30 and 90 min does not cause statistically different results, and the time could be minimized. Danso-Boateng et al. [34] developed a regression model for the solid yield of primary sewage sludge (SY = 118.49 − 0.26x_T − 0.04x_t, actual factors, range of 140–200 °C and 15–240 min) and it compares quite well with Equation (14) when set at a neutral pH of 7 (SY = 135.62 − 0.30x_T, actual factors). Though holding time is included, the main effect comes from temperature, as found in this study.

The regression model of the solid yield is displayed as a contour plot in Figure 3 as a representative for the other yields, as they follow the same trend. It shows that mild conditions (low reaction temperature and neutral pH) lead to higher solid, carbon, and energy yields. While Mäkelä et al. [18] found no effect of the addition of acid on the yields when investigating industrial mixed sludge from a pulp and paper mill, these experiments show a clear effect. This is likely to occur because a wider range of pH with a more concentrated acid was tested. Unlike Reza et al. [35], who found that mass yield was more affected by reaction temperature than initial pH when converting wheat straw, the opposite is indicated here with pH having 1.5 times the effect compared to that of reaction temperature.

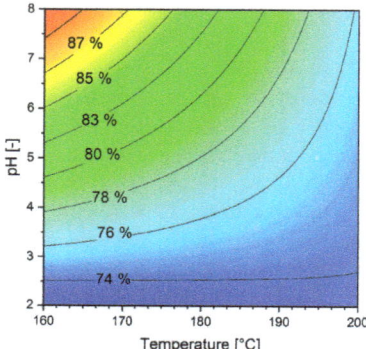

Figure 3. Regression model contour plot showing the influence of initial pH and reaction temperature on the solid yield (%$_{db}$).

The dry matter content after mechanical dewatering was used as a measure of the increase in sludge dewaterability. Analysis of variance has shown that only the influence of reaction temperature ($p < 0.001$) and initial pH ($p < 0.001$) was statistically relevant (see Table 2). Furthermore, their interaction ($p < 0.001$) as well as the quadratic term of temperature ($p < 0.01$) are included in the model due to their high significance level, resulting in the following equation with actual factors:

$$\text{Dry matter content } (\%) = -91.48 + 1.64 x_T - 7.74 x_{pH} + 0.036 x_T \times x_{pH} - 0.0047 x_T^2 \qquad (17)$$

The resulting response contour, plotted in Figure 4, shows the general trend that low pH and higher temperatures favor dewaterability. Initial pH has a clear and strong influence, as a more acidic initial pH yields increasing dry matter contents. A deviation from the expected trend exists at low pH levels, where an increase in reaction temperature does not result in ever increasing dry matter contents. This trend is caused by the quadratic term x_T^2 in the regression model, which causes the slight curvature. However, when determining to what extent this term contributes to the model by comparing the SS of the term x_T^2 to the total SS, it becomes clear that it contributes only with 5% to the model and should not be emphasized.

The regression model of the dry matter content is in agreement with Danso-Boateng et al. [8], Gao et al. [36], and Wang et al. [37], who also found a strong influence of temperature on dewaterability, but not of holding time. When the threshold time of 30 min is exceeded, there is no longer any holding time dependency according to Wang et al. [37]. At the selected temperatures, the typical HTC reactions already take place sufficiently and another increase in temperature will yield even faster reactions. Because dehydration and decarboxylation cause a decrease in hydrophilic functional groups, such as hydroxyl and carboxyl groups [38], the product is more easily dewatered already by a simple mechanical press. The effect of acidic pH on improved dewaterability can be explained by the enhancement of dehydration reactions, caused by sulfuric acid [39], which adds to the temperature effect. Dewaterability is additionally improved by the floc structure disintegration, which releases the bound water as free water [22]. Escala et al. [9], who conducted a comparable experiment, achieved similar dry matter contents. They were able to reach a dry matter content of 52 ± 5.5% after applying a pressure of 40 bar to HC produced at 205 °C and 24 min. This is higher than the 45.6 ± 2.24% achieved in this study at comparable reaction conditions (200 °C, 30 min, pH 2 and 7) and is very likely due to a much lower dewatering pressure.

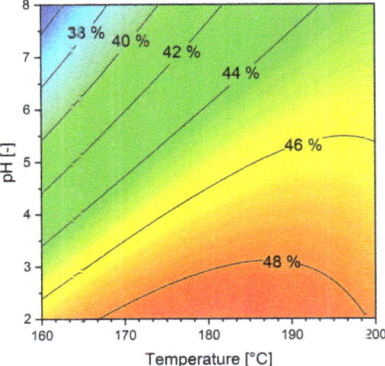

Figure 4. Regression model contour plot showing the influence of initial pH and reaction temperature on the dry matter content (%) after mechanical dewatering.

3.4. Phosphorus Release

For the phosphorus distribution, the pH value is the most significant driver and the variation of holding time showed no statistically relevant effect again, resulting in the following regression models with actual factors:

$$\text{Phosphorus content } (\text{mg L}^{-1}) = 12.97 - 0.021 x_T - 3.03 x_{pH} + 0.027 x_T \times x_{pH} + 0.18 x_T^2 \quad (18)$$

$$\text{Phosphorus share } (\%) = 204.32 - 0.42 x_T - 47.22 x_{pH} + 0.059 x_T \times x_{pH} + 2.63 x_T^2 \quad (19)$$

A lower initial pH generally leads to a lower pH value after the reaction (Figure 5a). This lower pH value favors the solution of phosphorus from the solid and the phosphorus concentration in the process water increases. The influence of the pH value is particularly evident in a range of the initial pH of 2–5 (Figure 5b).

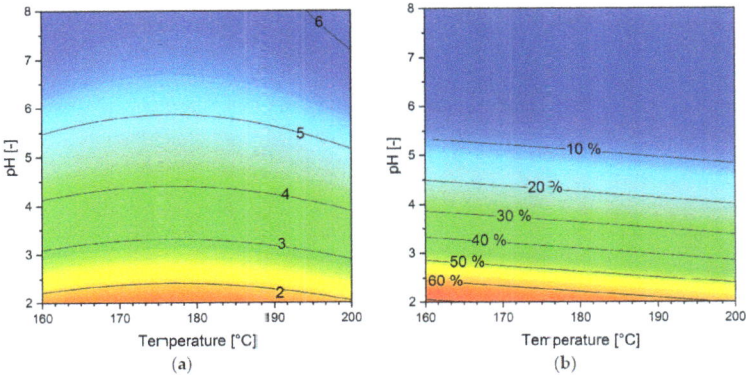

Figure 5. Regression model contour plots showing (**a**) the dependence of process water pH and (**b**) the dependence of liquid phase phosphorus release (% of total P) on reaction temperature and initial pH.

At neutral pH, not much phosphorus will be dissolved by varying holding time and reaction temperature. As others already found, most of the phosphorus is retained in the solid phase [9,13,22]. However, the variation of the initial pH has a major influence on the transfer of phosphorus into the liquid phase. When Shi et al. [14] reached an initial pH of 0.24 and applied HTC conditions of 170 °C and 30 min, they were able to transfer 83% of phosphorus into the liquid phase. The results of this study confirm their findings, as the maximum phosphorus share achieved in these experiments is 78% at 160 °C and an initial pH of 1.93. An even higher share of 88% was achieved in an experiment at 160 °C and an initial pH of 2.25 but was excluded as an outlier from the model as a result from the t-test according to Equation (12). It is, thus, possible to transfer phosphorus into the liquid phase to be potentially recovered by precipitation, e.g., as magnesium ammonium phosphate (struvite) [14], though this comes at the cost of high acid consumption. Additionally, precipitation of phosphorus in HTC process water brings additional challenges due to the huge variety of other components found in the process water.

The future upscaling of this process requires considerations regarding holding time, mixing behavior and heat transfer, especially because a continuous process is desired. While the results of this study show that the holding time can be minimized to 30 min for maximum dewaterability and phosphorus release in a batch process, this time needs to be assessed anew for the application in a continuous process. During the heating time in a batch process, the conversion of sewage sludge already begins and, therefore, the holding time of a batch cannot be directly transferred to a continuous process. More severe reaction conditions (higher reaction temperature and longer holding time) would be necessary, if energy densification and a further increase in HHV are desired [17].

4. Conclusions

The results of this study provide useful information for the design and optimization of HTC systems for sewage sludge treatment. It aimed at finding the optimal HTC reaction conditions for a maximized mechanical dewaterability and phosphorus release in the liquid phase. In the investigated design space (160–200 °C, 30–90 min, pH 1.93–8.09), low initial pH levels are desirable and holding time can be minimized as it had no statistically relevant effect. The influence of reaction temperature is not completely clear. While low temperatures are somewhat more advantageous for phosphorus release, medium temperatures are favorable for dewaterability.

It was investigated if using reaction time as an input parameter instead of holding time would cause different results. Though only minor changes were observed, some models were more refined. However, with regard to future experimental plans using batch reactors, aiming at identifying the impact of HTC process parameters on several product characteristics, it could be considered to exclude holding time from the parameter variation. Residence time is of course a main factor in the design of larger plants, but can only be addressed sufficiently in lab-scale using equipment allowing continuous operation to minimize actual heating and cooling times.

A numerical optimization to maximize mechanical dewaterability and phosphorus release yielded achievable dry matter contents of 48.6% and a phosphorus share of 70.3% in the liquid phase at 170 °C and an initial pH of 1.93. This would be accompanied by an HHV of 13.95 MJ kg_{db}^{-1}.

Author Contributions: Conceptualization, T.L. and B.W.; Formal analysis, T.L.; Funding acquisition, B.W.; Investigation, T.L.; Methodology, T.L. and B.W.; Project administration, T.L.; Supervision, B.W.; Validation, T.L. and B.W.; Visualization, T.L.; Writing—original draft, T.L.; Writing—review and editing, B.W. All authors have read and agreed to the published version of the manuscript.

Funding: This research received funding by the Federal Ministry of Food and Agriculture of Germany (BMEL) and by the Sächsische Aufbaubank (SAB) financed by the European Regional Development Fund (ERDF) under grant no. 100283030.

Acknowledgments: The authors would like to thank all of the laboratory and technical staff at DBFZ for conducting the experiments and analyses. Furthermore, the authors would like to thank L. Prokot for her contributions to the experimental part of this study.

Conflicts of Interest: The authors declare no conflict of interest.

Appendix A

Table A1. Experimental design (temperature, holding time, and pH) as well as reaction time and measured pH.

Experiment	T (°C)	t (min)	t_{140} (min)	pH, DoE (-)	pH, Measured (-)
1	160	90	275	7.0	7.8
2	180	60	251	4.5	5.6
3	200	90	411	7.0	8.1
4	160	30	130	2.0	2.1
5	160	90	274	7.0	7.7
6	200	30	247	2.0	2.3
7	160	30	144	7.0	7.7
8	200	30	207	7.0	7.8
9	200	90	365	7.0	7.1
10	160	30	147	7.0	7.4
11	180	60	221	7.0	7.3
12	180	30	178	4.5	5.3
13	160	90	274	2.0	2.2
14	200	30	236	2.0	2.2
15	180	60	264	2.0	2.1
16	160	90	280	2.0	1.9
17	180	90	352	4.5	5.5
18	200	60	324	4.5	5.5
19	160	60	194	4.5	5.3
20	200	90	398	2.0	2.1
21	180	60	251	4.5	5.4
22	180	60	272	4.5	5.4
23	180	60	241	4.5	5.4
24	160	30	155	2.0	2.3
25	200	30	242	7.0	8.1
26	200	90	413	2.0	2.0

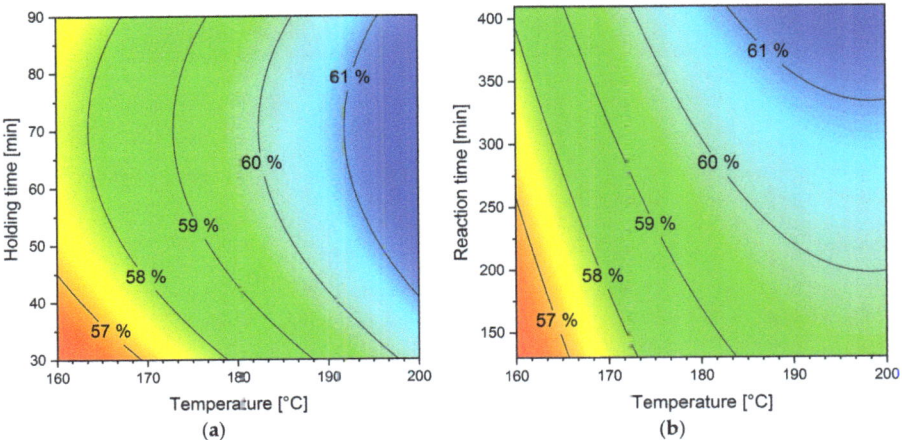

Figure A1. Regression model contour plots showing the dependence of carbon content (%$_{daf}$) on reaction temperature and (**a**) holding time and (**b**) reaction time.

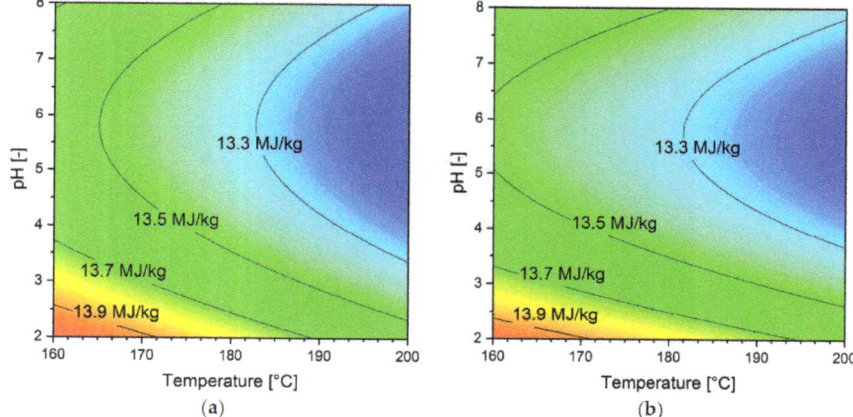

Figure A2. Regression model contour plots showing the dependence of the higher heating value (MJ kg_{TS}^{-1}) on reaction temperature and initial pH based on a model referring to (**a**) holding time and (**b**) reaction time (contours shown at the respective time at the corresponding center point).

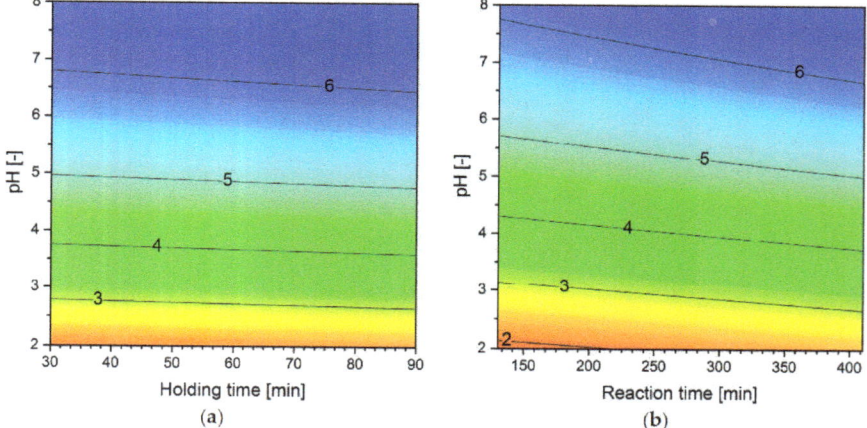

Figure A3. Regression model contour plots showing the dependence of the process water pH (-) on initial pH and (**a**) holding time and (**b**) reaction time.

References

1. Sterritt, R.M.; Lester, J.N. The value of sewage sludge to agriculture and effects of the agricultural use of sludges contaminated with toxic elements: A review. *Sci. Total Environ.* **1980**, *16*, 55–90. [CrossRef]
2. Beni, C.; Servadio, P.; Marconi, S.; Neri, U.; Aromolo, R.; Diana, G. Anaerobic Digestate Administration: Effect on Soil Physical and Mechanical Behavior. *Commun. Soil Sci. Plant Anal.* **2012**, *43*, 821–834. [CrossRef]
3. Hudcová, H.; Vymazal, J.; Rozkošný, M. Present restrictions of sewage sludge application in agriculture within the European Union. *Soil Water Res.* **2019**, *14*, 104–120. [CrossRef]
4. BMU. Verordnung über die Verwertung von Klärschlamm, Klärschlammgemisch und Klärschlammkompost-AbfKlärV. BGBl. I S. 3465. 2017. Available online: https://www.bmu.de/en/law/sewage-sludge-ordinance/ (accessed on 21 August 2020).
5. Raheem, A.; Sikarwar, V.S.; He, J.; Dastyar, W.; Dionysiou, D.D.; Wang, W.; Zhao, M. Opportunities and challenges in sustainable treatment and resource reuse of sewage sludge: A review. *Chem. Eng. J.* **2018**, *337*, 616–641. [CrossRef]

6. Werther, J.; Ogada, T. Sewage sludge combustion. *Prog. Energy Combust. Sci.* **1999**, *25*, 55–116. [CrossRef]
7. Wang, T.; Zhai, Y.; Zhu, Y.; Li, C.; Zeng, G. A review of the hydrothermal carbonization of biomass waste for hydrochar formation: Process conditions, fundamentals, and physicochemical properties. *Renew. Sust. Energ. Rev.* **2018**, *90*, 223–247. [CrossRef]
8. Danso-Boateng, E.; Holdich, R.G.; Wheatley, A.D.; Martin, S.J.; Shama, G. Hydrothermal carbonization of primary sewage sludge and synthetic faeces: Effect of reaction temperature and time on filterability. *Environ. Prog. Sustain. Energy* **2015**, *34*, 1279–1290. [CrossRef]
9. Escala, M.; Zumbühl, T.; Koller, C.; Junge, R.; Krebs, R. Hydrothermal Carbonization as an Energy-Efficient Alternative to Established Drying Technologies for Sewage Sludge: A Feasibility Study on a Laboratory Scale. *Energy Fuels* **2012**, *27*, 454–460. [CrossRef]
10. Meng, D.; Jiang, Z.; Yoshikawa, K.; Mu, H. The effect of operation parameters on the hydrothermal drying treatment. *Renew. Energy* **2012**, *42*, 90–94. [CrossRef]
11. Libra, J.; Ro, K.S.; Kammann, C.I.; Funke, A.; Berge, N.D.; Neubauer, Y.; Titirici, M.-M.; Fühner, C.; Bens, O.; Kern, J.; et al. Hydrothermal carbonization of biomass residuals: A comparative review of the chemistry, processes and applications of wet and dry pyrolysis. *Biofuels* **2011**, *2*, 71–106. [CrossRef]
12. Reza, M.T.; Andert, J.; Wirth, B.; Busch, D.; Pielert, J.; Lynam, J.G.; Mumme, J. Hydrothermal Carbonization of Biomass for Energy and Crop Production. *Appl. Bioenergy* **2014**, *1*, 11–29. [CrossRef]
13. Heilmann, S.M.; Molde, J.S.; Timler, J.G.; Wood, B.M.; Mikula, A.L.; Vozhdayev, G.V.; Colosky, E.C.; Spokas, K.A.; Valentas, K.J. Phosphorus reclamation through hydrothermal carbonization of animal manures. *Environ. Sci. Technol.* **2014**, *48*, 10323–10329. [CrossRef] [PubMed]
14. Shi, Y.; Luo, G.; Rao, Y.; Chen, H.; Zhang, S. Hydrothermal conversion of dewatered sewage sludge: Focusing on the transformation mechanism and recovery of phosphorus. *Chemosphere* **2019**, *228*, 619–628. [CrossRef]
15. Wang, T.; Zhai, Y.; Zhu, Y.; Peng, C.; Wang, T.; Xu, B.; Li, C.; Zeng, G. Feedwater pH affects phosphorus transformation during hydrothermal carbonization of sewage sludge. *Bioresour. Technol.* **2017**, *245*, 182–187. [CrossRef] [PubMed]
16. Ovsyannikova, E.; Arauzo, P.J.; Becker, G.C.; Kruse, A. Experimental and thermodynamic studies of phosphate behavior during the hydrothermal carbonization of sewage sludge. *Sci. Total Environ.* **2019**, *692*, 147–156. [CrossRef] [PubMed]
17. Mäkelä, M.; Benavente, V.; Fullana, A. Hydrothermal carbonization of lignocellulosic biomass: Effect of process conditions on hydrochar properties. *Appl. Energy* **2015**, *155*, 576–584. [CrossRef]
18. Mäkelä, M.; Benavente, V.; Fullana, A. Hydrothermal carbonization of industrial mixed sludge from a pulp and paper mill. *Bioresour. Technol.* **2016**, *200*, 444–450. [CrossRef]
19. Álvarez-Murillo, A.; Román, S.; Ledesma, B.; Sabio, E. Study of variables in energy densification of olive stone by hydrothermal carbonization. *J. Anal. Appl. Pyrolysis* **2015**, *113*, 307–314. [CrossRef]
20. Román, S.; Libra, J.; Berge, N.D.; Sabio, E.; Ro, K.S.; Li, L.; Ledesma, B.; Álvarez, A.; Bae, S. Hydrothermal Carbonization: Modeling, Final Properties Design and Applications: A Review. *Energies* **2018**, *11*, 216. [CrossRef]
21. Kieseler, S.; Neubauer, Y.; Zobel, N. Ultimate and Proximate Correlations for Estimating the Higher Heating Value of Hydrothermal Solids. *Energy Fuels* **2013**, *27*, 908–918. [CrossRef]
22. Wang, L.; Chang, Y.; Li, A. Hydrothermal carbonization for energy-efficient processing of sewage sludge: A review. *Renew. Sustain. Energ. Rev.* **2019**, *108*, 423–440. [CrossRef]
23. Deutsches Institut für Normung e. V. *Sludge, Treated Biowaste and Soil—Determination of pH (EN 15933:2012)*; German Version; Beuth Verlag GmbH: Berlin, Germany, 2012; DIN EN 15933:2012-11.
24. Deutsches Institut für Normung e. V. *Solid Biofuels—Sample Preparation (ISO 14780:2017)*; German Version; Beuth Verlag GmbH: Berlin, Germany, 2017; DIN EN ISO 14780:2017-08.
25. Deutsches Institut für Normung e. V. *Solid Biofuels—Determination of Moisture Content—Oven Dry Method—Part 3: Moisture in General Analysis Sample (ISO 18134-3:2015)*; German Version; Beuth Verlag GmbH: Berlin, Germany, 2015; DIN EN ISO 18134-3:2015-12.
26. Deutsches Institut für Normung e. V. *Solid biofuels—Determination of Total Content of Carbon, Hydrogen and Nitrogen—Instrumental Methods (EN 15104:2011)*; German Version; Beuth Verlag GmbH: Berlin, Germany, 2011; DIN EN 15104:2011-04.
27. Deutsches Institut für Normung e. V. *Solid Biofuels—Determination of Ash Content (EN 14775:2009)*; German Version; Verlag GmbH: Berlin, Germany, 2012; DIN EN 14775:2012-11.

28. Deutsches Institut für Normung e. V. *Solid Biofuels—Determination of Calorific Value (ISO 18125:2017)*; German Version; Verlag GmbH: Berlin, Germany, 2017; DIN EN ISO 18125:2017-08.
29. Deutsches Institut für Normung e. V. *Water quality—Determination of Selected Elements by Inductively Coupled Plasma Optical Emission Spectrometry (ICP-OES) (ISO 11885:2007)*; German Version; Beuth Verlag GmbH: Berlin, Germany, 2009; DIN EN ISO 11885:2009-09.
30. Channiwala, S.A.; Parikh, P.P. A unified correlation for estimating HHV of solid, liquid and gaseous fuels. *Fuel* **2002**, *81*, 1051–1063. [CrossRef]
31. Montgomery, D.C. *Design and Analysis of Experiments*, 8th ed.; John Wiley & Sons Inc.: Singapore, 2013; ISBN 978-1-118-09793-9.
32. Weisberg, S. *Applied Linear Regression*, 3rd ed.; Wiley-Interscience: Hoboken, NJ, USA, 2005; ISBN 9780471663799.
33. Wang, L.; Li, A. Hydrothermal treatment coupled with mechanical expression at increased temperature for excess sludge dewatering: The dewatering performance and the characteristics of products. *Water Res.* **2015**, *68*, 291–303. [CrossRef] [PubMed]
34. Danso-Boateng, E.; Shama, G.; Wheatley, A.D.; Martin, S.J.; Holdich, R.G. Hydrothermal carbonisation of sewage sludge: Effect of process conditions on product characteristics and methane production. *Bioresour. Technol.* **2015**, *177*, 318–327. [CrossRef]
35. Reza, M.T.; Rottler, E.; Herklotz, L.; Wirth, B. Hydrothermal carbonization (HTC) of wheat straw: Influence of feedwater pH prepared by acetic acid and potassium hydroxide. *Bioresour. Technol.* **2015**, *182*, 336–344. [CrossRef]
36. Gao, N.; Li, Z.; Quan, C.; Miscolczi, N.; Egedy, A. A new method combining hydrothermal carbonization and mechanical compression in-situ for sewage sludge dewatering: Bench-scale verification. *J. Anal. Appl. Pyrolysis* **2019**, *139*, 187–195. [CrossRef]
37. Wang, L.; Zhang, L.; Li, A. Hydrothermal treatment coupled with mechanical expression at increased temperature for excess sludge dewatering: Influence of operating conditions and the process energetics. *Water Res.* **2014**, *65*, 85–97. [CrossRef]
38. Funke, A.; Ziegler, F. Hydrothermal carbonization of biomass: A summary and discussion of chemical mechanisms for process engineering. *Biofuels Bioprod. Biorefining* **2010**, *4*, 160–177. [CrossRef]
39. Peterson, A.A.; Vogel, F.; Lachance, R.P.; Fröling, M.; Michael, J.; Antal, J.R.; Tester, J.W. Thermochemical biofuel production in hydrothermal media: A review of sub- and supercritical water technologies. *Energy Environ. Sci.* **2008**, *1*, 32–65. [CrossRef]

© 2020 by the authors. Licensee MDPI, Basel, Switzerland. This article is an open access article distributed under the terms and conditions of the Creative Commons Attribution (CC BY) license (http://creativecommons.org/licenses/by/4.0/).

Article

Comparative Studies on Water- and Vapor-Based Hydrothermal Carbonization: Process Analysis

Kyoung S. Ro [1],*, Judy A. Libra [2],* and Andrés Alvarez-Murillo [3]

1. USDA-ARS Coastal Plains Soil, Water & Plant Research Center, Florence, SC 29501, USA
2. Leibniz Institute for Agricultural Engineering and Bioeconomy, 14469 Potsdam, Germany
3. Department of Applied Physics, School of Industrial Engineering, University of Extremadura, 06006 Badajoz, Spain; andalvarez@unex.es
* Correspondence: kyoung.ro@usda.gov (K.S.R.); jlibra@atb-potsdam.de (J.A.L.); Tel.: +1-843-669-5203 (K.S.R.); +49-331-5699-866 (J.A.L.)

Received: 7 August 2020; Accepted: 29 October 2020; Published: 2 November 2020

Abstract: Hydrothermal carbonization (HTC) reactor systems used to convert wet organic wastes into value-added hydrochar are generally classified in the literature as liquid water-based (HTC) or vapor-based (VTC). However, the distinction between the two is often ambiguous. In this paper, we present a methodological approach to analyze process conditions for hydrothermal systems. First, we theoretically developed models for predicting reactor pressure, volume fraction of liquid water and water distribution between phases as a function of temperature. The reactor pressure model predicted the measured pressure reasonably well. We also demonstrated the importance of predicting the condition at which the reactor system enters the subcooled compression liquid region to avoid the danger of explosion. To help understand water–feedstock interactions, we defined a new solid content parameter $\%S(T)$ based on the liquid water in physical contact with feedstock, which changes with temperature due to changes in the water distribution. Using these models, we then compared the process conditions of seven different HTC/VTC cases reported in the literature. This study illustrates that a large range of conditions need to be considered before applying the label VTC or HTC. These tools can help in designing experiments to compare systems and understand results in future HTC research.

Keywords: hydrothermal carbonization (HTC); vapothermal carbonization (VTC); reactor pressure; process conditions; phase distribution of water; solid contents; hydrochar

1. Introduction

The process of hydrothermal carbonization (HTC) is used to carbonize organic residues and wastes for diverse applications ranging from fuels to soil amendments. In HTC, subcritical water is used as a solvent and reactant to transform a wide variety of organic feedstocks to solid carbonaceous products (hydrochar), which usually contain higher carbon contents, heating values and degrees of aromaticity than the original feedstocks [1–4]. Diverse types of reactors have been used, ranging from batch [3,5–12], semi-batch [2,9,10] to continuous reactors [13], with and without mixing, using direct heating through steam injection or through the reactor walls with controlled heating rates, or indirectly in muffle ovens. The pressurized reaction system usually consists of all three phases (gas, liquid, solid) and is heated to temperatures from 160 to 280 °C with pressures between 0.6 and 6.4 MPa due to the water vapor and gases produced in the reactions. Water is initially introduced into the reaction system via the moisture content of the feedstock and/or through the addition of water as a liquid or as steam. Common HTC process variations are differentiated according to how the water initially contacts the feedstock. When the feedstock is immersed in bulk liquid water, it is often called HTC. Where the feedstock is in direct contact with only steam, it is often called steam HTC, vapor HTC or vapothermal carbonization (VTC) [2,7,10,11]. Here, feedstock can be held in baskets away from liquid water, transported in and out on conveyor belts with little to no post-processing dewatering steps necessary. However, the dividing line between the two process types often does not remain sharp over the operating time, since the distribution of water between the liquid and vapor phases will change as temperature rises in batch systems, or as more steam is added in semi-batch systems. Usually we cannot see inside high-pressure reaction systems to gain visual insights, so the extent of the phase change is often unknown. Process comparisons between HTC vs. VTC based on the state of the initial water phase may be misleading. The phase changes may play an important role in causing structural variations in the hydrochars produced, their biodegradability, stability, and functionality in various applications [5].

Only a few studies comparing hydrochars produced by HTC vs. VTC have been reported in the literature. Cao et al. [5] found that biomass was more carbonized under liquid water in HTC than through steam in VTC. They determined that more aromatic and less alkyl groups were formed in the sugar beet and bark hydrochars made from HTC than in those from VTC under the same operating conditions (200 °C for 3 h). The hydrochars made from HTC also were less biodegradable than those from VTC as indicated by the lower values of the ratio BOD/COD for HTC-hydrochars. The lower biodegradability of hydrochar from HTC was probably due to its higher aromaticity produced during the reactions taking place in liquid water. In this case, the hydrochar made from HTC should be more stable in the environment than that made from VTC, which would have consequences for its use, e.g., as a soil amendment. On the other hand, comparisons between hydrochars from HTC vs. VTC for use as fuels have produced mixed results on how the phase of the reaction medium affects the important energy parameters: solid yield, higher heating value (HHV) and energy yield [10,11,14]. Comparing HTC vs. VTC for two feedstocks, digestate and straw, at 230 °C, 6 h, Funke et al. [11] found no clear trend in HHV: VTC-chars had a higher HHV than HTC-chars from digestate, but for straw, the order was reversed. However, VTC produced higher solid yields and, therefore, higher energy yields than HTC for both feedstocks. In contrast, Shafie et al. and Yeoh et al. reported mostly lower solid yields and higher HHV from the VTC process compared to HTC process (both at 220 °C, 1 h; two feedstocks [10,14], three feedstocks [14]). Again, as in [11], the combination of these two trends in the energy yield showed that the VTC process was more efficient for energy yield. We suspect that the reason for these conflicting trends in HHV and solid yield arise from the fact that a clear picture of how the water was distributed between gas and liquid phases was not given, nor to which extent the feedstocks were exposed to the liquid water phase. For instance, the VTC experimental setup used by Yeoh et al. (two concentric chambers with water filled in the outside chamber and the biomass in the inner chamber) was assumed to avoid the liquid water directly contacting the biomass feedstock. Yet it is not clear from their description that the liquid water was contained in the outside chamber throughout the reaction, since liquid water expands at higher temperature, possibly causing overflow into the biomass chamber. It appears that the transition between HTC/VTC reported in the literature is fuzzy at best because it can be changed by small variations in the same reactor system. How much water is present in each phase depends not only on the reactor temperature and pressure, but also on the total amount of water in the system relative to the volume of the reactor system.

As more knowledge is gained on the beneficial applications of hydrochars and HTC for waste and residue processing, more work on reactor designs for diverse settings (ranging from high to low tech systems) will be carried out. The variety of process variations may increase, with process configurations and conditions utilizing the unique transport properties of each medium, e.g., the higher thermal conductivity of bulk liquid water or the higher diffusivity of steam to penetrate the porous structure of the feedstock [7,15]. These changes in transport properties can affect reactions and product characteristics [15]. Therefore, knowledge on what influences the distribution of water between the reaction phases is essential for the production of the desired hydrochar quality. Especially if we want to replicate process conditions in various reactor types and scales to produce a desired hydrochar quality, we must be able to predict the distribution of water between the vapor and liquid phases at the design HTC reaction conditions.

Furthermore, reactor designs must consider how the HTC reactor pressure will change in response to operating conditions to ensure process safety. All reactors must be able to withstand the high temperatures and pressures that can develop during the process. As a rigid HTC reactor partially filled with water and feedstock is heated, the increase in the saturated water vapor and gases produced by the chemical reactions cause the reactor pressure to rise. At the same time, the density of the bulk liquid water decreases and consequently the volume of liquid water increases, decreasing the volume of the reactor headspace. When the liquid volume in the HTC reactor completely fills the headspace, it can no longer expand if the reactor temperature is increased further. The reactor water then enters a subcooled liquid compression region. In this region, pressure increases very rapidly with small increases in reactor temperature. To avoid the reactor pressure exceeding the tensile strength of reactor material, it is very important that the reactor system has a working safety disk or valve that can release pressure at a preset value. Without the use of proper rupture disks, the reactor can explode. Therefore, in order to maintain safe operating conditions, we need to predict the reactor pressure at the chosen process conditions. This requires understanding the relationship between the HTC reactor conditions (temperature, water volume, feedstock) and pressure.

The aim of this work is to present a methodological approach to analyze process conditions for hydrothermal systems in the framework of the hydrothermal carbonization reactions. In the paper, we first theoretically develop models for predicting reactor pressure, the distribution of water between phases, and the liquid water volume fractions as a function of reactor temperatures. Then, the evaluation is expanded to water and feedstock. Finally, using these new models, we analyze and compare process conditions for VTC and HTC systems reported in the literature.

2. Materials and Methods

2.1. Theoretical Development

For a reactor without any HTC reaction (i.e., without any feedstock inside reactor), we can estimate the HTC autogenic pressure with that of pure water at the HTC reaction temperature. This information is often visualized for hydrothermal systems with a pressure-temperature (P-T) phase diagram for water, showing the regions for the different types of processes, e.g., gasification, liquefaction, carbonization. However, to help us understand the process conditions during a hydrothermal reaction, the use of the temperature-volume (T-v) phase diagram for water is a powerful tool which provides information on P and T as well as the distribution of water between phases as a function of the overall specific volume of water (liquid and steam) in the reactor v_R (Figure 1). Using this diagram, one can understand the thermodynamic equilibrium at the chosen process conditions of the reactor system, e.g., temperature, pressure and mass of water in the system. In Figure 1, the saturation line represents the boundary condition for the phase change. For most HTC/VTC reactor systems, the reaction zone is usually located within the saturation curve, where steam and liquid phases coexist. The operating path for a batch system can be followed from the starting process conditions until the target conditions are met and the holding time begins. The closer the target point is to the steam or liquid saturation lines, the higher the amount of that phase. Since a log scale is used for the x-axis, the ratio between the two phases cannot easily be determined visually from the figure. The calculation procedure is developed in the following section.

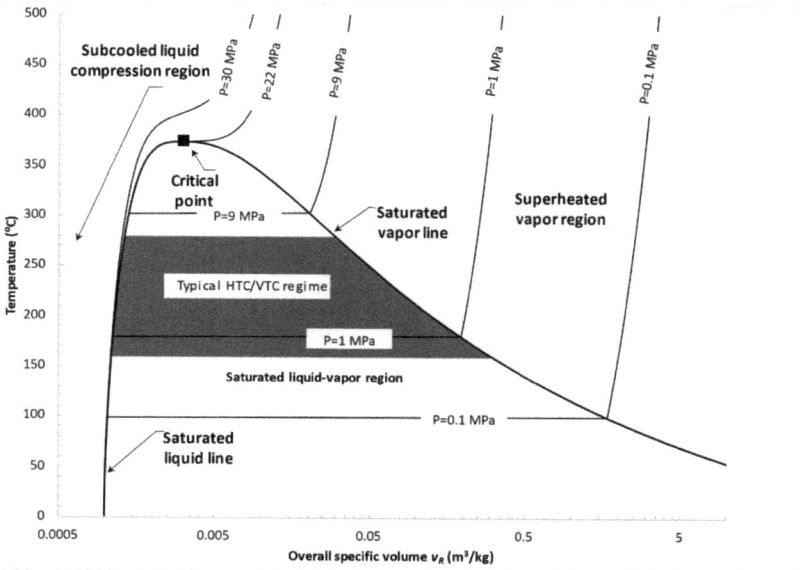

Figure 1. Temperature-volume (T-v) diagram for water showing the common operating region for vapor and hydrothermal carbonization reactions.

Total mass of water in the reactor is

$$M_{H2O} = x_L M_{H2O} + x_V M_{H2O} \quad (1)$$

where

M_{H2O} = total mass of liquid and vapor water in the reactor (kg);
x_L = mass fraction of liquid water;
x_V = mass fraction of vapor water (or steam quality);
$x_L + x_V = 1$.

As the HTC reactor is heated beyond the boiling temperature, the reactor volume is mostly filled with liquid water and steam, and the following relationship can be developed assuming both liquid water and steam are in equilibrium (i.e., for the saturated liquid–vapor region; Figure 1).

$$V_R = x_L M_{H2O} v_L + x_V M_{H2O} v_V \quad (2)$$

where

V_R = reactor volume (m³);
v_L = specific volume of saturated liquid water (m³/kg);
v_V = specific volume of saturated steam (m³/kg).

Combining Equations (1) and (2), the mass fraction of vapor water (x_v) can be calculated by knowing the thermophysical properties of water at those conditions, the mass of water in the reactor and the reactor volume:

$$x_V = \frac{v_R - v_L}{v_V - v_L} \quad (3)$$

where

$v_R = V_R/M_{H2O}$, overall specific volume of reactor water and steam mixture (m³/kg).

As the liquid-steam mixture in the reactor is heated, the volume of liquid water expands due to the decrease in water density ρ_L. Using Equation (3) along with values for saturated vapor and liquid specific volumes [16], the fraction of liquid-water occupying the reactor volume VF_w can be estimated at the HTC reaction temperature:

$$VF_w = \frac{V_w}{V_R} = \frac{(1-x_V)v_L}{v_R} \qquad (4)$$

where

V_w = volume of liquid water in the reactor at temperature T (m³);

VF_w = volume fraction of liquid water in the reactor at temperature T (-).

As long as the reactor volume is larger than the bulk liquid water volume (i.e., $VF_w < 1$), we can assume the liquid and vapor water phases are in equilibrium and the autogenic pressure can be estimated from the saturation properties of water using saturated steam tables [16–18].

If the temperature is further increased so that the liquid volume completely fills the reactor due to the decrease in its density (i.e., $VF_w = 1$, and $x_L = 1$), the liquid water will enter the subcooled liquid compression region. This region can be seen in the T-v phase diagram, left of the saturated vapor curve (Figure 1). There is no longer any headspace in the reactor and the water density in the reactor system at this point (also called overall reactor water density) becomes constant and can be calculated from $D = M_{H2O}/V_R$. As the rigid reactor walls are suppressing the tendency of the liquid volume to increase in response to the decrease in liquid water density, the reactor pressure increases rapidly as the water expands with the increase in temperature. When $VF_w \geq 1$ calculated from Equation (4) (i.e., physically impossible unless the reactor explodes), the reactor pressure in this range can be estimated with liquid compressibility factor for subcooled water:

$$P = Z_L \times D \times RT/MW_{H2O} \qquad (5)$$

where

P = reactor pressure (MPa);

Z_L = liquid compressibility factor for subcooled water (-);

D = overall reactor water density, M_{H2O}/V_R (kg/m³);

R = universal gas constant (8.31451 × 10⁻³ m³-MPa/kmol-K);

T = reactor temperature (K);

MW_{H2O} = molecular weight of water (kg/kmol).

To illustrate the danger of a potential reactor explosion if the liquid fills the reactor completely, example calculations to estimate the reactor pressure at three common HTC temperatures using Equation (5) are reported in Table 1. The values of liquid compressibility factor of the subcooled water reported by Lemmon et al. (2018) were used. A value for D was chosen that is slightly higher than the saturated liquid water density at 200 °C. This simulates the reactor pressure for the case when the liquid water fills the reactor completely at 200 °C. A further increase in T to 250 °C will rapidly increase P from 2 to 81.6 MPa, a pressure that many HTC reactors are not made to withstand. For instance, maximum allowable pressures for common laboratory reactors range from 13.3 to 34.5 MPa [19]. In contrast, if there is less liquid water added and more headspace in the reactor so that the liquid water-vapor equilibrium can exist at all operating temperatures, the pressure increase would follow the saturation pressure, increasing only from 1.6 to 4.0 MPa.

In order to avoid the subcooled compressible region, some manufacturers of pressure equipment recommend calculating the maximum allowable water mass using a safety factor and the ratio of ρ_L or its inverse v_L at the desired T to that at room temperature [20]. It is also important to note that the actual HTC pressure will be higher than that from the pure water because of gas production (predominantly CO_2) from HTC reactions.

Table 1. Example calculations for the effect of increasing the hydrothermal carbonization (HTC) temperature on reactor pressure for $D = 865$ kg/m^3 (values for saturated water properties taken from Lemmon et al. (2018).

Saturated Water (i.e., $VF_w < 1$)				
		200 °C	230	250
Specific volume of liquid water (v_l)	(m^3/kg)	0.00116	0.00121	0.00125
Specific volume of vapor water at (v_v)	(m^3/kg)	0.12721	0.07151	0.05008
Saturation pressure (P_{sat})	(MPa)	1.55	2.80	3.98
Subcooled Water (i.e., $VF_w > 1$)				
Liquid compressibility factor	(-)	0.010606	0.24601	0.39023
Reactor pressure (P) for $D = 865$ m^3/kg	(MPa)	2.00	49.44	81.55

2.2. HTC Reactor System

Three laboratory-scale HTC reactor systems were used to measure pressure change with temperature increase and validate the predicted values with experimental set-ups. Two sealed high pressure and temperature reactors made of Alloy C276 with valves and fittings made of T316 Stainless Steel (500-mL, Model 4575A and 1-L, Model 4680 HT, Parr Instrument Co., Moline, IL, USA) were used with various initial fillings with distilled water. A 1500-watts heater surrounding the outside reactor wall along with a programmable temperature controller was used to heat the reactants at a designed temperature. This reactor system was modified to improve control and data logging capability. In addition, a 18.75-L reactor system (Model 4555, T 316 Stainless Steel, Parr Instrument Co., Moline, IL, USA) with a similar heating system (6000 watts) and temperature controller (Model 4848BM) using SpecView data acquisition software was used to study the effect of initially pressurizing the system with nitrogen.

3. Results and Discussion

Understanding and replicating process conditions to produce a desired hydrochar quality require that we can estimate how much liquid water is in contact with the feedstock. Maintaining safe operating conditions requires that we can predict pressure increases during the reaction. In pressurized hydrothermal batch and semi-batch conversion systems, process conditions in the reaction system can be difficult to measure as well as to predict. The volume of liquid water and the distribution between the liquid and vapor phase change with temperature. Reactor pressure can increase with temperature due to (1) rising water vapor pressure, (2) the expansion of the liquid water, and (3) the production of process gas. In the following sections, the focus is on the effects caused by changes in the physical state of water. First, the relationships between temperature, pressure, the volume fraction of liquid water VF_w and the distribution of water between the liquid and vapor phase are shown for a reactor system filled only with water. Then, the evaluation is expanded to water and feedstock. Finally, the effect of these process conditions for VTC and HTC systems reported in the literature are discussed.

3.1. HTC Reactor Filled with Water Only

3.1.1. Estimating VF_w at Various Temperatures and VF_o

In HTC and VTC experiments, a wide range of initial water amounts can be used. For HTC experiments, high values of VF_o are commonly chosen. It is important, however, to choose process conditions so that the liquid water does not fill the reactor volume at the holding temperature (i.e., $VF_w = 1$) to avoid entering the subcooled liquid compression region. The higher the initial VF_o, the lower the reactor temperature at which VF_w becomes 1, because smaller headspace volumes cannot accommodate much expansion of liquid water as its density decreases with temperature. This behavior is shown in Figure 2 for a reactor filled with water only. Values for VF_w were estimated at various temperatures and VF_o using Equation (4). For a reactor initially filled with water at 90% (i.e., $VF_o = 0.9$), the liquid volume expands to the reactor volume (i.e., $VF_w = 1$) when the reactor temperature reaches 165 °C. Fortunately, this critical temperature, at which $VF_w = 1$, increases rapidly as VF_o is decreased, e.g., 305 °C for $VF_o = 0.7$ and 365 for $VF_o = 0.5$, so that process conditions can be chosen to remain well below the critical temperature. When the reactor is initially filled with water to less than half its volume (i.e., $VF_o < 0.5$), the liquid does not fill the reactor even when the temperature approaches the critical point of water around 374 °C. Interestingly, for experiments with low values of VF_o common to VTC operating conditions, VF_w can actually decrease with temperature. When the reactor is initially filled with a very low volume of water, such as $VF_o = 0.1$, the liquid volume decreases to zero at $T = 340$ °C. This happens when there is so much headspace that the liquid water completely vaporizes, i.e., the molecular collision frequency of H_2O molecules in the headspace is so small that condensation does not happen in this high headspace situation.

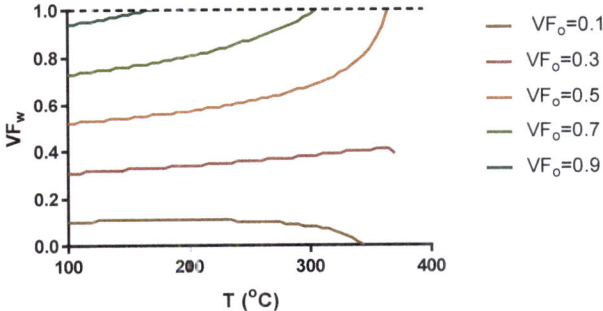

Figure 2. Change in volume fraction of reactor filled with liquid water VF_w (V_w/V_R) as a function of temperature for various initial liquid water volume fractions (VF_o).

3.1.2. Estimating Pressure and VF_w under Process Conditions

For a batch reactor system starting with only water at atmospheric pressure, the reactor pressure at the holding temperature can be easily estimated using the simple saturation water vapor pressure at T as long as $0 < VF_w < 1$. However, when $VF_w = 1$ or higher, if the temperature is further increased, the reactor pressure will now follow the subcooled water pressure, which rises rapidly. The pressure increase must then be estimated using Equation (5) with the liquid compressibility factor Z and the overall reactor water density D. This approach can then be used to predict the increase in reactor pressure as a function of the reactor temperature at various VF_o. The results are illustrated in Figure 3. When $VF_o = 0.3$, the liquid volume does not reach the reactor volume (i.e., $VF_w < 1$) even at the highest temperature simulated at 370 °C. Therefore, the reactor pressure follows the saturation water vapor pressure shown as the lower curve in Figure 3. For a reactor system with a high initial volume of water, e.g., $VF_o = 0.8$, the reactor pressure also follows the saturation pressure line below 250 °C, similar to the behavior at $VF_o = 0.3$. However, at $T = 250$ °C, VF_w becomes unity and the water enters the subcooled compression region. In this region, a small increase in temperature of only 5° can cause a rapid increase in pressure from 4.9 to 11.7 MPa (Figure 3).

The application of this approach to predict VF_w and pressure for experimental runs at various temperatures was evaluated by comparing the predicted pressures with the observed pressures in HTC reactors initially filled with three different amounts of water. As most researchers have discovered, determining the actual volume of an HTC reactor with cavity volume in the reactor head due to connections to the pressure gauge and sampling ports is difficult. Based on reactor volume estimations from simple geometric dimensions, the added water resulted in $VF_o = 0.3$, 0.63, and 0.67. For $VF_o = 0.3$, the observed pressures as the reactor was heated followed the saturation water vapor line at temperatures up to 349 °C (Figure 3). However, for $VF_o = 0.67$ at temperatures higher than 349 °C, the pressure increased much more rapidly than the saturation vapor pressure, indicating that the water entered the subcooled compression region. According to the predicted pressure line for the estimated VF_o, it should have entered the subcooled region at a lower temperature of 320 °C. Instead, the observed pressure followed the predicted pressure line of $VF_o = 0.6$ (Figure 3). We suspect that this discrepancy can be attributed to the fact that the actual reactor volume was larger than the estimated volume, (e.g., 1.1 L instead of 1 L for 0.67 L water initially filled). These results show that the approach is adequate to estimate VF_w at various reactor temperatures in practical applications, however, if the actual reactor volume is not known accurately, there will be some deviation from the predicted values.

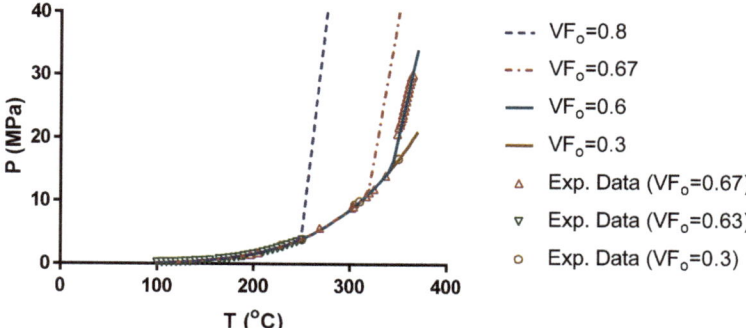

Figure 3. Comparison of measured and estimated reactor pressure at different initial liquid water volume fractions (VF_o).

Another important question to consider in deciding upon operating conditions and evaluating experimental results is: How does a higher initial pressure affect the pressure development and phase distribution of water in the reactor? For example, some experimenters pressurize the system initially using an inert gas such as N_2 or Ar. The answer in short is that the addition of pressure to the reactor headspace does not change the behavior of water. If enough water and time are available, water will vaporize to the gas phase to reach the saturation water vapor pressure at which liquid water and vapor water are in equilibrium. This pressure is a function of temperature only and independent of the presence of other gases. The added inert gas does not change the relationships for VF_w and the distribution of water between the liquid and vapor phases. However, the total reactor pressure will be higher in the reactor initially filled with N_2 or Ar than that without the initial inert gases. The total pressure $P(total)$ can be estimated by summing the partial pressures of all individual non-reacting gases as stated in Dalton's law. The increase in the partial pressure for each component with temperature can be calculated independently and added together. This can be seen in Figure 4 for two experimental runs in an 18.75-L Parr reactor in which water was heated to 220 °C ($VF_o = 0.63$): one starting at atmospheric pressure and the second one with N_2 addition to achieve an initial pressure of 1.4 MPa. The measured values from the nonpressurized run ($P_o = 0.1$ MPa) are compared to the saturation water vapor pressure $P(sat)$ from [2] in the lower curve. For the run at $P_o = 1.4$ MPa, the partial pressure increase for N_2 $P(N_2)$ was estimated using the ideal gas law, combined with Equation (4) to calculate the changes in headspace volume ($1-VF_w$) as temperature increases.

Comparison of the measured and theoretical values shows clearly that the contribution of the saturated water vapor to the total pressure is not affected by the initial addition of N_2 gas. The small deviation between the calculated pressure and the measured can be due to inaccuracies in the pressure measurement or in estimating the reactor volume, and the assumption that N_2 behaves as an ideal gas with no solubility in the liquid. Nevertheless, the difference does not mask that the fact that the addition of pressure to the reactor headspace does not change the behavior of water.

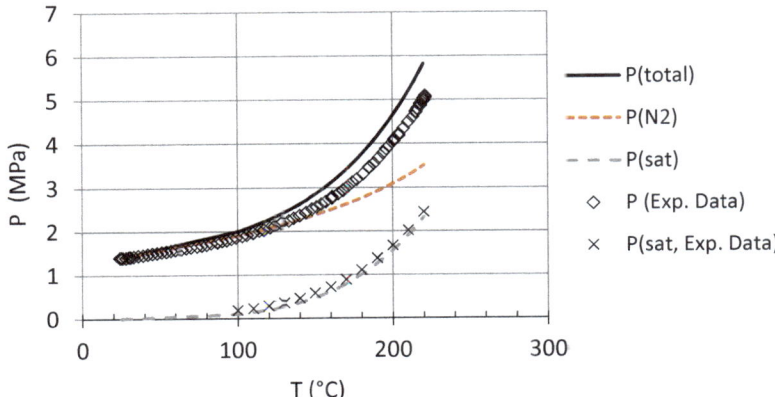

Figure 4. Comparison of the theoretical total pressure *P(total)* calculated from the partial pressures for N$_2$ *P(N$_2$)* and saturated water vapor *P(sat)* to the measured values for pressure *P* and *P(sat)* in the reactor for $VF_o = 0.63$.

3.2. HTC Reactor Filled with Water and Feedstock

3.2.1. Estimating the Distribution of Water between Phases as a Function of Temperature and Its Effect on Solid Content

In their comparison of hydrochars from VTC and from HTC systems, Cao et al. (2013) postulated that the amount of liquid water in contact with the feedstock in the reaction system may determine the degree of carbonization and influence which reactions take place and their sequence [5]. However, they did not quantify how much liquid water was in contact with the feedstock in their reaction systems. This is a common problem in most of the literature on HTC/VTC systems. Often the label used for the system is defined by the initial conditions. For instance when the feedstock is initially completely submerged in bulk liquid water, it is commonly called an HTC system. Whereas, when dry or wet feedstock is placed separately from the bulk liquid water, it is called a VTC system. However, the volume of liquid water and the distribution of water between the liquid and vapor phase change with temperature, which can change the amount of water contacting the feedstock. In addition, the feedstock characteristics such as moisture content, particle size, bulk density, as well as structural changes during the reaction can affect how water interacts with the feedstock. In VTC, carbonization reactions can take place between a wet feedstock and water within its cells or present as a film on its surface [10,11]. Ref. [11] Even with completely dried feedstock, the feedstock can be wetted during the process by absorbing water vapor or water vapor condensing on its surface.

The parameters often used to describe the relationship between water and feedstock in a reaction system do not differentiate between the bulk liquid water added to the process and the liquid water in contact with the feedstock. The nominal solid content at the start of the run is usually reported in published studies as:

$$\%S_o = \left(\frac{M_{biomass}}{M_{H2O} + M_{biomass}} \right)_{T=T_o} \times 100 \tag{6}$$

where

$\%S_o$ = nominal solid content;
$M_{biomass}$ = initial feedstock dry mass;
M_{H2O} = total mass of water in the reactor.

A similar parameter *R* which describes the initial ratio of feedstock dry mass to total mass of water is also often used. These parameters only describe the initial conditions based on the initial filling masses of water and feedstock, but do not provide critical information on the extent to which feedstock is exposed to liquid water in the HTC or VTC systems to promote important hydrothermal carbonization reactions. In order to provide useful information on the degree of physical contact between the feedstock and liquid water throughout the process, we propose reporting the following solid content parameter:

$$\%S(T) = \left(\frac{M_{biomass}}{m_{H2O} + M_{biomass}}\right)_{T=T} \times 100 \qquad (7)$$

where

$\%S(T)$ = actual solid content based on liquid water in contact with feedstock;
m_{H2O} = mass of liquid water in contact with feedstock;
T = reactor temperature.

With these new definitions, one can systematically distinguish various HTC/VTC process conditions in terms of fraction of liquid water physically in contact with feedstock. For HTC systems, where the feedstock is assumed to be completely submerged in the bulk liquid water over the whole reaction time, $m_{H2O} = x_L \cdot M_{H2O}$. Using Equations (3) and (4), these assumptions can be checked for the reaction temperature and the solid content values adjusted with Equation (7). For example, the change in the distribution of water between the two phases can be seen in Figure 5a. For temperatures below 250 °C and VF_o larger than 0.3, less than 4% of the water will be vaporized. The expansion of VF_w, as seen in Figure 2, should offset the small loss of liquid water to the vapor phase and submerged feedstocks should remain submerged at these conditions. Therefore, the actual solid content will be approximately the same as the nominal solid content at the initial reactor temperature T_o (i.e., $\%S(T) = \%S_o$). Only for systems with VF_o closer to 0.1, more common to VTC systems, will approximately 20% of the water be present as vapor at 250 °C.

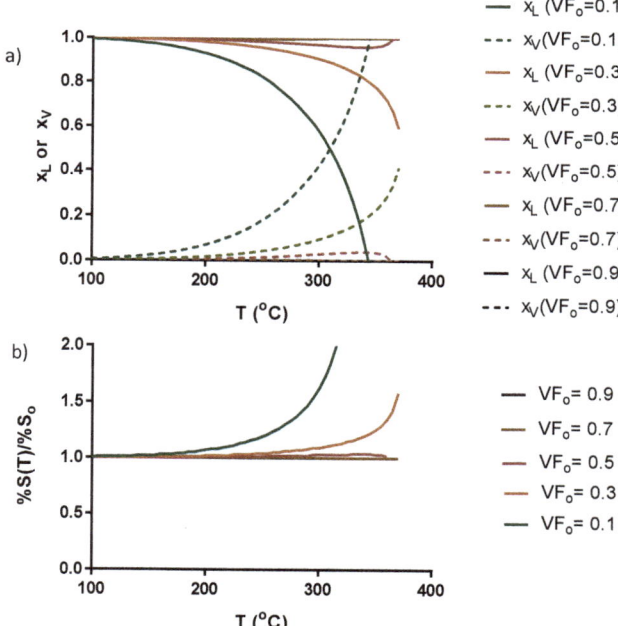

Figure 5. Changes in process parameters at various initial VF_o as the operating temperature increases for (**a**) mass fractions of water in liquid x_L and vapor x_V, (**b**) the ratio of actual to nominal solid content in the reactor system, $\%S(T)$ at T to $\%S_o$ at the initial temperature.

In VTC systems, wetted or completely dried feedstock can be suspended without any physical contact with bulk liquid water. The bulk liquid water can be placed either at the bottom of the reactor or in a separate interconnected chamber, or steam can be injected to heat the reactor. The value reported for $\%S_o$ for such systems often includes the bulk water. However, this can be misleading, especially for dried feedstock, where the actual initial solid content $\%S(T_o) = 1$ because $m_{2O} = 0$. Although $\%S(T_o) = 1$ initially, $\%S(T)$ will become less than one over time because water vapor will be volatilized from the physically separated bulk liquid as the VTC reactor temperature increases and will condense on the surface of the dry feedstock. The extent to which the vaporized water condenses onto the feedstock depends on the kinetics of condensation and vaporization at the reaction temperature, but $\%S(T)$ will rarely reach $\%S_o$. The condensed water will promote typical hydrolysis and other important carbonization reactions as in HTC systems. For VTC systems with initially wetted feedstock, bulk liquid water may or may not be added to the reactor. If no bulk liquid water is added similar to that of Funke et al. [11], $\%S(T_o) = \%S_o$ where the moisture content (MC) determines the value of the initial solid content. As the temperature increases, liquid water is lost to vaporization, reducing the water content of the feedstock. $\%S(T)$ can then be calculated with Equation (7) and:

$$m_{H2O} = x_L \times M_{H2O} = x_L \times \frac{M_{biomass} \times MC}{1 - MC} \qquad (8)$$

Using these equations for the new solid content parameters, the ratios of actual to nominal solid content are plotted against the reactor temperatures in Figure 5b for various initial volume fractions of liquid water. For HTC systems with VF_o larger than 0.3 and temperatures below 250 °C, there is little difference between the two values. At 250 °C, only a 4% increase is seen in $\%S(T)$, and after the reactor is half-filled (i.e., $VF_o > 0.5$), no differences are noticeable. In contrast, for systems with low values of VF_o (e.g., $VF_o = 0.1$), $\%S(T)$ becomes 20% higher than $\%S_o$. Less liquid water is in contact with the solids to participate in reactions. For VTC systems with wet feedstocks where the liquid water is associated in or on the feedstock, the transfer of water to the vapor would change the $\%S(T)$ in the reactor significantly. For wetted feedstock suspended over bulk liquid water, the situation is more complicated, since water can vaporize from the wet feedstock or bulk liquid water, mass transfer within and in-between feedstock, and the kinetics of condensation and vaporization all play roles in the location of the liquid water. This is beyond the scope of this paper. Moreover, the implications of the reduction in the mass of bulk liquid water on the potential of reducing physical contact between feedstock with a bulk volume larger than that of liquid water and subsequent carbonization reactions need to be further studied.

3.2.2. Estimating VF_w and Pressure under Process Conditions with Feedstock

Up until now, we have analyzed the changes in volume fractions due to changes in the physical properties of water. The addition of feedstock to the reactor can reduce the headspace volume available to accommodate the expansion of liquid water. To adjust the volume fractions for the presence of feedstock, the reactor volume must be reduced by the volume occupied by the feedstock. To strictly determine this volume, we need to know the true density of the material, i.e., the ratio between the feedstock mass and its volume excluding the cavities, pores and gaps in the material where water and air could be trapped. In addition, the loss of solid mass and structure during HTC reactions would have to be taken into consideration. For practical purposes, simple estimates of the initial volume of feedstock can be made with liquid displacement methods and used to adjust the calculation of VF_w.

HTC reactions with the feedstock can also change the gas composition and pressure of the headspace. The composition and amount of the produced gases is closely tied to the process conditions and feedstock material. In general, most HTC reactions with biomass produce predominantly CO_2 (~>80%) with minor percentages of N_2, H_2S, O_2, CH_4, H_2, etc., in the gas besides water vapor. Explosive gas mixtures are not expected in HTC, unlike hydrothermal gasification where reactor temperatures are near or above the critical temperature of water (i.e., ~374 °C) and produce approximately 1:1 of CH_4

and CO_2. However, an in-depth analysis about gas production and compositions is beyond the scope of this paper. The impact of the product gas on the total reactor pressure and its partition between water and gas phases need to be further investigated as gas solubility changes with temperature and pressure. The results of this study provide a theoretical framework for further experimental and modeling research on this aspect of HTC.

3.3. Comparison of Process Conditions for Hydrothermal Treatment (HTC and VTC) Reported in the Literature

The results from published HTC/VTC studies that have been made at various scales, ranging from 1 L to 10 m^3, and with different modes of operation, e.g., batch and semi-batch with respect to the steam, are analyzed in this section. As summarized in the introduction, few studies comparing HTC and VTC systems have been published and some results are contradictory. The goal here is to identify the effect of process conditions on the distribution of the water between the phases to understand what is behind the labels—HTC and VTC—and develop criteria on how to label systems, either HTC or VTC. This is necessary especially for cases in which we want to replicate process conditions used to produce a desired hydrochar quality in other reactor types and/or scales. The results are structured into seven cases for the discussion here and an overview of the process conditions and feedstocks is given in Table 2.

Table 2. Overview of the process conditions and feedstock for the seven cases with VTC/HTC processes.

Case	Reactor					Feedstock			Water in System		Literature
	Type	T (°C)	V_R (L)	Heating	Mixed	Type	MC (%)	$M_{biomass}$ (gDM)	M_{H2O} (g)	v_R (m^3/kg)	
1	Batch HTC	200	1	heating band	No	bark	7	23.5	101.6	0.010	Cao et al. [5]
						sugar beet	7	37.6	162.9	0.006	
2	Semi-batch VTC	200	70	steam, condensate removal	No	bark	40–70	n.r.	n.r.	0.013 *	
						sugar beet	40–70	n.r.	n.r.	0.012 *	
3	Semi-batch VTC	200	10,000	steam, no condensate removal	Yes	MSW	53	1771.9	start 1998.1	start 0.005	Safril et al. [9]
									end 3398.1	end 0.0029	
4	Batch HTC	230	18.75	heating band	No	wheat straw	0	450	8550	0.00219	Funke et al. [11]
						digestate	0	630	11,970	0.00157	
5	Batch VTC	230	18.75	heating band	No	wheat straw	25	450	1350	0.01390	
						digestate	25	630	1890	0.00995	
6	Batch HTC	220	4.6	heating band	No	bagasse	67.2	16.6–28.8	1495.5	0.003	Shafie et al. [10]
						lime peel	78	14.4–44.0	1495.5	0.003	
7	Semi-batch VTC	220	4.6	heating band, condensate separated	No	bagasse	67.2	n.r.	n.r.	0.041–0.053 @	
						lime peel	78	n.r.	n.r.	0.022–0.044 @	

n.r.—not reported. MC – moisture content. * Assumed 50% reactor volume filled with bark or sugar beet feedstock (MC = 55%) suspended in baskets with bulk density of 0.267 kg/L for bark [21] and 0.298 kg/L for sugar beet pulp [22].
@ Assumed the same amount of feedstock as in Case 6 with biomass bulk density of 0.616 kg/L for bagasse [23] and 0.490 kg/L for lime peel [24].

In Cases 1 and 2, a comparison of batch HTC (Case 1) with semi-batch VTC (Case 2) using the same feedstocks was made in different reactor systems [5]. For Case 1, the feedstock was dried, ground, and water added to the 1 L reactor, submerging the feedstock. In Case 2, the wet feedstock was suspended in baskets and steam (1.6 MPa) was injected to heat the 70 L reactor. As water condensed over the heating up and holding time, it was removed except for that which remained on the feedstock. Therefore, there was no increase in the mass of condensed water in the reactor over time (Revatec GmbH, DE 10 2009 010 233.7). The feedstock was reported to have a moisture content between 40 and 70%. For the calculations here, the mass of water in the system was estimated from the overall specific volume of saturated water vapor at 200 °C in the 70 L reactor and an average moisture content of the feedstock (55%) during the HTC reaction. As feedstock loading information is not available, we assumed 50% of the reactor volume was filled with bark or sugar beet pulp suspended in baskets inside the reactor. The mass of water was estimated using bulk densities of bark and sugar beet pulp reported in the literature [21,22].

In Case 3, a commercial-scale hybrid system with municipal solid waste (MSW) in a 10 m^3 reactor system was used [9]. The semi-batch system was fed saturated steam (2.5 MPa) to heat the feedstock and held at 180–230 °C for 30 min. The system started similar to a VTC system with only wet feedstock, but as the condensed steam was mixed with the feedstock over time, the system became more similar to HTC. A mid-range temperature of 200 °C was assumed for further analyses here. In Cases 4 and 5, Funke et al. compared VTC and HTC for two feedstocks in the same batch reactor system [11]. For VTC, the dried feedstock was first soaked in water and then suspended in a basket without any additional liquid water added to the reactor. For HTC, the dried feedstock was submerged in water. In Cases 6 and 7, Shafie et al used bagasse with MC of 67.16% (cut into less than 10 mm) and lime peel with MC of 78.04% (size as received) as feedstock [10]. For HTC, the feedstock was fully submerged inside the water in a reactor. For VTC, saturated steam was supplied to the reactor with the feedstock suspended in order to avoid contact with condensed liquid water accumulated at the bottom of the reactor. As feedstock loading information was not reported for the VTC experiments, we assumed the mass of feedstock was the same as that used in HTC experiments, along with bulk densities from the literature (bagasse [23]; lime peel (value for lemon peel was used [24]). The process conditions and feedstock for each case are summarized in Table 2, along with the respective v_R, overall specific volume of reactor liquid water and steam mixture.

3.3.1. Change in Process Conditions in Batch HTC (Cases 1, 4, 6) or VTC Processes (Case 5)

In batch reactors, the solids and liquids are introduced at the beginning of the run and the reactor is sealed before heating starts. The initial VF_0 and $\%S_0$ can be easily calculated and are usually reported (Table 3). Feedstock initially submerged in water can unequivocally be called HTC when VF_w at the holding temperature remains as large or larger than VF_0. This is true for all batch HTC cases (1, 4, 6) analyzed here. Each case includes results for two feedstocks under slightly differing conditions. In Case 4, with a relatively large amount of initial water (VF_0 = 0.46 and 0.64, for wheat straw and digestate, respectively), the expansion of water at 230 °C causes VF_w to increase by approximately 20%. As only 0.5 to 1.4% (m/m) of the initial liquid water is transferred to the vapor phase, there is little to no change in $\%S(T)$. Similarly, very little increase in solid content is observed in Case 6. In contrast, Case 1 at 200 °C has a low degree of initial water filling for both feedstocks (i.e., VF_0 = 0.1 and 0.16), and VF_w is very similar to VF_0. Between 4 and 7% of the water is transferred to the vapor, causing a corresponding increase in the value of $\%S(T)$. The values for $\%S(T)$ ranged from 1.0 to 19.9% for all batch HTC cases, ensuring adequate contact with liquid water to promote HTC reactions. Despite the loss of liquid water due to vaporization, the filling volume (VF_w) slightly increases because the volume of water expands with the reactor temperature, guaranteeing that the feedstock is completely submerged in the liquid water throughout the reaction period. Therefore, hydrothermal reactions will take place between the feedstock and liquid water and the process can be called batch HTC in Cases 1, 4, and 6.

If the solids are suspended in the reactor in baskets or on trays so that they are not submerged in water, the process is commonly called VTC (Cases 5). If the feedstock has a high moisture content such as the dried feedstock soaked in water (Case 5, MC = 75% or $\%S_0 = \%S(T_0) = 25\%$), or it is made up of intact microorganisms or fresh plant material, the actual $\%S(T)$ slightly increases compared to $\%S_0$ (27.6 or 28.9 vs. 25%) due to the small loss of liquid water in the feedstock to vapor (Table 3).

Table 3. Overview of process conditions and water distribution for the seven cases of VTC/HTC.

Case	Reactor Type	Feedstock Type	v_R (m^3/kg)	Mass Fraction in Vapor x_v (-)	VF_o (-)	VF_w (-)	$\%S_o$ (%)	$\%S(T)$ (%)
1	Batch HTC	bark	0.010	0.069	0.1	0.11	18.8%	19.9%
		sugar beet	0.006	0.04	0.16	0.18	18.8%	19.4%
2	Semi-batch VTC	bark	0.013 *	0.051 *	n.r.	-	45%#	45% #
		sugar beet	0.012 *	0.046 *	n.r.	-	45%#	45% #
3	Semi-batch VTC	MSW	start 0.005	start 0.03	0.2		47.0%	
			end 0.0029	end 0.0142		0.34		34.6%
4	Batch HTC	wheat straw	0.00219	0.014	0.46	0.54	5.0%	5.1%
		digestate	0.00157	0.0051	0.64	0.77	5.0%	5.0%
5	Batch VTC	wheat straw	0.01390	0.1804	0.07	0.07	25.0%	28.9%
		digestate	0.00995	0.1239	0.1	0.11	25.0%	27.6%
6	Batch HTC	bagasse	0.003	0.03	0.33	0.378	1.1–1.9%	1.1–2.0%
		lime peel	0.003	0.03	0.33	0.378	1.0–2.9%	1.0–3.0%
7	Semi-batch VTC	bagasse	0.041–0.053 @	0.473–0.609 @	n.r.	-	32.8%	32.8%
		lime peel	0.022–0.044 @	0.251–0.510 @	n.r.	-	22.0%	22.0%

n.r.—not reported. * Assumed 50% of reactor volume filled with bark or sugar beet suspended in baskets. # Averaged value for feedstock. @ Assumed the same amount of feedstock as in Case 6.

3.3.2. Change in Process Conditions for Semi-Batch VTC Process with Steam Injection (Cases 2 and 7 with Condensate Removal or Separation, and Case 3 without Condensate Removal)

For reactors in Cases 2 and 3 with a semi-batch mode of operation where saturated steam is introduced over time to first heat the reactor and then to maintain the desired operating temperature, the calculations for how much mass of the water is present as liquid or vapor are not as straightforward. The steam condenses as it heats the feedstock to the targeted operating temperature, and more will condense over the targeted holding time. As steam is introduced, the reactor pressure will remain constant at the saturation pressure if there are no reaction products entering the vapor phase. However, gases are normally produced by the hydrothermal reactions and the pressure rises as the gases, mainly CO_2, enter the headspace.

For systems with condensate removal, as in Case 2, or condensate separation, as in Case 7, the mass of bulk liquid water in contact with feedstock comes from moisture already present within the feedstock and water condensation on the surface of feedstock. Wet feedstocks will retain most of their moisture. For dried feedstock, the majority of the water in the system will be in the vapor, with some steam condensing on the feedstock surface, especially in the heating phase. Assuming that the amount of steam condensed on the feedstock surface is negligible, the overall specific volume v_R is mostly that of the saturated water vapor and the moisture content of the feedstock. In such systems, the process can be labeled VTC without much ambiguity. The amount of liquid water that can react with the feedstock for VTC systems mainly depends on the moisture content of the original feedstock and the condensed water on the feedstock surface. It is very difficult to quantify this amount of water. For these two cases, $\%S(T)$ was assumed to remain the same as the initial value. The condensed water is sometimes flashed off at the end of the run (e.g., for energy recovery (Revatech, 2012)), so that the solids come out about as wet as they went in. This is helpful in reducing dewatering requirements, but this hinders easily assessing how much water was in contact with the feedstock.

In systems without condensate removal, as in Case 3, the continuous injection of steam will build up the total mass of water in the system, with the majority present in the form of liquid water. The VTC process then approaches the HTC process. In this hybrid VTC–HTC commercial-scale unit, the reaction system is well-mixed and liquid water is mixed into the feedstock, gradually lowering the value of $\%S(T)$ in the reactor from the initial $\%S_o$ value 47% to 34.6%. The volume fraction of vapor water (x_V) changes somewhat. Starting with 3% of the water present as vapor, it reduces to 1.4% at the end of the run (Table 3). In general, it is important to measure the mass of steam introduced in systems without condensate removal, so that the mass of accumulated condensed water can be monitored as a safety precaution. Steam injection must stop before VF_w approaches 1 to avoid rupture of the reactor.

It is interesting to note that the solid content $\%S(T)$ of feedstock for all seven cases was less than 45% at the reaction temperature. The solid content for HTC systems ranged from 1.0 to 19.9%, while that for VTC systems from 27.6 to 45%. It means that 55% or more of the total water mass was present as liquid water and had direct physical contact with feedstock promoting carbonization reactions. According to Cao et al., the lower the $\%S(T)$, the more the product was carbonized. The highest solids content for VTC was 45% in Case 2 because of the water already present within the raw feedstock even though additional water was not supplied. This leads to questions on what will happen if we conduct VTC with completely dried feedstock, such as: Is it possible to carbonize the dried feedstock with steam alone? For such reaction systems, the initial value of $\%S(T_o)$ equals one. The value of $\%S(T)$ will become less than one as some of the steam condenses on the surface of feedstock promoting the HTC reactions. In such a system, the extent of carbonization will be determined by the extent of the wetting of the feedstock by steam. More detailed study is needed to understand the relationship between the degree of wetting by steam and carbonization.

3.4. Comparison of the Processes Using the T–v Diagram

To graphically illustrate the process conditions for each case at its reaction temperature, the values for the seven cases are plotted on a T-v phase diagram (Figure 6). Their locations in relation to the saturation curve show whether steam or liquid water predominates at the specific process conditions. Due to the log scale for the x-axis, the ratio between the two phases cannot easily be determined visually from the figure. Nevertheless, this visualization may help us to understand why some results from these studies comparing HTC and VTC systems are contradictory. The operating conditions in the HTC vs. VTC comparative studies are very different. The thermodynamic conditions in Cases 4 and 6 result in water being present mainly as a liquid for the HTC reactions as expected (i.e., toward the left side of the dome), while Case 5 is mid-range and Case 7 is located nearer the vapor saturation curve with a predominant steam phase. Process conditions for Case 1 (HTC) are to the left of Case 2 (VTC) on the 200 °C and 1.5 MPa isobaric line, suggesting more HTC reactions in Case 1, but the locations are closer together than the other pairs. Thus, this diagram visualizes the differences in the reaction phases, and allows us to subsequently interpret whether the system can be characterized more as HTC or VTC. When the conditions result in the same overall specific volume but with different process temperatures, the amount of water present as steam will change. This is true for Case 1-bark and Case 5-digestate (Table 3 and Figure 6). Both have a similar v_R, but Case 5-digestate at 230 °C has almost double the amount of water present as steam ($x_v = 0.1239$) than that for Case 1-bark at 200 °C ($x_v = 0.069$).

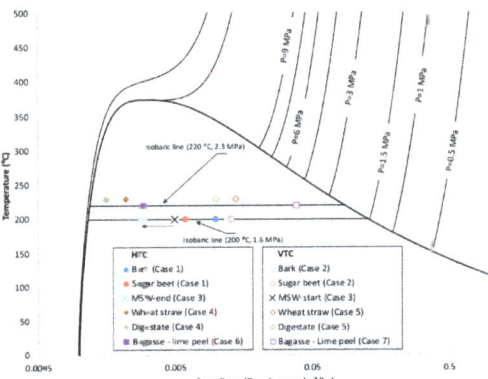

Figure 6. Comparison of the process conditions for the seven cases in the T–v diagram for water.

Furthermore, the diagram helps visualize the safety aspects. It is easy to see that the target conditions in Cases 1, 2, 5, and 7 are well away from entering the subcooled liquid compression region, where pressure increases rapidly with an increase in reactor temperature. For the semi-batch system in Case 3, the overall specific volume v_R decreases from 0.005 to 0.0029 m^3/kg due to the increase in the total mass of water as steam is injected into the reactor, and we move from the right to the left on the isobaric line at 1.6 MPa (Figure 6). For such semi-batch systems, it is important to make sure the start and end points remain far enough away from the subcooled compression region. Temperature increases above the initial target conditions due to use of superheated steam or exothermic reactions could move the system diagonally upwards towards the subcooled compression region and high pressure as steam is added. In the subcooled compression region, if a safety rupture disk valve is not present to release at a preset pressure, the reactor pressure can exceed the tensile strength of reactor material, and the reactor can explode.

4. Conclusions

There are many types of hydrothermal reactor systems being used with many process variations in the literature. The analysis presented in this paper illustrates that a large range of conditions need to be considered before labeling a reactor system VTC and HTC. The analysis of the process conditions of seven different HTC/VTC cases reported in the literature through the use of the models developed in this paper and a T-v phase diagram showed that the distinction between HTC and VTC is often ambiguous. The models developed in this study for predicting pressure, the volume fraction of liquid water and the distribution of water between phases as a function of reactor temperature can be used to systematically analyze various HTC/VTC process conditions. Furthermore, this study also demonstrates the importance of predicting the condition at which the reactor system enters the subcooled compression liquid region to avoid the danger of explosion. Comparison of the reactor pressures predicted by the models to the actual pressure for reactors filled with varying amounts of water with and without initial pressurization showed reasonable agreement. However, higher pressures can be expected with the addition of feedstock due to the production of CO_2 and other gases by the hydrothermal reactions and the decrease in headspace volume occupied by the feedstock. In order to describe the amount of liquid water in physical contact with feedstock, we defined a new solid content parameter $\%S(T)$ which changes with reaction temperature due to changes in the water distribution between phases. This parameter is more useful in describing the solid content than the nominal parameter $\%S_0$ typically reported in the literature. While the models developed here can help determine whether steam or liquid water predominates at the specific process conditions, more research and modeling on hydrothermal systems with feedstock present are required to understand the effect of the water phase on the hydrothermal reactions. The tools presented here can help in designing experiments to compare systems and in understanding results in future HTC research.

Author Contributions: K.S.R. conceived the original idea, conducted W-HTC experiments, analyzed the data, and participated in manuscript writing. J.A.L. conducted experiments using an 18.75-L reactor system, and both J.A.L. and A.A.-M. participated in all phases of the research, analyzed the data, simulated T-v diagrams, and participated in manuscript writing. All authors have read and agreed to the published version of the manuscript.

Funding: Financial help for A.A.M. came from the Junta de Extremadura and FEDER (Fondo Europeo de Desarrollo Regional "Una manera de hacer Europa") project IB16108, and also from the program "Ayudas a grupos de la Junta de Extremadura" GR18150. The open access journal fee was supported by the Leibniz Association's Open Access Publishing Fund.

Acknowledgments: The authors would like to acknowledge the technical support provided by Melvin Johnson of the USDA-Agricultural Research Service (ARS), Coastal Plains Soil, Water and Plant Research Center, Florence, SC, and Marcus Fischer of the Leibniz Institute of Agricultural Engineering and Bioeconomy. This research was supported by the USDA-ARS National Program 212 Soil and Air. Mention of trade names or commercial products is solely for the purpose of providing specific information and does not imply recommendation or endorsement by the U.S. Department of Agriculture.

Conflicts of Interest: The authors declare no conflict of interest.

Glossary

Symbol	Description	Unit
D	overall reactor water density, M_{H2O}/V_R	kg/m³
$M_{biomass}$	mass of feedstock in the reactor as dry matter (DM)	
M_{H2O}	total mass of water as liquid and vapor water in the reactor	kg
MC	moisture content	%
MW_{H2O}	molecular weight of water	kg/kmol
P	reactor pressure	MPa
P_o	initial reactor pressure	MPa
P_{sat}	saturated vapor pressure	-
R	universal gas constant (8.31451 × 10⁻³ m³-MPa/kmol-K)	-
$\%S$	% solid in reactor—ratio of mass of feedstock in DM to (total mass of water + mass of feedstock in DM in reactor), $M_{biomass}/(M_{H2O} + M_{biomass})$	-
T	reactor temperature	°C
VF_o	volume fraction of liquid water in the reactor at initial temperature T_o	-
VF_w	volume fraction of liquid water in the reactor at temperature T	-
V_R	reactor volume	m³
V_w	volume of liquid water in the reactor at temperature T	m³
v_R	overall specific volume of reactor liquid water and steam mixture, V_R/M_{H2O}	m³/kg
v_L	specific volume of saturated liquid water	m³/kg
v_v	specific volume of saturated steam	m³/kg
x_L	mass fraction of liquid water	-
x_v	mass fraction of vapor water or steam (or also called steam quality)	-
Z	compressibility factor for liquid or vapor	-

References

1. Libra, J.A.; Ro, K.S.; Kammann, C.; Funke, A.; Berge, N.D.; Neubauer, Y.; Titirici, M.M.; Fuhner, C.; Bens, O.; Kern, J.; et al. Hydrothermal carbonization of biomass residuals: A comparative review of the chemistry, processes and applications of wet and dry pyrolysis. *Biofuels* **2011**, *2*, 71–106. [CrossRef]
2. Cao, X.; Ro, K.S.; Chappell, M.; Li, Y.; Mao, J. Chemical structures of swine-manure chars produced under different carbonization conditions investigated by advanced solid-state 13C nuclear magnetic resonance (NMR) spectroscopy. *Energy Fuels* **2011**, *25*, 388–397. [CrossRef]
3. Ro, K.S.; Flora, J.R.V.; Bae, S.; Libra, J.A.; Berge, N.D.; Álvarez-Murillo, A. Properties of animal-manure-based hydrochars and predictions using published models. *ACS Sustain. Chem. Eng.* **2017**, *5*, 7317–7324. [CrossRef]
4. Roman, S.; Libra, J.A.; Berge, N.D.; Sabio, E.; Ro, K.S.; Li, L.; Ledesma, B.; Alvarez, A.; Bae, S. Hydrothermal carbonization: Modeling final properties design and applications: A review. *Energies* **2018**, *11*, 216. [CrossRef]
5. Cao, X.; Ro, K.S.; Libra, J.A.; Kammann, C.; Lima, I.M.; Berge, N.D.; Li, L.; Li, Y.; Chen, N.; Yang, J.; et al. Effects of biomass types and carbonization conditions on the chemical characteristics of hydrochars. *J. Agric. Food Chem.* **2013**, *61*, 9401–9411. [CrossRef] [PubMed]
6. Ro, K.S.; Novak, J.M.; Johnson, M.G.; Szogi, A.A.; Libra, J.A.; Spokas, K.A.; Bae, S. Leachate water quality of soils amended with different swine manure-based amendements. *Chemosphere* **2016**, *142*, 92–99. [CrossRef] [PubMed]
7. Minaret, J.; Dutta, A. Comparison of liquid and vapor hydrothermal carbonization of corn husk for the use as a solid fuel. *Bioresour. Technol.* **2016**, *200*, 804–811. [CrossRef] [PubMed]
8. Álvarez-Murillo, A.; Roman, S.; Ledesma, B.; Sabio, E. Study of variables in energy densification of oliver stone by hydrothermal carbonization. *J. Anal. Appl. Pyrolysis* **2015**, *113*, 307–314. [CrossRef]
9. Safril, T.S.; Safril, B.I.; Yoshikawa, K. Commercial demonstration of solid fuel production from municipal solid waste employing the hydrothermal treatment. *Int. J. Environ. Sci.* **2017**, *2*, 316–323. [CrossRef]

10. Shafie, S.A.; Al-attab, K.A.; Zainal, Z.A. Effect of hydrothermal and vapothermal carbonization of wet biomass waste on bound moisture removal and combustion characteristics. *Appl. Therm. Eng.* **2018**, *139*, 187–195. [CrossRef]
11. Funke, A.; Reebs, F.; Kruse, A. Experimental comparison of hydrothermal and vapothermal carbonization. *Fuel Process. Technol.* **2013**, *115*, 261–269. [CrossRef]
12. Berge, N.D.; Ro, K.S.; Mao, J.D.; Flore, J.R.V.; Chappell, M.A.; Bae, S. Hydrothermal carbonization of municipal waste streams. *Environ. Sci. Technol.* **2011**, *45*, 5696–5703. [CrossRef] [PubMed]
13. Gomez, J.; Corsi, G.; Pino-Cortes, E.; Diaz-Robles, L.A.; Campos, V.; Cubillos, F.; Pelz, S.K.; Paczkowsk, S.; Carrasco, S.; Silva, J.; et al. Modeling and simulation of a continuous biomass hydrothermal carbonization process. *Chem. Eng. Commun.* **2020**, *207*, 751–768. [CrossRef]
14. Yeoh, K.-H.; Shafie, S.A.; Al-attab, K.A.; Zainal, Z.A. Upgrading agricultural wastes using three different carbonization methods: Thermal, hydrothermal and vapothermal. *Bioresour. Technol.* **2018**, *265*, 365–371. [CrossRef] [PubMed]
15. Akiya, N.; Savage, P.E. Roles of water for chemical reactions in high-temperature water. *Chem. Rev.* **2002**, *102*, 2725–2750. [CrossRef] [PubMed]
16. Lemmon, E.W.; Bell, I.H.; Huber, M.L.; McLinden, M.O. *NIST Standard Seference Database 23: Reference Fluid Thermodynamic and Transport Properties-REFPROP, Version 10.0*; National Institute of Standards and Technology (NIST): Gaithersburg, MD, USA, 2018.
17. Lemmon, E.W.; McLinden, M.O.; Friend, D.G. Thermophysical properties of fluid systems. In *NIST Chemistry Webbook*; NIST Standard Reference Database No. 69; National Institute of Standards and Technology (NIST): Gaithersburg, MD, USA, 2017.
18. IAPWS Thermodynamic Property Formulations. Available online: http://www.iapws.org/newform.html (accessed on 1 October 2020).
19. Parr Stirred Reactors and Pressure Vessels. Chapter 2. Bulletin 4500, Volume 12. Available online: https://www.parrinst.com/de/files/downloads/2013/08/4500MB-v12.0_Ch2_Parr_Stirred-Reactors-Literature.pdf (accessed on 15 May 2020).
20. Parr Safety in the Operation of Laboratory Reactors and Pressure Vessels. No. 230M, 8; Moline, IL. 2009. Available online: https://www.ccmr.cornell.edu/wp-content/uploads/sites/2/2015/11/ParrReactorSafetyInfo-230m.pdf (accessed on 15 May 2020).
21. Corder, S.E. Properties and uses of bark as an energy source. In Proceedings of the XVI IUFRO World Congress, Oslo, Norway, 20 June–2 July 1976; Volume 31.
22. Karpaky, H.; Maalouf, C.; Bliard, C.; Gacoin, A.; Lachi, M.; Polidori, G. Mechanical and thermal characteriziation of a beet pulp-starch composite for building applications. *E3S Web Conf.* **2019**, *85*, 1–8. [CrossRef]
23. Oliveira, S.L.; Mendes, R.F.; Mendes, L.M.; Freire, T.P. Particleboard panels made from sugarcane bagasse: Characterization for use in the furniture industry. *Mater. Res.* **2016**, *19*, 914–922. [CrossRef]
24. Pathak, P.D.; Mandavgane, S.A.; Kulkarni, B.D. Fruit peel waste: Characterization and its potential uses. *Curr. Sci.* **2017**, *113*, 444–454. [CrossRef]

Publisher's Note: MDPI stays neutral with regard to jurisdictional claims in published maps and institutional affiliations.

© 2020 by the authors. Licensee MDPI, Basel, Switzerland. This article is an open access article distributed under the terms and conditions of the Creative Commons Attribution (CC BY) license (http://creativecommons.org/licenses/by/4.0/).

Article

Formation of Carbon Quantum Dots via Hydrothermal Carbonization: Investigate the Effect of Precursors

Md Rifat Hasan [1], Nepu Saha [2], Thomas Quaid [2] and M. Toufiq Reza [2,*]

[1] Department of Chemical and Biomolecular Engineering, Ohio University, Athens, OH 45701, USA; mh919116@ohio.edu

[2] Department of Biomedical and Chemical Engineering and Sciences, Florida Institute of Technology, Melbourne, FL 32901, USA; nsaha2019@my.fit.edu (N.S.) tquaid2018@my.fit.edu (T.Q.)

* Correspondence: treza@fit.edu; Tel.: +1-321-674-8578

Citation: Hasan, M.R.; Saha, N.; Quaid, T.; Reza, M.T. Formation of Carbon Quantum Dots via Hydrothermal Carbonization: Investigate the Effect of Precursors. *Energies* **2021**, *14*, 986. https://doi.org/10.3390/en14040986

Academic Editor: David Chiaramonti

Received: 17 January 2021
Accepted: 11 February 2021
Published: 13 February 2021

Publisher's Note: MDPI stays neutral with regard to jurisdictional claims in published maps and institutional affiliations.

Copyright: © 2021 by the authors. Licensee MDPI, Basel, Switzerland. This article is an open access article distributed under the terms and conditions of the Creative Commons Attribution (CC BY) license (https://creativecommons.org/licenses/by/4.0/).

Abstract: Carbon quantum dots (CQDs) are nanomaterials with a particle size range of 2 to 10 nm. CQDs have a wide range of applications such as medical diagnostics, bio-imaging, biosensors, coatings, solar cells, and photocatalysis. Although the effect of various experimental parameters, such as the synthesis method, reaction time, etc., have been investigated, the effect of different feedstocks on CQDs has not been studied yet. In this study, CQDs were synthesized from hydroxymethylfurfural, furfural, and microcrystalline cellulose via hydrothermal carbonization at 220 °C for 30 min of residence time. The produced CQDs showed green luminescence behavior under the short-wavelength UV light. Furthermore, the optical properties of CQDs were investigated using ultraviolet-visible spectroscopy and emission spectrophotometer, while the morphology and chemical bonds of CQDs were investigated using transmission electron microscopy and Fourier-transform infrared spectroscopy, respectively. Results showed that all CQDs produced from various precursors have absorption and emission properties but these optical properties are highly dependent on the type of precursor. For instance, the mean particle sizes were 6.36 ± 0.54, 5.35 ± 0.56, and 3.94 ± 0.50 nm for the synthesized CQDs from microcrystalline cellulose, hydroxymethylfurfural, and furfural, respectively, which appeared to have similar trends in emission intensities. In addition, the synthesized CQDs experienced different functionality (e.g., C=O, O-H, C-O) resulting in different absorption behavior.

Keywords: carbon quantum dots; hydrothermal carbonization; hydroxymethylfurfural; furfural; microcrystalline cellulose

1. Introduction

Carbon quantum dots (CQDs) are a new class of carbon nanomaterials sized below 10 nm [1,2]. CQDs have attracted tremendous attention in the research community due to their unique photoluminescence (PL) properties, biocompatibility, electrochemical luminescence properties, and low toxicity [3–5]. These properties enable them to be used in bio-imaging, biosensor, drug delivery, and photo-catalysis applications [6–13]. CQDs are mainly synthesized via two approaches: top-down and bottom-up. The top-down approach refers to the breakdown of larger carbon particles by laser ablation, electrochemical oxidation, chemical oxidation, and ultrasonic synthesis [14–17], while the bottom-up approach synthesizes the CQDs from molecular precursors through microwave synthesis, thermal decomposition, and hydrothermal treatment [4,18–20]. Among all synthesis methods, hydrothermal carbonization (HTC) has been considered as the most promising method due to high quantum efficiency, lower cost, environmentally friendly nature, and non-toxicity [4,5,20–22].

HTC is an emerging technology that converts carbohydrates into high-value materials, fuels, and chemicals [23–26]. HTC is typically performed at 180 to 260 °C for 5 min to 12 h under water saturation pressure, depending on the application [27]. In this conditions, water is more reactive and behaves as a non-polar solvent because of high ionic product and low dielectric constant [28]. Therefore, carbohydrates undergo for hydrolysis and the hydrolyzed products then undergo for simultaneous dehydration, decarboxylation, condensation, and polymerization to make cross-linked polymeric materials [29]. The particle sizes of the solid product, often referred as hydrohcar, are generally between 10 nm to 100 µm [30]. As HTC is a bottom-up approach, the particles smaller than 10 nm remain in the liquid phase (known as process liquid). HTC process liquid is referred to as the waste product from a HTC process; thus far, it is considered as a liability for the HTC development, as it requires expensive treatment. Various treatment technologies including anaerobic digestion (AD), wet air oxidation (WAO), and membrane distillation (MD) have been proposed to treat HTC process liquid [31–33]. However, the lack of value-added product separation along with the additional cost of treatment often prohibits the adoption of these aforementioned technologies for HTC process liquid.

In the recent past, various researchers have tried to separate the CQDs from the liquid phase of the HTC process [3,4,34–36]. For instance, Mehta et al. produced highly fluorescent CQDs from sugarcane via hydrothermal treatment at 120 °C for 3 h [4]. The obtained the CQDs were about 3.0 nm in size with highly blue fluorescence. On the other hand, Sahu et al. reported highly photoluminescent CQDs with sizes 1.5 to 4.5 nm from orange juice via HTC at 120 °C for 2.5 h [34]. Papaioannou et al. studied the effect of HTC residence time from 2 to 12 h at 200 °C on the properties of CQDs produced from D-(+)-glucose [35]. They reported that the sizes of CQDs decreased and their level of crystallinity increased with an increase in reaction time.

From the above discussed literature, it is clear that various parameters such as reaction time, residence time, etc. have been examined to understand their effect on the properties of CQDs. Although researchers studied different feedstocks at various conditions, to the best of the authors' knowledge, no study has been reported about how the properties of CQDs change with the variation of feedstock at the same experimental conditions (temperature and residence time). The hypothesis was that simultaneous evolutions of amorphous hydrochar and semi-crystalline CQDs occur during HTC reactions. The amorphous hydrochar agglomerates into micro-meter-sized supramolecules, whereas, nano-sized CQDs remain in the liquid phase due to their high aqueous solubility. As various biopolymers react differently under HTC conditions, the authors' expectation was that the presence and properties of CQDs will be different as well. Therefore, the objectives of the project were to investigate the effect of feedstock on the properties of CQDs. To achieve this goal, HTC of three organic precursors (i.e., furfural, 5-hydroxymethyl furfural (HMF), and microcrystalline cellulose) have been performed at 220 °C for 30 min of residence time. In this study, the HTC experiments were conducted at a relatively low temperature and residence time as they are reported to be favorable for CQD production. HMF and furfural might react at temperatures lower than 220 °C under an HTC environment, but the literature indicates that microcrystalline cellulose starts to react at around 220 °C [29,37]. As the purpose of this study was to investigate the effect of precursor, the authors wanted to choose the lowest temperature where all the precursors could react under the HTC environment. Multi-staged filtration was performed on the process liquids to remove supramolecules. The presence of CQDs was confirmed by investigating the optical (luminescence, ultraviolet-visible absorption, and emission), morphology (transmission electron microscopy), and chemical (Fourier Transform Infrared Spectroscopy) properties.

2. Materials and Methods

2.1. Materials

In this study, CQDs were produced from three different biomass precursors. Of them, 5-hydroxymethylfurfural (HMF) was purchased from Carbosynth LLC. (San Diego, CA, USA). Meanwhile, furfural was purchased from Sigma-Aldrich (St. Louis, MO, USA). On the other hand, microcrystalline cellulose (extra pure, avg. particle size 90 µm) was purchased from Acros Organics (Fair Lawn, NJ, USA). Analytical-grade ethanol was purchased from Fisher Scientific (Waltham, MA, USA).

2.2. Methods

2.2.1. Hydrothermal Carbonization and Separation of CQDs from HTC Process Liquid

HTC experiments of different precursors were performed in a glass-lined 100 mL Parr batch reactor (reactor series 4590, Moline, IL, USA). A Parr proportional-integral-derivative (PID) controller (model 4590) with an accuracy of ±2 °C was used to control the reaction temperature. The pressure was not controlled but was monitored during the experiment with a pressure transducer and a gauge. HTC experiments were conducted at 220 °C for a residence time of 30 min. The reactor was loaded with 45 mL solution contained 10% (w/v) of precursor and the rest was deionized (DI) water. The reactor was closed and heated at a constant rate of 10 °C/min until it reached the set temperature and then it was maintained at isothermal conditions for 30 min. At the end of the residence time, the heater was turned off, heating elements were removed from the reactor, and the reactor was rapidly cooled to room temperature (~30 °C) by placing it in an ice-water bath. Once the reactor reached room temperature, the gaseous products produced during the reaction were vented in a fume hood by opening the vent valve. Finally, the lid of the reactor was opened and HTC process liquid was filtered by Whatman 41 filter paper. Dark brown HTC process liquid samples were collected in centrifuge tubes and stored in the refrigerator for further synthesis. As the major goal of this study was to evaluate the variation of optical, morphological, and chemical properties of CQDs with precursors, the hydrochar and gaseous products were not further characterized. However, the mass balance and physico-chemical characterization of hydrochars can be found elsewhere [23]. All the experiments were completed in duplicate to check reproducibility.

The dark brown HTC process liquid was centrifuged at 10,000 rpm for 15 min by a Sorvall BIOS 16-series centrifuge from Thermo Fisher Scientific (Waltham, MA, USA) to separate the larger particles from the liquid phase. The fluorescent CQD containing liquid phase was then filtered with a standard syringe filter (0.22 µm). The filtered CQD containing solution was then evaporated under reduced pressure by vacuum distillation. The concentrated product from vacuum distillation was kept in the high vacuum freeze dryer for 24 h to obtain powdered CQDs. The powdered CQDs were collected and refrigerated in glass vials for further investigations. A schematic of the CQD synthesis and characterization strategy is shown in Figure 1.

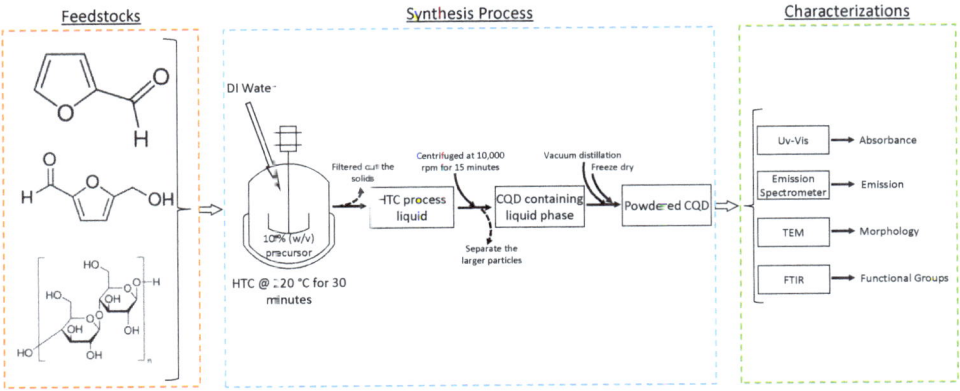

Figure 1. A schematic of the CQD synthesis and characterization strategy.

2.2.2. Optical, Morphological, and Chemical Characterizations of CQDs

To investigate the fluorescent properties of the synthesized CQDs, short-wavelength (254 nm) UV light was used. Although the produced CQDs appeared as light brown in color under the day light, it showed a green luminescence under UV light (see graphical abstract). The green luminescence under the UV light provided hints about the presence of quantum particles in the produced solution. This finding encouraged the authors to further investigate the optical properties of CQDs.

A Perkin Elmer Lambda LS 35 UV-visible spectrometer (Waltham, MA, USA) was used to observe the absorption behavior of the CQDs. The syringe-filtered CQD containing liquid phase was used in this analysis. The absorption capacity of those CQDs containing liquid phase were beyond the maximum limit (2.5) of the UV-vis spectrometer. As a result, all the original samples were diluted 5000 times with DI water. For each run, 10 mL of diluted sample was taken in a quartz cuvette and absorption was monitored for a wavelength range of 200 to 700 nm. The absorption capability of the CQDs was observed relative to the DI water blank run.

A Horiba FluoroMax-4 emission spectrometer (Irvine, CA, USA) was used to investigate the fluorescence emission capability of the synthesized CQDs at a certain excitation wavelength. For a clear comparison of the emission capacity of different CQDs, all the CQD samples were diluted to an absorbance value of 0.2 at 350 nm wavelength. A quartz cuvette (considered for better light transmission) was used for obtaining the fluorescence emission spectra. A slit width of 3 mm was used for the excitation and fluorescence emission spectra of aqueous CQD solution. For 360 nm of excitation wavelength, emission spectra were recorded for a wavelength range of 375 to 700 nm. Emission spectra were compared to DI water blank run at the same condition.

To observe the morphologies and obtain size images of CQDs, transmission electron microscopy (TEM) was performed on a Tecnai F20 system (Hillsboro, OR, USA). The powdered CQDs were dispersed into ethanol and sonicated for 6 h. One drop of the ethanol dispersed CQDs was placed on a copper grid coated with amorphous carbon, dried at room temperature, and analyzed in TEM. The instrument was operated at 200 kV with an X-TWIN lens and high-brightness field emission electron gun (FEG). The TEM images were then processed with Image J software to determine the particle size distribution of the CQDs.

To observe the changes in functional groups, Fourier-transform infrared spectroscopy (FTIR) analysis was performed in Bruker Optics Vertex 80 FTIR (Billerica, MA, USA). For the analysis, CQDs particles were dispersed in DI water. FTIR transmittance spectra were obtained for the wavenumber range of 4000 to 500 cm^{-1} with respect to the reference of DI water run.

All the above-mentioned characterizations were completed in duplicate to check reproducibility.

3. Results and Discussions

3.1. Absorption Properties

The optical characteristics of CQDs were investigated in terms of their absorption and emission properties. From the UV-vis spectra (absorption properties) shown in Figure 2, it was observed that all CQDs exhibit a broad range of absorption in the UV region, with the tails of the spectra in the near-visible region. Although all of the CQD samples started to show absorption behavior at the same wavelength (240 nm), their peak intensities were different. For instance, the cellulose peak showed the highest intensity of 2.35 while HMF and furfural showed peaks of 1.45 and 0.7, respectively. Additionally, the cellulose derived CQDs showed normal distribution while the furfural showed average distribution. This phenomenon indicates that the CQDs produced from furfural can absorb wide wavelengths while cellulose CQDs can adsorb more specific wavelengths among the three studied precursors. This variation of the absorption properties could be further supported by the FTIR spectra (see Figure 3). In the cellulose spectrum, only a carbonyl (C=O) peak was observed at 1670 cm^{-1}. On the other hand, both HMF and furfural showed various other peaks, such as medium alcohol (O-H) peaks between 1330 and 1420 cm^{-1}, aromatic ester (C-O) peaks between 1200 and 1300 cm^{-1}, and a sharp alkyl ether (C-O) peak at 1027 cm^{-1}. Due to the presence of additional functionality, HMF and furfural could exhibit broader absorbance compared to cellulose. In addition to the broad peaks, CQDs showed tails in the visible region. These tails are typically related to nanoparticle functionalization and are reported as lower energy surface centers [38]. These tails are also attributed the presence of various π→π* (C=C) and n→π* (C=O and/or others) transitions [2]. As the aromatic rings increase with the hydrothermal treatment, the energy gaps between π states gradually decreased [15]. On the other hand, functional groups (i.e., carbonyl) with electron lone pairs could be bonded with aromatic carbon that allows electron transition from n states [39].

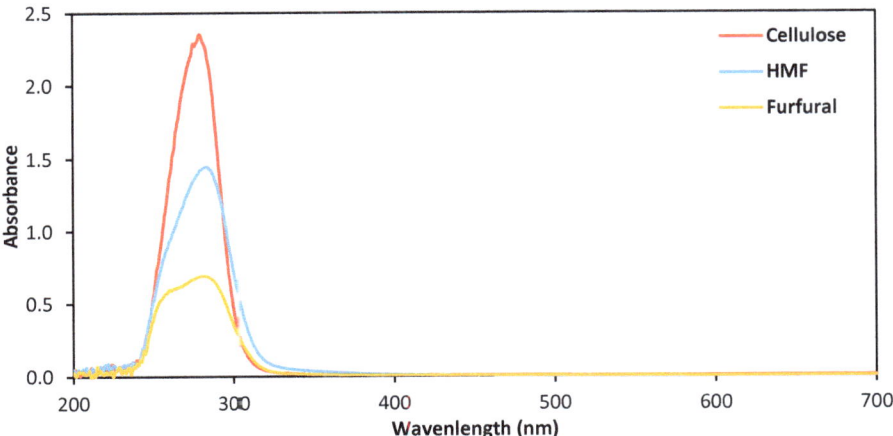

Figure 2. UV-Vis spectra for carbon quantum dots derived from various precursors.

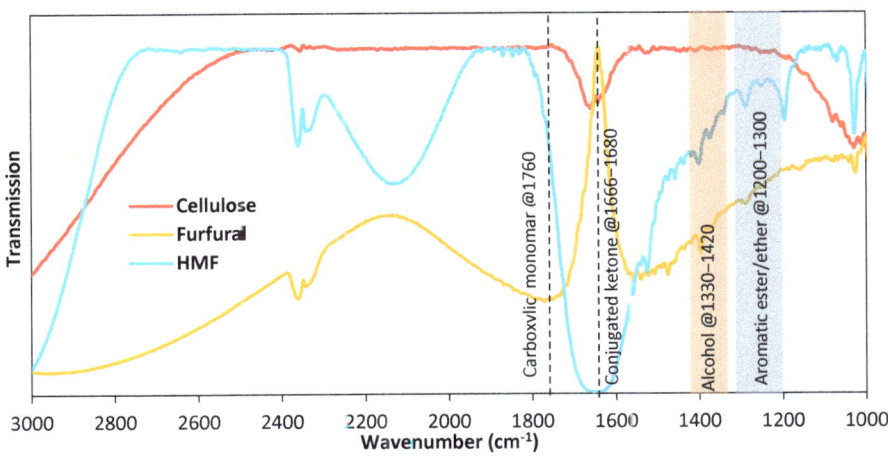

Figure 3. FTIR spectra of the CQDs synthesized from various precursors.

3.2. Emission Properties

In addition to the adsorption properties, fluorescence emission properties of the CQDs were investigated. The emission spectra of the CQDs are shown in Figure 4, where it can be seen that the emission wavelength (360 nm) is longer than the excitation wavelength (240 nm) shown in the absorbance spectra (Figure 2). This observation is in line with previous literature, where the researchers reported the reason behind this is mainly due to bandgap of the conjugated π domains and/or the presence of defects in the structures [39–41]. The spectra also show that the emission capacity of cellulose-based CQDs is the highest among the studied feedstocks, where furfural-derived CQDs show the least emission capacities. As the absorbance intensity of the furfural-derived CQDs was the least, it was expected to have the highest emission properties. However, the emission properties showed the similar trend as absorption capacity. This could be due to the combined effect of functional groups, ligand chain length, surface defects, and morphology of the CQDs [42].

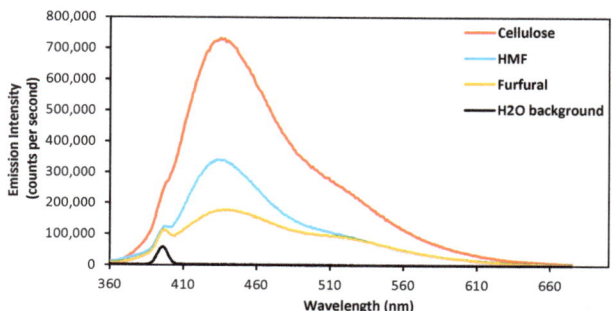

Figure 4. Emission spectra of carbon quantum dots derived from various precursors.

3.3. Morphology of the CQDs

The morphology and particle size of the produced CQDs were obtained from TEM analysis which are shown in Figure 5. It is observed that these CQD particles are spherical in shape and most of the particles are separated from each other. Size distribution results revealed that the spherical nanoparticles diameter ranged from 2 to 9 nm. Although the CQDs were synthesized under the same hydrothermal conditions, the sizes of the CQDs were different. This could be due to the various degradation temperatures of the precursors allowing them to nucleate at different conditions, thus forming different-sized CQDs [43,44]. In addition, the agglomeration of the particles could be the reason for getting various mean sizes [43]. A general trend between CQD emission capacity and mean size was observed. For instance, the highest emission capacity was attributed to the mean particle diameter of 6.36 ± 0.54 nm for the cellulose-derived CQDs, while the lowest emission intensity occurred with the mean particle diameter of 3.94 ± 0.60 nm from the furfural-derived CQDs. This could be attributed to the increased surface area of the particles, allowing more light-emitting functional groups to be present and active.

The CQDs produced in this study contain remarkably similar characteristics, as found by the Zhao et al. and Gao at al., where they produced CQDs from pine wood and alkali lignin, respectively, via HTC treatment [45,46]. The similarities include but are not limited to the absorption range of 225 to 300 nm, the emission range of 435 to 450 nm, and the size 2 to 5 nm. With these characteristics, the CQDs were successfully used as a nanosensor to detect the iron (Fe^{3+}) and ascorbic acid. It has been determined that the functionalization of the CQDs with groups containing oxygen are crucial to the success of the detection of the ions as they are chelated by these groups to induce a fluorescent quenching [45,46]. The CQDs produced in this study all share this vital characteristic as well. Although the dots in this study fall within those ranges in every respect except for the size of the cellulose dots, these CQDs could be used for similar purposes as well.

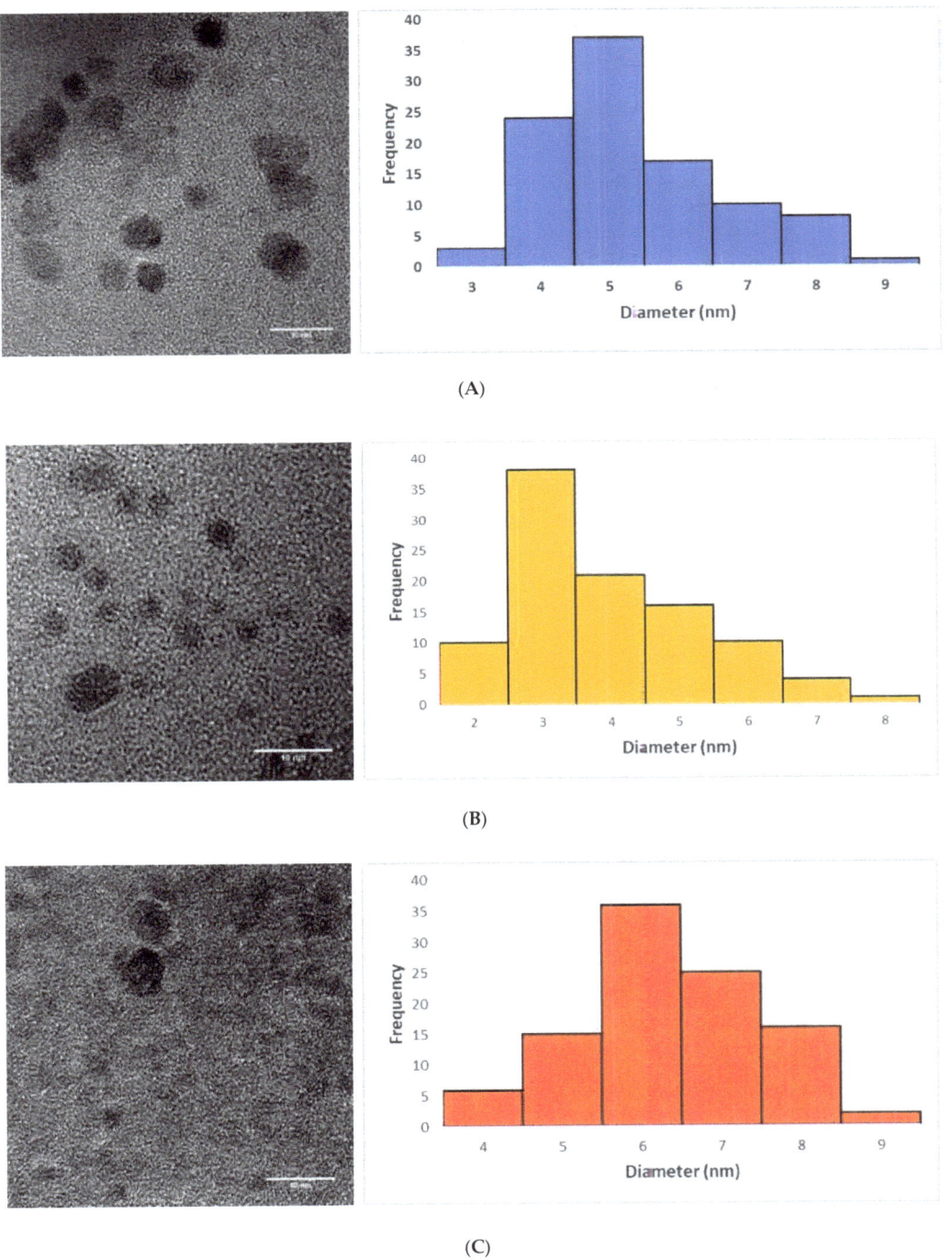

Figure 5. TEM size images along with the particle size distribution of the different types of CQDs synthesized from: (**A**) HMF; (**B**) Furfural; (**C**) Cellulose.

4. Conclusions

In this study, CQDs were produced from HMF, furfural, and microcrystalline cellulose via HTC at 220 °C for 30 min. The optical and morphological properties of the produced CQDs were investigated. It was observed that CQDs showed green luminescence under short-wavelength UV light (254 nm). Additionally, they can absorb light in a broad range. The variation of the absorption range was justified from the FTIR spectra where CQDs from different precursors showed different chemical bonds (e.g., alcohol, ketone, etc.). Although all the studied CQDs were within two to nine nm in diameter, the mean diameter varied with the precursor. Depending on the particle size, the CQDs showed different emission capacities. Finally, this study provides an insight into the effect of precursor on the CQD properties. So, the precursors for the CQDs should be selected based on the targeted applications.

Author Contributions: Conceptualization, M.R.H., N.S., M.T.R.; methodology, M.R.H., N.S.; formal analysis, M.R.H., N.S., T.Q.; investigation, M.R.H., N.S., M.T.R.; writing—original draft preparation, M.R.H., N.S., T.Q.; writing—review and editing, M.R.H., N.S., T.Q., M.T.R. All authors have read and agreed to the published version of the manuscript.

Funding: The research is partially funded by the United States Department of Agriculture (USDA) grant (2019-67019-2928) through M.T.R.

Institutional Review Board Statement: Not applicable.

Informed Consent Statement: Not applicable.

Data Availability Statement: Not applicable.

Acknowledgments: The authors acknowledge Travis A. White from the Department of Chemistry and Biochemistry at Ohio University for allowing to use emission spectrofluorometer. The authors are also thankful to Kyle McGaughy from the biofuels lab at the Florida Institute of Technology for his laboratory efforts in this project.

Conflicts of Interest: The authors declare no conflict of interest.

References

1. Liu, W.; Li, C.; Ren, Y.; Sun, X.; Pan, W.; Li, Y.; Wang, J.; Wang, W. Carbon dots: Surface engineering and applications. *J. Mater. Chem. B* **2016**, *4*, 5772–5788. [CrossRef]
2. Wang, Y.; Hu, A. Carbon quantum dots: Synthesis, properties and applications. *J. Mater. Chem. C* **2014**, *2*, 6921–6939. [CrossRef]
3. Zhu, S.; Meng, Q.; Wang, L.; Zhang, J.; Song, Y.; Jin, H.; Zhang, K.; Sun, H.; Wang, H.; Yang, B. Highly photoluminescent carbon dots for multicolor patterning, sensors, and bioimaging. *Angew. Chem.* **2013**, *125*, 4045–4049. [CrossRef]
4. Mehta, V.N.; Jha, S.; Kailasa, S.K. One-pot green synthesis of carbon dots by using Saccharum officinarum juice for fluorescent imaging of bacteria (Escherichia coli) and yeast (*Saccharomyces cerevisiae*) cells. *Mater. Sci. Eng. C* **2014**, *38*, 20–27. [CrossRef] [PubMed]
5. Qu, D.; Zheng, M.; Du, P.; Zhou, Y.; Zhang, L.; Li, D.; Tan, H.; Zhao, Z.; Xie, Z.; Sun, Z. Highly luminescent S, N co-doped graphene quantum dots with broad visible absorption bands for visible light photocatalysts. *Nanoscale* **2013**, *5*, 12272–12277. [CrossRef] [PubMed]
6. Choudhary, R.; Patra, S.; Madhuri, R.; Sharma, P.K. Equipment-free, single-step, rapid, "on-site" kit for visual detection of lead ions in soil, water, bacteria, live cells, and solid fruits using fluorescent cube-shaped nitrogen-doped carbon dots. *ACS Sustain. Chem. Eng.* **2016**, *4*, 5606–5617. [CrossRef]
7. Gao, X.; Lu, Y.; Zhang, R.; He, S.; Ju, J.; Liu, M.; Li, L.; Chen, W. One-pot synthesis of carbon nanodots for fluorescence turn-on detection of Ag+ based on the Ag+-induced enhancement of fluorescence. *J. Mater. Chem. C* **2015**, *3*, 2302–2309. [CrossRef]
8. Kim, S.; Choi, Y.; Park, G.; Won, C.; Park, Y.-J.; Lee, Y.; Kim, B.-S.; Min, D.-H. Highly efficient gene silencing and bioimaging based on fluorescent carbon dots in vitro and in vivo. *Nano Res.* **2017**, *10*, 503–519. [CrossRef]
9. Li, H.; Kang, Z.; Liu, Y.; Lee, S.-T. Carbon nanodots: Synthesis, properties and applications. *J. Mater. Chem.* **2012**, *22*, 24230–24253. [CrossRef]
10. Liu, J.; Liu, Y.; Liu, N.; Han, Y.; Zhang, X.; Huang, H.; Lifshitz, Y.; Lee, S.-T.; Zhong, J.; Kang, Z. Metal-free efficient photocatalyst for stable visible water splitting via a two-electron pathway. *Science* **2015**, *347*, 970–974. [CrossRef] [PubMed]

11. Loukanov, A.; Sekiya, R.; Yoshikawa, M.; Kobayashi, N.; Moriyasu, Y.; Nakabayashi, S. Photosensitizer-conjugated ultrasmall carbon nanodots as multifunctional fluorescent probes for bioimaging. *J. Phys. Chem. C* 2016, *120*, 15867–15874. [CrossRef]
12. Wang, Z.; Fu, B.; Zou, S.; Duan, B.; Chang, C.; Yang, B.; Zhou, X.; Zhang, L. Facile construction of carbon dots via acid catalytic hydrothermal method and their application for target imaging of cancer cells. *Nano Res.* 2016, *9*, 214–223. [CrossRef]
13. Wu, Y.-F.; Wu, H.-C.; Kuan, C.-H.; Lin, C.-J.; Wang, L.-W.; Chang, C.-W.; Wang, T.-W. Multi-functionalized carbon dots as theranostic nanoagent for gene delivery in lung cancer therapy. *Sci. Rep.* 2016, *6*, 21170. [CrossRef] [PubMed]
14. Zhou, J.; Booker, C.; Li, R.; Zhou, X.; Sham, T.-K.; Sun, X.; Ding, Z. An electrochemical avenue to blue luminescent nanocrystals from multiwalled carbon nanotubes (MWCNTs). *J. Am. Chem. Soc.* 2007, *129*, 744–745. [CrossRef] [PubMed]
15. Li, H.; He, X.; Kang, Z.; Huang, H.; Liu, Y.; Liu, J.; Lian, S.; Tsang, C.H.A.; Yang, X.; Lee, S.T. Water-soluble fluorescent carbon quantum dots and photocatalyst design. *Angew. Chem.* 2010, *122*, 4532–4536. [CrossRef]
16. Qiao, Z.-A.; Wang, Y.; Gao, Y.; Li, H.; Dai, T.; Liu, Y.; Huo, Q. Commercially activated carbon as the source for producing multicolor photoluminescent carbon dots by chemical oxidation. *Chem. Commun.* 2010, *46*, 8812–8814. [CrossRef]
17. Park, S.Y.; Lee, H.U.; Park, E.S.; Lee, S.C.; Lee, J.-W.; Jeong, S.W.; Kim, C.H.; Lee, Y.-C.; Huh, Y.S.; Lee, J. Photoluminescent green carbon nanodots from food-waste-derived sources: Large-scale synthesis, properties, and biomedical applications. *ACS Appl. Mater. Interfaces* 2014, *6*, 3365–3370. [CrossRef]
18. Tang, L.; Ji, R.; Cao, X.; Lin, J.; Jiang, H.; Li, X.; Teng, K.S.; Luk, C.M.; Zeng, S.; Hao, J. Deep ultraviolet photoluminescence of water-soluble self-passivated graphene quantum dots. *ACS Nano* 2012, *6*, 5102–5110. [CrossRef] [PubMed]
19. Chen, B.; Li, F.; Li, S.; Weng, W.; Guo, H.; Guo, T.; Zhang, X.; Chen, Y.; Huang, T.; Hong, X. Large scale synthesis of photoluminescent carbon nanodots and their application for bioimaging. *Nanoscale* 2013, *5*, 1967–1971. [CrossRef] [PubMed]
20. Wang, W.; Ni, Y.; Xu, Z. One-step uniformly hybrid carbon quantum dots with high-reactive TiO2 for photocatalytic application. *J. Alloy. Compd.* 2015, *622*, 303–308. [CrossRef]
21. Bian, J.; Huang, C.; Wang, L.; Hung, T.; Daoud, W.A.; Zhang, R. Carbon dot loading and TiO2 nanorod length dependence of photoelectrochemical properties in carbon dot/TiO2 nanorod array nanocomposites. *ACS Appl. Mater. Interfaces* 2014, *6*, 4883–4890. [CrossRef]
22. Wang, J.; Gao, M.; Ho, G.W. Bidentate-complex-derived TiO2/carbon dot photocatalysts: In situ synthesis, versatile heterostructures, and enhanced H 2 evolution. *J. Mater. Chem. A* 2014, *2*, 5703–5709. [CrossRef]
23. Saha, N.; Saba, A.; Reza, M.T. Effect of hydrothermal carbonization temperature on pH, dissociation constants, and acidic functional groups on hydrochar from cellulose and wood. *J. Anal. Appl. Pyrolysis* 2019, *137*, 138–145. [CrossRef]
24. Saha, N.; Volpe, M.; Fiori, L.; Volpe, R.; Messineo, A.; Reza, M.T. Cationic Dye Adsorption on Hydrochars of Winery and Citrus Juice Industries Residues: Performance, Mechanism, and Thermodynamics. *Energies* 2020, *13*, 4686. [CrossRef]
25. Saha, N.; Saba, A.; Saha, P.; McGaughy, K.; Franqui-Villanueva, D.; Orts, W.J.; Hart-Cooper, W.M.; Reza, M.T. Hydrothermal carbonization of various paper mill sludges: An observation of solid fuel properties. *Energies* 2019, *12*, 858. [CrossRef]
26. Reza, M.T. Upgrading biomass by hydrothermal and chemical conditioning. Ph.D. Dissertation, University of Nevada, Reno, NV, USA, 2013.
27. Berge, N.D.; Ro, K.S.; Mao, J.; Flora, J.R.V.; Chappell, M.A.; Bae, S. Hydrothermal Carbonization of Municipal Waste Streams. *Environ. Sci. Technol.* 2011, *45*, 5696–5703. [CrossRef] [PubMed]
28. Bandura, A.V.; Lvov, S.N. The ionization constant of water over wide ranges of temperature and density. *J. Phys. Chem. Ref. Data* 2006, *35*, 15–30. [CrossRef]
29. Sevilla, M.; Fuertes, A.B. The production of carbon materials by hydrothermal carbonization of cellulose. *Carbon* 2009, *47*, 2281–2289. [CrossRef]
30. Sevilla, M.; Fuertes, A.B. Chemical and structural properties of carbonaceous products obtained by hydrothermal carbonization of saccharides. *Chem.–A Eur. J.* 2009, *15*, 4195–4203. [CrossRef] [PubMed]
31. Reza, M.T.; Freitas, A.; Yang, X.; Coronella, C.J. Wet air oxidation of hydrothermal carbonization (HTC) process liquid. *ACS Sustain. Chem. Eng.* 2016, *4*, 3250–3254. [CrossRef]
32. Silva, N.A. Evaluation of Membrane Distillation for Treating Hydrothermal Carbonization Aqueous Product (HAP) from Dairy Manure. Master's Thesis, University of Nevada, Reno, NV, USA, 2020.
33. Wirth, B.; Reza, T.; Mumme, J. Influence of digestion temperature and organic loading rate on the continuous anaerobic treatment of process liquor from hydrothermal carbonization of sewage sludge. *Bioresour. Technol.* 2015, *198*, 215–222. [CrossRef]
34. Sahu, S.; Behera, B.; Maiti, T.K.; Mohapatra, S. Simple one-step synthesis of highly luminescent carbon dots from orange juice: Application as excellent bio-imaging agents. *Chem. Commun.* 2012, *48*, 8835–8837. [CrossRef]
35. Papaioannou, N.; Titirici, M.-M.; Sapelkin, A. Investigating the Effect of Reaction Time on Carbon Dot Formation, Structure, and Optical Properties. *ACS Omega* 2019, *4*, 21658–21665. [CrossRef]
36. Yang, Z.-C.; Wang, M.; Yong, A.M.; Wong, S.Y.; Zhang, X.-H.; Tan, H.; Chang, A.Y.; Li, X.; Wang, J. Intrinsically fluorescent carbon dots with tunable emission derived from hydrothermal treatment of glucose in the presence of monopotassium phosphate. *Chem. Commun.* 2011, *47*, 11615–11617. [CrossRef]
37. Saha, N.; McGaughy, K.; Reza, M.T. Elucidating hydrochar morphology and oxygen functionality change with hydrothermal treatment temperature ranging from subcritical to supercritical conditions. *J. Anal. Appl. Pyrolysis* 2020, *152*, 104965. [CrossRef]

38. Carbonaro, C.M.; Corpino, R.; Salis, M.; Mocci, F.; Thakkar, S.V.; Olla, C.; Ricci, P.C. On the emission properties of carbon dots: Reviewing data and discussing models. *C—J. Carbon Res.* **2019**, *5*, 60. [CrossRef]
39. Wang, R.; Lu, K.-Q.; Tang, Z.-R.; Xu, Y.-J. Recent progress in carbon quantum dots: Synthesis, properties and applications in photocatalysis. *J. Mater. Chem. A* **2017**, *5*, 3717–3734. [CrossRef]
40. Baker, S.N.; Baker, G.A. Luminescent carbon nanodots: Emergent nanolights. *Angew. Chem. Int. Ed.* **2010**, *49*, 6726–6744. [CrossRef] [PubMed]
41. Gokus, T.; Nair, R.; Bonetti, A.; Bohmler, M.; Lombardo, A.; Novoselov, K.; Geim, A.; Ferrari, A.; Hartschuh, A. Making graphene luminescent by oxygen plasma treatment. *Acs Nano* **2009**, *3*, 3963–3968. [CrossRef] [PubMed]
42. Spirin, M.G.; Brichkin, S.B.; Gak, V.Y.; Razumov, V.F. Influence of photoactivation on luminescent properties of colloidal InP@ZnS quantum dots. *J. Lumin.* **2020**, *226*, 117297. [CrossRef]
43. Zhang, Y.; Wang, Y.; Feng, X.; Zhang, F.; Yang, Y.; Liu, X. Effect of reaction temperature on structure and fluorescence properties of nitrogen-doped carbon dots. *Appl. Surf. Sci.* **2016**, *387*, 1236–1246. [CrossRef]
44. Wongso, V.; Sambudi, N.S.; Sufian, S. The effect of hydrothermal conditions on photoluminescence properties of rice husk-derived silica-carbon quantum dots for methylene blue degradation. *Biomass Convers. Biorefin.* **2020**, 179. [CrossRef]
45. Zhao, S.; Song, X.; Chai, X.; Zhao, P.; He, H.; Liu, Z. Green production of fluorescent carbon quantum dots based on pine wood and its application in the detection of Fe^{3+}. *J. Clean. Prod.* **2020**, *263*, 121561. [CrossRef]
46. Gao, X.; Zhou, X.; Ma, Y.; Qian, T.; Wang, C.; Chu, F. Facile and cost-effective preparation of carbon quantum dots for Fe^{3+} ion and ascorbic acid detection in living cells based on the "on-off-on" fluorescence principle. *Appl. Surf. Sci.* **2019**, *469*, 911–916. [CrossRef]

Article

Integration of Air Classification and Hydrothermal Carbonization to Enhance Energy Recovery of Corn Stover

Md Tahmid Islam [1], Nepu Saha [1], Sergio Hernandez [2], Jordan Klinger [2] and M. Toufiq Reza [1,*]

[1] Department of Biomedical and Chemical Engineering and Sciences, Florida Institute of Technology, 150 W University Boulevard, Melbourne, FL 32901, USA; islamm2019@my.fit.edu (M.T.I.); nsaha2019@my.fit.edu (N.S.)

[2] Biomass Characterization Department, Idaho National Laboratory, 2525 Fremont Ave, Idaho Falls, Idaho, ID 83402, USA; sergio.hernandez@inl.gov (S.H.); jordan.klinger@inl.gov (J.K.)

* Correspondence: treza@fit.edu; Tel.: +1-321-674-8578

Abstract: Air classification (AC) is a cost-effective technology that separates the energy-dense light ash fraction (LAF) from the inorganic-rich high ash fraction (HAF) of corn stover. HAF could be upgraded into energy-dense solid fuel by hydrothermal carbonization (HTC). However, HTC is a high-temperature, high-pressure process, which requires additional energy to operate. In this study, three different scenarios (i.e., AC only, HTC only, and integrated AC–HTC) were investigated for the energy recovery of corn stover. AC was performed on corn stover at an 8 Hz fan speed, which yielded 84.4 wt. % LAF, 12.8 wt. % HAF, and 2.8 wt. % below screen particles. About 27 wt. % ash was reduced from LAF by the AC process. Furthermore, HTC was performed on raw corn stover and the HAF of corn stover at 200, 230, and 260 °C for 30 min. To evaluate energy recovery, solid products were characterized in terms of mass yield, ash yield, ultimate analysis, proximate analyses, and higher heating value (HHV). The results showed that the energy density was increased with the increase in HTC temperature, meanwhile the mass yield and ash yield were decreased with the increase in HTC temperature. Proximate analysis showed that fixed carbon increased 18 wt. % for original char and 27 wt. % for HAF char at 260 °C, compared to their respective feedstocks. Finally, the hydrochar resulting from HAF was mixed with LAF and pelletized at 180 bar and 90 °C to densify the energy content. An energy balance of the integrated AC–HTC process was performed, and the results shows that integrated AC with HTC performed at 230 °C resulted in an additional 800 MJ/ton of energy recovery compared to the AC-only scenario

Keywords: corn stover; air classification; hydrothermal carbonization; pelletization; energy recovery

1. Introduction

More than one billion tons of biomass will be available for bioenergy production in the USA by 2030 [1–3]. Among other resources, corn stover (CS) has a huge potential to contribute to the renewable energy portfolio, as more than 250 million dry tons of CS are produced annually in the USA [1,4]. With proper feedstock handling and preprocessing, CS could become an abundant source of bioenergy [5,6]. However, a significant amount of agricultural waste, including CS, is burnt or left unprocessed in the field due to high logistical costs. This results in serious environmental problems and economic forfeiture [7,3]. Therefore, the utilization of potential bioenergy from waste biomass sources such as CS needs to be considered with urgency.

Air classification (AC) is a low-cost preprocessing technology that has been developed particularly in agricultural processing and mining applications. In AC, the separations are performed based on a combination of material size, density, and drag properties, utilizing screens and air streams to effectively separate high-density fractions from low-density fractions. In general, large amounts of soil, rocks, and other foreign matters are incorporated into the feedstock during harvest and collection. The inorganics, dirt, rocks,

and foreign materials are mostly dense materials compared to the lignocellulosic structure of biomass. Therefore, AC separates a significant amount of these exogenous inorganics and creates an enriched stream of biomass (often called the light ash fraction, LAF) and a soil-laden reject stream (also known as the high ash fraction, HAF) [9–11]. Lacey et al. [11] showed such separations for chipped forest residues after AC and reported remarkable ash (wt. %) reductions (>32 wt. %) in the final throughputs. Emerson et al. [9] also found quite reasonable reductions in ash (wt. %) for hybrid poplar (from 2.34 wt. % to 1.67 wt. %) and shrub willow (from 2.60 wt. % to 2.14 wt. %). Recently, Thompson et al. [2] showed that AC can effectively upgrade CS by removing 30 wt. % ash at a mild 7.5 Hz fan speed, or approximately 2.1 m/s counter-stream air flow of AC. It has also been reported that a large fraction of HAF byproduct is produced during AC, accounting for more than 12 wt. % of the initial CS. This HAF of CS contains more than 80 wt. % organic content that can be further utilized [2]. Instead, an estimated HAF disposal cost of USD 28.86 per ton has been suggested by Reza et al. [12] and Humbird et al. [13], accounting for 2.5 cents of the USD 2.15 per gallon minimum ethanol selling price. In order to make AC more economically viable, the further utilization of HAF needs to be implemented in conjunction with AC technologies and system-wide economics and sustainability for full material utilization. However, to the best of the authors' knowledge, no work has been done to date to utilize the waste energy in value-added fuel. Therefore, to make use of the vital unrecovered energy from this solid waste and integrate it with the AC technology, the HAF could be converted into essential products.

Hydrothermal carbonization (HTC) is a promising thermochemical process that transforms biomass, such as CS, into a carbon-rich solid product called hydrochar [5,14,15]. The reaction temperature varies between 180 and 260 °C, and the pressures are maintained above the saturation pressure. The ionic products of water increase three orders of magnitude ($K_{H2O\ (573K)}/K_{H2O\ (298K)} = 10^3$) [5,15,16] and the dielectric constant of water reduces from 78.5 (298 K, 0.1 MPa) to 27.1 (523 K, 5 MPa) [15,17]. As a result, biomass undergoes a series of reactions, i.e., hydrolysis, decarboxylation, dehydration, condensation polymerization, and aromatization, in subcritical water due to the increased reactivity and solvent-like properties [18,19]. For agricultural residues, the HTC reaction products can be divided into three streams: 41–90 wt. % solids with 80–95 wt. % of the original calorific value; 2–10 wt. % gas consisting mainly of CO_2; the rest is process liquid [20–24]. The energy-dense solid hydrochar can be pelletized to improve the mass and energy density of the feedstock and reduce the cost of transportation, handling and storage [25–27].

During HTC, the low density and low viscosity of subcritical water, in combination with the acidic process liquid and the modification of the biomass structure, aid in the leaching of inorganics from biomass structures [18,28–31]. The reduction of ash is particularly important for hydrochars' application as fuel, as this can significantly reduce the ash-fusion temperatures, leading to slagging in fuel boilers [1]. Therefore, HTC could be used to recover the energy in the HAF stream while reducing ash from the hydrochar. However, HTC is an energy-intensive process as it requires high temperatures and high pressure. Therefore, an energy balance in the integrated AC–HTC is warranted to evaluate the net energy recovery from the integrated process compared to the individual AC and HTC processes. Therefore, the main goal of this work was to study the feasibility of three different scenarios, to determine the energy recovery from CS: 1. AC-only process; 2. HTC-only process; 3. Integrated AC–HTC process. First, the AC of CS was performed to separate LAF and HAF. Then HTC on HAF was performed at various temperatures and the hydrochars were pelletized with LAF. The energy recovery from the integrated AC–HTC process was compared with AC-only and HTC-only, wherein the original (ORG) CS was used for both technologies.

2. Materials and Methods

2.1. Materials

A raw corn stover bale from Kadolph Farms in Hardin, Iowa was deconstructed in the process development unit (PDU) of the Idaho National Laboratory's Biomass Feedstock National User Facility (BFNUF). The bale was harvested in the fall of 2018 using an 8-row stripper header and raked using rotary tedder prior to baling. Bales were stored immediately after baling in stacks at the field edge with a six-high by three-length stacking position. Size reduction in the BFNUF was accomplished using a Vermeer prototype horizontal bale grinder (BG480, Pella, Iowa) with a 15.24 cm screen. This mill was powered by two 200 HP electric motors with 96 swinging hammers on each of the two grinding drums. The sample, referred to as ORG CS hereafter, was dried below 10 wt. % moisture content prior to AC and HTC treatments.

2.2. Methods

2.2.1. Air Classification of Corn Stover

The ORG CS intended for HTC were sampled using mechanical sampling procedures and equipment according to Solid Biofuels Sampling ISO 18135:2017. A super sack of the deconstructed CS bale was divided using a custom rotary splitter that consists of a conveyor and eight bins mounted on a rotating table. In this unit, a feed hopper accommodates approximately 120 L of sample and a live-bottom style belt feeder slowly dispenses the samples to 8 sample bins below. The 45 L bins rotate over 100 times during each splitting operation and make at least one full rotation for each belt flight to ensure the samples are representative of the original bulk solid. Both the belt feeder and the rotary table are equipped with speed control devices (variable frequency drive and/or DC potentiometer) to adjust the processing parameters according to the sample feed behavior to ensure analytical splitting (an image of AC can be found in the supplementary information (Figure S1)). AC separates the samples into two fractions as the material is passed over a screen-covered fan: LAF and HAF. The HAF is blown upward and removed while the LAF (air classified material) remains. Air classification was performed on CS feedstocks using a 2× Air Cleaner equipped with an Iso-flo dewatering infeed shaker (Key Technologies, Walla Walla, WA, USA). CS was air classified using a fan speed of 1.8 m/s. Both light and heavy fractions were collected after classification for downstream processing and analyses.

2.2.2. Hydrothermal Carbonization (HTC)

HTC was carried out in the 300 mL Parr reactor (Moline, IL, USA) for 30 min at three different temperatures 200, 230, and 260 °C (an image of the HTC reactor system is shown in the supplementary section (Figure S2)). In a typical HTC run, 20 g of dry CS (HAF or ORG CS) was mixed with 180 mL of deionized (DI) water inside the reactor vessel. The reactor was stirred at 150 ± 5 rpm throughout the HTC experiment. The temperature of the reactor was controlled using a proportional integral derivative (PID) controller. The heating rate was maintained at 10 °C min^{-1}. The pressure was not controlled but monitored throughout the HTC, which was found to be 210 ± 20 psi, 380 ± 20 psi and 620 ± 20 psi for the 200 °C, 230 °C and 260 °C operating temperatures, respectively. When the reactor reached the desired HTC temperature, it was kept isothermal for 30 min. At the end of the HTC reaction time, the heater was turned off and the reactor was quenched by an ice water bath. It took 15 ± 5 min to cool down the reactor below 50 °C. The gas produced during HTC was not the focus of this work and was vented in a fume hood. The slurry was then filtered using Whatman 41 filter paper in a vacuum filtration system. The hydrochar was first washed with approx. 200 mL deionized (DI) water to remove the HTC liquid products that adhered to the hydrochar, and then dried at 105 °C for 24 h in an oven. The dried hydrochar was stored in a ziplock bag for further analysis. The solid mass yield (MY) was calculated using Equation (1). Each hydrochar referred to in this work is labeled according to the HTC temperature. For instance, CS ORG HTC200 represents the hydrochar produced from ORG CS at 200 °C.

$$MY(wt.\ \%) = \frac{\text{Mass of dried post}(g) - \text{process solid}(g)}{\text{Mass of untreated dry feedstock }(g)} \times 100\% \qquad (1)$$

2.2.3. Pelletization of Hydrochar with Corn Stover

Samples (LAF, ORG hydrochar, and hydrochar–LAF mix) were pelletized using a 24 ton benchtop single-press pellet press (Across International, model # MP24A, Livingston, NJ) (an image of the pellet press can be found in the supplementary information (Figure S3)). About 1.5 g of dry sample was powdered using mortar pestle to make it homogeneous. From the homogeneous powdered sample, 0.50 g of dry was taken out and put in the die sleeve of the 13 mm circular pressing die. A compressive force of 7.5 MT was applied at the top surface of the sample by the lever. The temperature was set at 90 °C (beyond the glass transition temperature for CS) [32] using a band heater and the holding time was set at 30 s for each pellet. Pellets were stored in sealed bags for further characterizations.

2.3. Product Characterization

2.3.1. Higher Heating Value (HHV)

An IKA C 200 bomb calorimeter (Staufen, BW, Germany) was used to determine higher heating values (HHV) of the raw and hydrochar pellets by following the ASTM D240 method. Energy yield (EY) was calculated by Equation (2). Triplicates were performed for each sample to report reproducibility.

$$EY(\%) = MY(wt.\ \%) \times \frac{\text{HHV of dried hydrochar }\left(\frac{MJ}{kg}\right)}{\text{HHV of untreated dry feedstock }\left(\frac{MJ}{kg}\right)} \qquad (2)$$

2.3.2. Ash

The ASTM D1102 method was followed to determine the ash content of the dry solid samples by using a muffle furnace (Thermo Scientific, Model # FB1415M, Waltham, MA) at 575 °C for 5 h 30 min. The ash yield (AY) was calculated by using Equation (2). Triplicates were performed for each sample to report reproducibility.

$$AY\ (wt.\ \%) = MY(wt.\ \%) \times \frac{\text{Ash in hydrochar (wt.\ \%)}}{\text{Ash in raw feedstock (wt.\ \%)}} \qquad (3)$$

2.3.3. Thermogravimetric Analysis (TGA)

The volatile matter (VM) and fixed carbon (FC) of samples were determined by TGA using a TGA Q5000 (TA instruments, New Castle, DE, USA). The experimental procedure was taken from the literature [33]. Experiments were carried out under inert atmosphere using a constant flowrate (10 mL/min) of nitrogen to avoid any possible oxidation and to continuously purge the VM. The sample was first heated to 105 °C and kept isothermal for 5 min. The mass loss at this temperature accounts for the moisture content (MC). The sample was then heated to 900 °C at a ramp rate of 20 °C/min and kept isothermal for 5 min. The mass loss from 105 °C to 900 °C accounts for VM. Finally, the FC (wt. %) was then calculated from Equation (3):

$$FC\ (wt.\ \%) = 100\ wt.\ \% - MC\ (wt.\ \%) - VM\ (wt.\ \%) - ash\ (wt.\ \%) \qquad (4)$$

2.3.4. CHNS/O Analysis

A FLASH EA 1112 Series (Thermo Scientific, Waltham, MA, USA) elemental analyzer was used to quantify the elemental carbon (C), hydrogen (H), nitrogen (N), and sulfur (S) content in the sample using the method described in the literature [34]. For the analysis, 2, 5-Bis (5-tert-butyl-benzoxazol-2-yl) thiophene (BBOT) was used as a calibration standard and vanadium oxide (V$_2$O$_5$) as a conditioner for the samples, which were combusted around 950 °C in ultra-high purity oxygen with helium carrier gas and passed over copper oxide pellets and then electrolytic copper. The produced gases were then analyzed by a thermal conductivity detector (TCD), with the peak areas of detection being compared to those of BBOT standards. The oxygen (O) content was found by subtraction method. The following equation was used to find the oxygen wt. %.

$$O~(wt.~\%) = 100~wt.~\% - C~(wt.~\%) - H~(wt.~\%) - N~(wt.~\%) - S~(wt.~\%) - ash~(wt.~\%) \quad (5)$$

3. Results and Discussion

3.1. Characterization of Air Classified CS

The AC of ORG CS led to an LAF or "clean" fraction and an HAF or "dirty" fraction. All the separations were conducted under the fan speed and frequency of 1.8 m/s and 8 Hz, respectively. These conditions were chosen for optimum separation as reported in the literature [2,9,11,35]. As seen in Figure 1, the separated LAF contains more low ash plant tissue fractions such as cobs, stalks, and husks, while the HAF contains more of the undesired plant fractions (leaves, pith, rind) that usually contain higher ash values as well as soil entrained during plant growth or picked up and baled with the CS during harvest. Table 1. Proximate and ultimate analysis of ORG and separated CS. Measurements made using ASTM D7582, ASTM D5373, and ASTM D4239.shows the results from the three materials for comparison, detailing the proximate and ultimate analysis of each fraction. The elemental carbon and moisture content increased more in the LAF than the HAF because of the tissue types and sizes that are preferentially separated into the heavy stream (cob and stalk) after AC. Meanwhile, the elemental oxygen content and volatile content also increased in LAF after AC. On the other hand, the nitrogen and ash content were increased in the HAF after AC. In fact, the LAF fraction from the air classification resulted in a 27 wt. % reduction (from 7.13 wt. % to 5.24 wt. %) in total ash compared with the original as-received CS material, whereby the trend is also in compliance with Lacey et al. [11] and Thompson et al. [2]. The HAF resulted in a higher ash fraction, with 19.10 wt. % total ash that resulted from the higher portion of fines, leaves, and contaminants such as plastic, twine, and introduced dirt typical of harvesting methods. Similar result was also found in other reports in the literature [2]. This indicates a significant reduction in inorganics that are non-convertible and lead to excessive equipment wear [2]. This material beneficiation creates an enhanced feedstock for conversion, but also generates an alternate biomass fraction that requires utilization. As discussed in this work, we investigate HTC as a method of utilization. The conversion and analysis results further considered in this study are focused around the HAF.

As-received corn stover.

Air classified LAF "clean" fraction with fan speed of 1.8 m/s air velocity.

Air classified HAF "dirty" fraction with fan speed of 1.8 m/s air velocity.

Figure 1. Photo of the original (ORG) prepared corn stover (CS) and the air classified separated CS with high ash fraction (HAF) and light ash fraction (LAF) divisions.

Table 1. Proximate and ultimate analysis of ORG and separated CS. Measurements made using ASTM D7582, ASTM D5373, and ASTM D4239.

Sample Name	CS ORG	CS LAF 8 Hz	CS HAF 8 Hz
Moisture (wt. %)	6.5 ± 0.0	6.7 ± 0.0	3.1 ± 0.0
VM (wt. %)	77.2 ± 0.2	78.9 ± 0.1	67.6 ± 0.2
Ash (wt. %)	7.1 ± 0.2	5.2 ± 0.0	19.1 ± 0.1
FC (wt. %)	15.7 ± 0.2	15.9 ± 0.2	13.3 ± 0.1
H (wt. %)	5.5 ± 0.0	5.3 ± 0.4	4.9 ± 0.0
C (wt. %)	46.2 ± 0.1	47.2 ± 0.3	40.4 ± 0.2
N (wt. %)	0.6 ± 0.0	0.5 ± 0.0	1.1 ± 0.0
O (wt. %)	40.4 ± 0.2	41.7 ± 0.2	34.4 ± 0.2
S (wt. %)	0.2 ± 0.0	0.2 ± 0.0	0.2 ± 0.0

3.2. Characterization of Hydrochars Prepared from HAF of CS

HTC experiments were conducted to determine the effect of HTC temperature on the physiochemical properties of hydrochar. Table 2 shows the MY, EY, proximate and ultimate analyses of the ORG CS and hydrochar samples. The MY decreased with the increase in HTC temperature for both ORG and HAF hydrochars. For ORG, the MY was 61.7 ± 3.9 wt. % at HTC200, while it was decreased to 43.1 ± 2.8 wt. % at HTC260. For HAF, the MY was 62.2 ± 1.9 wt. % at HTC200 and then it decreased to 50.1 ± 0.4 wt. % at HTC 260. It was previously reported that MY decreased with increasing HTC temperature [20,22,24]. The attenuation in MY between the lower and higher temperature HTC treatments could be primarily due to the degradation of different components at successively higher temperatures [28]. For example, hemicellulose starts degrading at a lower temperature at approximately 180 °C, cellulose starts breaking down at approximately 230 °C, and the lignin starts decaying significantly with temperatures at and above 260 °C. All these are responsible for the lower MY at higher temperatures [19,24,28,36–38]. Previous studies also suggest that the hydrolysis reaction, which requires the lowest activation energy compared to other decomposition reactions, occurs below 200 °C and results in high MY [36,39]. On the other hand, HTC temperature elevation releases volatile matters, which enhance the dehydration and elimination reactions. This phenomenon takes part in dropping the MY with the increase in HTC temperature [39]. It might be anticipated that the transformation from CS to hydrochars entailed a high proportion of organic degradation, and inorganics solubilization and removal from the solution. It could be possible that when the HTC temperature increased from 200 °C to 260 °C, the dehydration reaction became more prominent than hydrolysis reaction (see Figure 2). The dehydration reaction might synthesize more organic acids (lowering the pH of process liquid), which could potentially catalyze the decomposition of biomass, meaning that more process liquid and gas were being produced and resulted in a low mass yield at high temperatures. Although the MY values at 230 °C for both feedstocks were similar, at 260 °C, HAF showed the highest value. This could be due to the relatively higher content of inorganic dirt and foreign materials in the HAF compared to the ORG.

Table 2. Proximate and ultimate analysis of air classified CS hydrochars. DB is dry basis.

Sample Type	Sample Condition	MY (wt. %)	LY (%)	HHV$_{DB}$ (MJ/kg)	AY (wt. %)	MC (wt. %)	Ash (wt. %)	VM (wt. %)	FC (wt. %)	C (wt. %)	H (wt. %)	N (wt. %)	S (wt. %)	O (wt. %)
								Thermogravimetric Analysis				Ultimate Analysis		
ORG	raw	100.0 ± 0.0	100.0 ± 0.0	18.3 ± 0.2	100.0 ± 0.0	5.9 ± 1.2	5.7 ± 0.9	76.9 ± 0.8	13.3 ± 1.6	45.3 ± 0.9	5.5 ± 0.1	0.6 ± 0.04	BD**	42.8 ± 1.0
	HTC 200	61.7 ± 3.9	66.4 ± 4.2	19.7 ± 0.2	50.4 ± 4.5	1.8 ± 0.0	4.7 ± 0.4	75.5 ± 0.7	17.9 ± 0.7	48.1 ± 0.4	5.7 ± 0.0	0.3 ± 0.1	BD**	41.2 ± 0.4
	HTC 230	54.1 ± 0.6	60.9 ± 0.7	20.7 ± 0.2	72.1 ± 6.4	1.4 ± 0.3	7.6 ± 0.7	69.2 ± 0.8	21.8 ± 1.0	50.9 ± 0.1	5.4 ± 0.1	0.4 ± 0.0	BD**	35.6 ± 0.1
	HTC 260	43.1 ± 2.8	52.0 ± 3.3	22.1 ± 0.2	85.9 ± 11.6	1.2 ± 0.1	11.3 ± 1.6	55.9 ± 0.5	31.6 ± 0.4	55.5 ± 0.0	2.6 ± 2.6	0.5 ± 0.2	BD**	30.0 ± 2.4
LAF	raw	100.0 ± 0.0	100.0 ± 0.0	18.8 ± 0.2	100.0 ± 0.0	5.6 ± 1.6	4.5 ± 0.1	71.2 ± 3.9	17.5 ± 2.2	51.3 ± 0.0	5.5 ± 0.0	0.1 ± 0.0	BD**	38.51 ± 0.0
HAF	raw	100.0 ± 0.0	100.0 ± 0.0	16.8 ± 0.1	100.0 ± 0.0	7.9 ± 0.2	10.3 ± 1.1	64.9 ± 0.4	17.1 ± 0.7	40.6 ± 0.3	5.1 ± 0.2	0.9 ± 0.1	BD**	42.9 ± 0.4
	HTC 200	62.2 ± 1.9	68.9 ± 2.8	18.6 ± 0.2	64.1 ± 3.8	2.3 ± 0.1	10.6 ± 0.6	69.5 ± 0.4	17.6 ± 0.3	43.9 ± 0.2	5.1 ± 0.1	0.9 ± 0.2	BD**	39.4 ± 0.2
	HTC 230	64.9 ± 1.9 *	76.7 ± 2.6	19.9 ± 0.1	67.1 ± 1.3	1.9 ± 0.4	10.6 ± 0.2	66.9 ± 1.5	20.5 ± 1.9	47.5 ± 1.9	4.7 ± 0.3	1.1 ± 0.2	BD**	35.9 ± 1.9
	HTC 260	50.1 ± 0.4 *	66.4 ± 0.4	22.2 ± 0.2	69.7 ± 1.7	1.1 ± 0.2	14.3 ± 0.4	40.3 ± 0.2	44.4 ± 0.0	51.1 ± 0.5	3.5 ± 0.1	1.7 ± 0.2	BD**	29.4 ± 0.5

* values refer to duplicate experiments. All other experiments were triplicated; ** below detection limit.

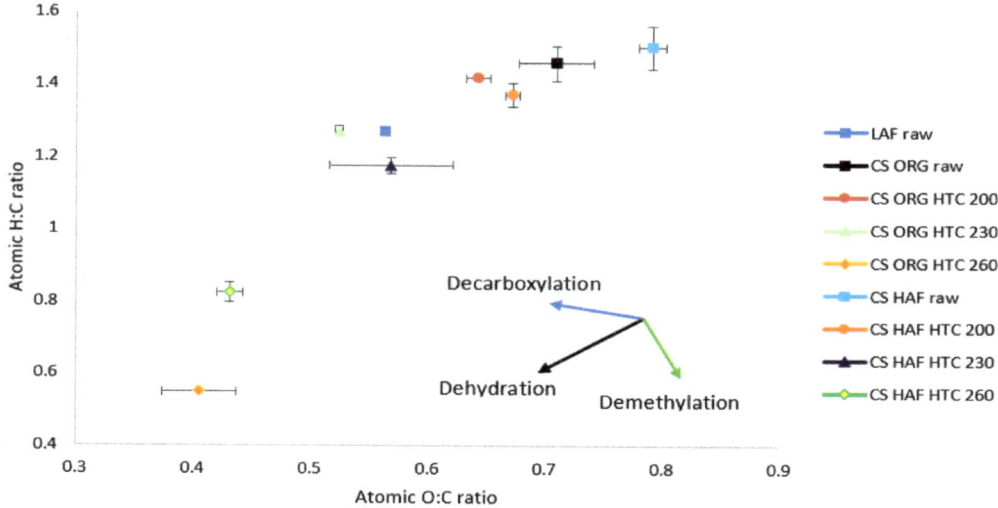

Figure 2. Van-Krevelen diagram.

The AY showed slight changes in their values (Table 2) for HAF hydrochars, but these changes were significant for ORG hydrochars. AY showed remarkable reductions of ~50 wt. % and ~36 wt. % from raw ORG to ORG HTC200 and raw HAF to HAF HTC200, respectively, but it increased for successive HTC temperatures. Qadi et al. [40] and Chen et al. [41] found that HTC aids in the removal of loose minerals from the biomass. One possible explanation is that CS can be accompanied by large amounts of loose dirt during harvesting, depending on method, and this portion was removed significantly during the HTC200 process. It is likely that this agronomic practice contributed to the trapped soil in this sample, as discussed on the previous section. On the other hand, the ORG HTC260 and HAF HTC260 showed the highest AY values of ~86 wt. % and ~70 wt. %, respectively. This could be due to the adsorption of the inorganics from the liquid phase to the solid phase (hydrochar) at elevated temperatures. The oversaturation with minerals in the liquid phase at high temperatures could be responsible for this precipitation phenomenon. Several researchers found that HTC enhances the degradation process with higher temperatures, and produces sugar monomers, furfurals, and organic acids, which leave porous structures of hydrochar. As such, some entrapped/loosely bonded inorganics in the crosslinked matrix might have been adsorbed into the pores of hydrochar during HTC [18,24,40]. Since cellulose and lignin start degrading at around 230 °C and 260 °C, respectively, and create porous structures, they might permit the insoluble inorganics to be absorbed from the process liquid into the hydrochar surface, resulting in higher AY.

The ultimate analysis shown in Table 2 indicates that with the increase in HTC temperature, the carbon content was increased by ~10 wt. % for ORG HTC260 and ~11 wt. % for HAF HTC260 from raw ORG and HAF, respectively, wherein the oxygen content dropped about 13 wt. % and 14 wt. %. Regarding hydrogen, although ORG chars showed essentially no change, HAF chars showed a slight drop (~1.5 wt. %) for HTC 260 from raw HAF. The nitrogen showed insignificant changes in the hydrochars with respect to the feedstock. The higher carbon content and lower oxygen content rationalized the rising HHV with the increase in HTC process temperature. The fuel quality was also analyzed with the van-Krevelen diagram (Figure 2). The van-Krevelen diagram depicts that the closer to the origin of the data the points are, the better the fuel is [33]. The HAF fuel quality was very low compared to the LAF quality, as expected due to the higher concentration of non-combustible species. Even considering this on an inorganics-free basis, there is a small discrepancy with a lower calorific value in the HAF. As mentioned above, this is attributed to the partitioning of tissues during AC. In addition to having higher inorganics content, these HAF tissues (larger portions of leaf, sheaths) have higher amounts of extractives (cutin resin, protein, etc.). With the increase in HTC temperature, the hydrochar fuel quality (calorific value, and carbon, oxygen, and hydrogen content) increases (up to 32 wt. % in the HAF). Hydrochars produced at 260 °C showed the best fuel quality among all the chars, and this agrees with prior studies undertaken in these condition ranges [5]. As the rise in HTC temperature favored the dehydration and decarboxylation reactions (see Figure 2), this could justify the increase in elemental C and decrease in elemental O content during the carbonization process.

The TGA of the ORG chars showed a decreasing trend (Table 2) of VM and a rising trend of FC. The ORG HTC260 showed ~21 wt. % VM decrease and ~19 wt. % FC increase from the raw ORG, whereas HAF HTC260 showed a significant decrease (~25 wt. %) in VM and increase (~27 wt. %) in FC compared to raw HAF. However, a minimum change in both FC and VM was observed for both feedstocks at 200 °C. Previous studies suggested that the degradation of cellulose at around 220 °C might be responsible for this phenomenon [36,42,43]. Sharma et al. [39] and Titirici et al. [44] further explained cellulose degradation via two routes: (1) cellulose > glucose > 5-hydroxymethyl furfural > carbonized structure; (2) cellulose > aromatic structure. In this study, a possible explanation for receiving higher FC for both ORG and HAF HTC260 is the cellulose carbohydrate's degradation into more carbonaceous particles at successively higher temperatures.

The highest energy content was observed at the 260 °C hydrochars for both ORG and HAF (about 22 MJ/kg), as shown in Table 2, which was almost 4 MJ/kg and 5 MJ/kg higher than with the ORG CS (~18 MJ/kg) and HAF (~17 MJ/kg), respectively. Earlier studies demonstrated that with the increasing HTC temperature, the HHV and corresponding EY showed upward and downward trends, respectively [20,22,24]. The EY reported in Table 2 showed a descending trend for ORG chars, but a more complex pattern for HAF hydrochars. A possible explanation for the discrepancies between the hydrochar EY values is that the HAF HTC230 had higher MY, which could be due to the retention of cellulose in the hydrochar due to partial degradation. In addition, the enriched energy content and mass yield is convolved with the reduction in inorganics to create a potentially complex optimization for the energy yield of the HAF.

3.3. Energy Recovery by Integrating AC–HTC Process

Figure 3 shows three different proposed scenarios: the AC-only process, the HTC-only process, and the integrated AC–HTC process. The main objective of these processes was to determine the energy recovery from CS. For all processes, the basis was arbitrarily taken as 100 kg dry ORG feed. In the AC-only process, the throughputs were LAF and HAF, as the below-screen particles were assumed to be negligible. The HAF was considered as waste and discarded from the process for AC. The LAF was the only throughput that was pelletized and the HHV was determined. In the HTC-only process, dry ORG was directly fed into the HTC reactor at three different temperatures (200, 230, and 260 °C), and then corresponding hydrochars were pelletized to find out the energy recovery of the HTC process. The last process was considered as the integrated AC–HTC process, which comprised the AC followed by the HTC process. As mentioned earlier, the AC process had two main throughputs: LAF and HAF. The HAF waste was fed into the HTC process at three different temperatures (200, 230, and 260 °C). The hydrochars were then mixed with LAF at their corresponding weight ratios and pelletized to find out the energy recovery from the integrated process.

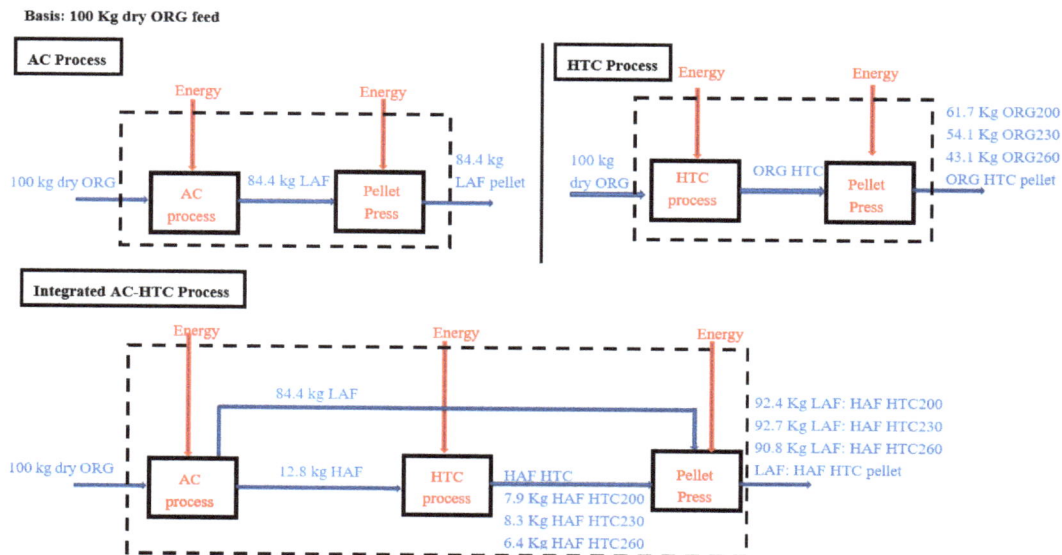

Figure 3. AC, HTC, and integrated AC–HTC process.

The overall energy balances for the three scenarios were calculated using Equation (6)–(11). For process units, the specific energy consumptions (SECs) were found from [45,46] and shown in Table 3. Specific energy consumption for each process unit. The SEC of AC was calculated as 0.0036 MJ/kg by INL and the basis was taken as per kg dry LAF. The pellet press SEC was found to be 0.4 MJ/kg dry feed for the raw LAF and LAF–HTC mixtures and 0.2 MJ/kg dry hydrochar for the ORG chars [45]. Earlier, it was demonstrated that the pellet press SEC increases with a higher feed rate, feed moisture (wt. %), and length-to-diameter (L/D) ratio of pellet die, etc. [47]. For example, the moisture (wt. %) in the biomass feed is higher than in the hydrochar feed due to the higher hydrophilicity [48], which can increase the overall SEC of biomass pelletization. Here, the SEC for biomass and biomass–hydrochar mixtures under pelletization (Table 3. Specific energy consumption for each process unit) was considered higher than the hydrochar SEC due to the higher feed, moistures (wt. %), and/or L/D ratio. For the case of SEC of HTC, the required SEC was found to be 5.9, 6.3, and 6.7 MJ/kg dry feed for HTC200, HTC230, and HTC260, respectively (for both ORG and HAF chars) [46]. As higher temperatures require higher energy consumption [46], HTC260 has the highest SEC (6.7 MJ/kg dry feed) among all the chars. Table 4 shows the overall energy balance of the processes.

As per the AC process (Table 4), the overall net energy in was 1833.6 MJ (Equations (6), (7), and (11) and Table 4) and the energy out was 1588.3 MJ (Equation (8), Table 4). The recoverable energy of the HAF was found by the difference between the ORG and LAF energies, which was ~245 MJ (Equation (9), Table 4). Since the HAF waste was not processed and eliminated after AC, the recoverable energy was 0 MJ for the AC-only

In the case of the HTC-only process (Table 4), the overall energy in was 2439.6 MJ, 2474.1 MJ, and 2507.9 MJ for ORG HTC200, 230 and 260, respectively (Equations (6), (7) and Table 4). The overall energy out was 1217.6 MJ, 1116.4 MJ and 953.8 MJ for HTC200, 230 and 260, respectively (Equation (8) and Table 4). The energy consumptions for the bulk quantity of HTC of ORG significantly elevated the process' overall energy demand. Since the HTC process was highly energy-intensive, the energy recovery was not feasible. process.

Table 3. Specific energy consumption for each process unit.

Process Units	Specific Energy Consumption (SEC)	Unit	Reference Paper
AC	0.0036	MJ/kg dry LAF	This study
Pellet press for LAF	0.4	MJ/kg dry feed	[45]
HTC for ORG200	5.9	MJ/kg dry feed	[46]
HTC for ORG230	6.3	MJ/kg dry feed	[46]
HTC for ORG260	6.7	MJ/kg dry feed	[46]
Pellet press for ORG200	0.2	MJ/kg dry hydrochar	[45]
Pellet press for ORG230	0.2	MJ/kg dry hydrochar	[45]
Pellet press for ORG260	0.2	MJ/kg dry hydrochar	[45]
HTC for HAF200	5.9	MJ/kg dry feed	[46]
HTC for HAF230	6.3	MJ/kg dry feed	[46]
HTC for HAF260	6.7	MJ/kg dry feed	[46]
Pellet press for LAF:HAF HTC200	0.4	MJ/kg dry feed	[45]
Pellet press for LAF:HAF HTC230	0.4	MJ/kg dry feed	[45]
Pellet press for LAF:HAF HTC260	0.4	MJ/kg dry feed	[45]

Table 4. Energy extracted from HAF waste.

Process Type	HTC	Energy in (MJ)				Energy out (MJ)			Potential Energy (MJ) in HAF Waste	Energy Recovery (MJ) from HAF Waste
		ORG Feed	AC	HTC of HAF	Pellet Press	LAF Pellet	ORG Pellet	LAF:HAF HTC Pellet		
AC Process		1833.3	0.3	-	37.1	1588.3	-	-	245.0	0
HTC Process	ORG HTC 200	1833.3	-	594.0	12.3	-	1217.6	-	245.0	-
	ORG HTC 230	1833.3	-	630.0	10.8	-	1116.4	-	245.0	-
	ORG HTC260	1833.3	-	666.0	8.6	-	953.8	-	245.0	-
Integrated AC-HTC Process	LAF: HAF HTC200	1833.3	0.3	76.0	40.6	-	-	1736.5	245.0	68.7
	LAF: HAF HTC230	1833.3	0.3	80.6	40.8	-	-	1752.6	245.0	80.0
	LAF: HAF HTC260	1833.3	0.3	85.3	40.0	-	-	1727.4	245.0	51.0

As per the integrated AC–HTC process (Table 4), the HAF waste was hydrothermally carbonized to utilize the recoverable energy present (~245 MJ) in the product, and further mixed with the LAF stream for taking advantage of AC preprocessing (Equation (10), and Table 4). The results in Table 4 show that the energy recovery was the maximum (~80 MJ) for LAF:HAF under HTC230, which was about ~11 MJ and ~29 MJ higher than LAF:HAF HTC200 and LAF:HAF HTC260, respectively (using Equations (6)–(8), (10), (11), and Table 4). The higher FC and elemental carbon for HAF HTC230 than HAF HTC200 could contribute as a better solid fuel compared to LAF: HAF HTC200. On the other hand, although HAF HTC260 has higher FC and elemental carbon than HAF HTC230, due to the higher volatile matters as well as the elemental oxygen in HAF HTC230, this could enhance the energy content of the LAF:HAF HTC230. Since volatiles are mainly made of short chain hydrocarbons, long chain hydrocarbons and aromatic hydrocarbons that are easy to distill off, it might be possible that the heavier hydrocarbons break into lighter gases during combustion and react with limited oxygen via partial oxidation, releasing more energy as heat [49]. In this context, integrating the LAF and HAF HTC230 process streams might give a better energy recovery out of the AC reject stream. Moreover, in this mode, the HTC process energy consumptions were significantly less than the HTC-only mode, which might impact the overall energy recovery in a positive dimension. On the other hand, the final product energy value was much higher than the AC-only mode, which bears the importance of combining the AC and HTC processes together. Overall, the integrated AC–HTC could become a better option to take the advantage of the potential waste energy as valuable utilizable energy, and minimize the overall energy lost.

Equations used:

$$\text{Energy in (MJ) of feed} = \text{HHV of feed} \times \text{Total mass of feed} \qquad (6)$$

$$\text{Energy in (MJ) of process unit} = \text{Specific energy consumption of the unit} \times \text{Total mass of feed or product} \qquad (7)$$

$$\text{Energy out (MJ) of product} = \text{HHV} \times \text{Total mass of product} \qquad (8)$$

$$\text{Potential energy (MJ) in HAF waste} = \text{LAF pellet-ORG feed} \qquad (9)$$

$$\text{Energy recovery (MJ) from HAF waste for the integrated AC-HTC process} = \text{LAF:HAF HTC pellet-HTC of HAF-pellet press for LAF:HAF HTC pellet-net energy of LAF pellet} \qquad (10)$$

$$\text{Net energy (MJ) of LAF pellet} = \text{LAF pellet-Pellet press for LAF} \qquad (11)$$

4. Conclusions

This study investigated an integrated AC–HTC process to recover the energy from corn stover. An air classifier was used to beneficiate CS into a purified biomass stream with 27 wt. % reduction in ash content from LAF with 84.4 wt. % mass recovery. The waste from this fractionation technique was used as a feedstock for HTC to produce a high-energy fuel pellet (19–22 MJ/kg). When CS was hydrothermally carbonized, both MY and AY were decreased with the HTC temperature; meanwhile, the energy content of the hydrochar was increased. This study showed that the AC and HTC processes cannot recover any energy from the HAF stream individually. On the other hand, along with the high-energy densified hydrochar, the integrated AC–HTC process further showed significant energy recovery (~800 MJ/tonne) from the HAF. Therefore, this study provides evidence of an HTC that can be integrated with AC to reduce the inorganic content and recover energy. Adopting HTC with AC could potentially transform CS into an advanced biorefinery feedstock, while still utilizing the otherwise lost organics from the AC process. Further sustainability and technoeconomic assessment of AC and HTC are needed to justify the economic viability of the integrated process. However, the energy consumption of this study was calculated based on the laboratory-scale data, which could vary at a large scale. Process economics with large-scale data could reveal additional data that might assist technology maturation.

Supplementary Materials: The following are available online at https://www.mdpi.com/1996-1073/14/5/1397/s1, Figure S1: Air classification technology, Figure S2: Hydrothermal Carbonization (HTC) reactor system, Figure S3: Pellet press for pelletization.

Author Contributions: Conceptualization, M.T.I. and M.T.R.; methodology, M.T.I., N.S., S.H., J.K.; formal analysis, M.T.I., N.S., S.H.; investigation, M.T.I., S.H.; resources, J.K., M.T.R.; data curation, M.T.I., M.T.R.; writing—original draft preparation, M.T.I., S.H.; writing—review and editing, N.S., M.T.R., J.K.; supervision, M.T.R., J.K.; funding acquisition, M.T.R. All authors have read and agreed to the published version of the manuscript.

Funding: This research was funded by United States Department of Agriculture, grant number 2019-67019-31594. This work was authored in part by the Idaho National Laboratory under USA Department of Energy (DOE) Idaho Operations Office with Contract no. DE-AC07-05ID14517.

Institutional Review Board Statement: Not applicable.

Informed Consent Statement: Not applicable.

Data Availability Statement: Not applicable.

Acknowledgments The authors are grateful for the air classification technology of the Biomass Feedstock National User Facility (BFNUF) at Idaho National Laboratory. The authors acknowledge laboratory assistances from Kyle McGaughy, Cadianne Chambers, and Travis Rembrandt from Biofuels Lab at Florida Institute of Technology for their laboratory efforts in this project.

Conflicts of Interest: The authors declare no conflict of interest.

References

1. *U.S. Billion-Ton Update: Biomass Supply for a Bioenergy and Bioproducts Industry*; U.S Department of Energy, Energy Efficiency and Renewable Energy, Office of the Biomass Program: Washington, DC, USA, 2011; p. 235.
2. Thompson, V.S.; Lacey, J.A.; Hartley, D.; Jindra, M.A.; Aston, J.E.; Thompson, D.N. Application of air classification and formulation to manage feedstock cost, quality and availability for bioenergy. *Fuel* **2016**, *180*, 497–505. [CrossRef]
3. Perlack, R.D.; Wright, L.L.; Turhollow, A.F.; Graham, R.L.; Stokes, B.J.; Erbach, D.C. *Biomass as Feedstock for a Bioenergy and Bioproducts Industry: The Technical Feasibility of a Billion-Ton Annual Supply*; Oak Ridge National Laboratory: Oak Ridge, TN, USA, 2005; p. 1216415. [CrossRef]
4. Zhichao, W.; Dunn, J.B.; Wang, M.Q. *Updates to the Corn Ethanol Pathway and Development of an Integrated Corn and Corn Stover Ethanol Pathway in the GREET™ Model*; No. ANL/ESD-14/11; Argonne National Lab. (ANL): Argonne, IL, USA, 2014.
5. Machado, N.; Castro, D.; Queiroz, L.; Santos, M.; Costa, C. Production and Characterization of Energy Materials with Adsorbent Properties by Hydrothermal Processing of Corn Stover with Subcritical H$_2$O. *J. Appl. Solut. Chem. Model.* **2016**, *5*, 117–130. [CrossRef]
6. Kim, S.; Dale, B.E. Global potential bioethanol production from wasted crops and crop residues. *Biomass Bioenergy* **2004**, *26*, 361–375. [CrossRef]
7. Zhang, Y.; Jiang, Q.; Xie, W.; Wang, Y.; Kang, J. Effects of temperature, time and acidity of hydrothermal carbonization on the hydrochar properties and nitrogen recovery from corn stover. *Biomass Bioenergy* **2019**, *122*, 175–182. [CrossRef]
8. Biswas, B.; Pandey, N.; Bisht, Y.; Singh, R.; Kumar, J.; Bhaskar, T. Pyrolysis of agricultural biomass residues: Comparative study of corn cob, wheat straw, rice straw and rice husk. *Bioresour. Technol.* **2017**, *237*, 57–63. [CrossRef]
9. Emerson, R.M.; Hernandez, S.; Williams, C.L.; Lacey, J.A.; Hartley, D.S. Improving bioenergy feedstock quality of high moisture short rotation woody crops using air classification. *Biomass Bioenergy* **2018**, *117*, 56–62. [CrossRef]
10. Williams, C.L.; Emerson, R.M.; Hernandez, S.; Klinger, J.L.; Fillerup, E.P.; Thomas, B.J. Preprocessing and hybrid biochemical/thermochemical conversion of short rotation woody coppice for biofuels. *Front. Energy Res.* **2018**, *6*, 74. [CrossRef]
11. Lacey, J.A.; Aston, J.E.; Westover, T.L.; Cherry, R.S.; Thompson, D.N. Removal of introduced inorganic content from chopped forest residues via air classification. *Fuel* **2015**, *160*, 265–273. [CrossRef]
12. Reza, M.T.; Emerson, R.; Uddin, M.H.; Gresham, G.; Coronella, C.J. Ash reduction of corn stover by mild hydrothermal preprocessing. *Biomass Convers. Biorefinery* **2014**. [CrossRef]
13. Humbird, D.; Davis, R.; Tao, L.; Kinchin, C.; Hsu, D.; Aden, A. *Process Design and Economics for Biochemical Conversion of Lignocellulosic Biomass to Ethanol: Dilute-Acid Pretreatment and Enzymatic Hydrolysis of Corn Stover*; Technical Report NREL/TP-5100-47764; National Renewable Energy Lab. (NREL): Golden, CO, USA, 2011; p. 1013269.
14. Möller, M.; Nilges, P.; Harnisch, F.; Schröder, U. Subcritical Water as Reaction Environment: Fundamentals of Hydrothermal Biomass Transformation. *ChemSusChem* **2011**, *4*, 566–579. [CrossRef]
15. Brunner, G. Heat Transfer. In *situ Spectroscopic Techniques at High Pressure*; Elsevier: Amsterdam, The Netherlands, 2014; Volume 5, pp. 227–263. [CrossRef]
16. Öztürk, I.; Irmak, S.; Hesenov, A.; Erbatur, O. Hydrolysis of kenaf (*Hibiscus cannabinus* L.) stems by catalytical thermal treatment in subcritical water. *Biomass Bioenergy* **2010**, *34*, 1578–1585. [CrossRef]
17. Machado, N.; De Castro, D.; Santos, M.; Araújo, M.; Lüder, U.; Herklotz, L.; Werner, M.; Mumme, J.; Hoffmann, T. Process analysis of hydrothermal carbonization of corn Stover with subcritical H$_2$O. *J. Supercrit. Fluids* **2018**, *136*, 110–122. [CrossRef]
18. Funke, A.; Ziegler, F. Hydrothermal carbonization of biomass: A summary and discussion of chemical mechanisms for process engineering. *Biofuels Bioprod. Biorefin.* **2010**, *4*, 160–177. [CrossRef]
19. Miraret, J.T. *Hydrothermal Carbonization of Corn Residuals to Produce a Solid Fuel Replacement for Coal*; The University of Guelph: Keiu Lake, ON, Canada, 2015; p. 133.
20. Kobayashi, N.; Okada, N.; Hirakawa, A.; Sato, T.; Kobayashi, J.; Hatano, S.; Itaya, Y.; Mori, S. Characteristics of Solid Residues Obtained from Hot-Compressed-Water Treatment of Woody Biomass. *Ind. Eng. Chem. Res.* **2009**, *48*, 373–379. [CrossRef]
21. Lynam, J.G.; Coronella, C.J.; Yan, W.; Reza, M.T.; Vasquez, V.R. Acetic acid and lithium chloride effects on hydrothermal carbonization of lignocellulosic biomass. *Bioresour. Technol.* **2011**, *102*, 6192–6199. [CrossRef] [PubMed]
22. Yan, W.; Acharjee, T.C.; Coronella, C.J.; Vásquez, V.R. Thermal pretreatment of lignocellulosic biomass. *Environ. Prog. Sustain. Energy* **2009**, *28*, 435–440. [CrossRef]
23. Hoekman, S.K.; Broch, A.; Robbins, C. Hydrothermal Carbonization (HTC) of Lignocellulosic Biomass. *Energy Fuels* **2011**, *25*, 1802–1810. [CrossRef]

24. Reza, M.T.; Lynam, J.G.; Uddin, M.H.; Coronella, C.J. Hydrothermal carbonization: Fate of inorganics. *Biomass Bioenergy* **2013**, *49*, 86–94. [CrossRef]
25. Tu, R.; Sun, Y.; Wu, Y.; Fan, X.; Wang, J.; Cheng, S.; Jia, Z.; Jiang, E.; Xu, X. Improvement of corn stover fuel properties via hydrothermal carbonization combined with surfactant. *Biotechnol. Biofuels* **2019**, *12*, 1–19. [CrossRef]
26. Wang, T.; Zhai, Y.; Zhu, Y.; Li, C.; Zeng, G. A review of the hydrothermal carbonization of biomass waste for hydrochar formation: Process conditions, fundamentals, and physicochemical properties. *Renew. Sustain. Energy Rev.* **2018**, *90*, 223–247. [CrossRef]
27. Wu, Q.; Yunqiao, P.; Hao, N.; Wells, T.; Meng, X.; Li, M.; Pu, Y.; Liu, S.; Ragauskas, A.J. Characterization of products from hydrothermal carbonization of pine. *Bioresour. Technol.* **2017**, *244*, 78–83. [CrossRef] [PubMed]
28. Smith, A.M.; Singh, S.; Ross, A.B. Fate of inorganic material during hydrothermal carbonisation of biomass: Influence of feedstock on combustion behaviour of hydrochar. *Fuel* **2016**, *169*, 135–145. [CrossRef]
29. Wagner, W.; Pruß, A. The IAPWS Formulation 1995 for the Thermodynamic Properties of Ordinary Water Substance for General and Scientific Use. *J. Phys. Chem. Ref. Data* **2002**, *31*, 387–535. [CrossRef]
30. Archers, D.G.; Wang, P. The Dielectric Constant of Water and Debye-HOckel Limiting Law Slopes. *J. Phys. Chem. Ref. Data* **1990**, *19*, 41.
31. Bandura, A.V.; Lvov, S.N. The Ionization Constant of Water over Wide Ranges of Temperature and Density. *J. Phys. Chem. Ref. Data* **2006**, *35*, 15–30. [CrossRef]
32. Kaliyan, N.; Morey, R.V. Densification Characteristics of Corn Stover and Switchgrass. *Trans. ASABE* **2009**, *52*, 907–920. [CrossRef]
33. Saba, A.; Saha, P.; Reza, M.T. Co-Hydrothermal Carbonization of coal-biomass blend: Influence of temperature on solid fuel properties. *Fuel Process. Technol.* **2017**, *167*, 711–720. [CrossRef]
34. Saha, N.; Xin, D.; Chiu, P.C.; Reza, M.T. Effect of Pyrolysis Temperature on Acidic Oxygen-Containing Functional Groups and Electron Storage Capacities of Pyrolyzed Hydrochars. *ACS Sustain. Chem. Eng.* **2019**, *7*, 8387–8396. [CrossRef]
35. Thompson, V.S.; Aston, J.E.; Lacey, J.A.; Thompson, D.N. Optimizing Biomass Feedstock Blends with Respect to Cost, Supply, and Quality for Catalyzed and Uncatalyzed Fast Pyrolysis Applications. *BioEnergy Res.* **2017**, *10*, 811–823. [CrossRef]
36. Libra, J.A.; Ro, K.S.; Kammann, C.; Funke, A.; Berge, N.D.; Neubauer, Y.; Titirici, M.-M.; Führer, C.; Bens, O.; Kern, J.; et al. Hydrothermal carbonization of biomass residuals: A comparative review of the chemistry, processes and applications of wet and dry pyrolysis. *Biofuels* **2011**, *2*, 71–106. [CrossRef]
37. Pastor-Villegas, J.; Pastor-Valle, J.; Rodríguez, J.M.; García, M.G. Study of commercial wood charcoals for the preparation of carbon adsorbents. *J. Anal. Appl. Pyrolysis* **2006**, *76*, 103–108. [CrossRef]
38. Kumar, S.; Gupta, R.; Lee, Y.; Gupta, R.B. Cellulose pretreatment in subcritical water: Effect of temperature on molecular structure and enzymatic reactivity. *Bioresour. Technol.* **2010**, *101*, 1337–1347. [CrossRef] [PubMed]
39. Sharma, R.; Jasrotia, K.; Singh, N.; Ghosh, P.; Srivastava, S.; Sharma, N.R.; Singh, J.; Kanwar, R.; Kumar, A. A Comprehensive Review on Hydrothermal Carbonization of Biomass and Its Applications. *Chem. Afr.* **2019**, *3*, 1–19. [CrossRef]
40. Qadi, N.; Takeno, K.; Mosqueda, A.; Kobayashi, M.; Motoyama, Y.; Yoshikawa, K. Effect of Hydrothermal Carbonization Conditions on the Physicochemical Properties and Gasification Reactivity of Energy Grass. *Energy Fuels* **2019**, *33*, 6436–6443. [CrossRef]
41. Chen, S.-F.; Mowery, R.A.; Scarlata, C.J.; Chambliss, C.K.; Chambliss, K. Compositional Analysis of Water-Soluble Materials in Corn Stover. *J. Agric. Food Chem.* **2007**, *55*, 5912–5918. [CrossRef]
42. Saha, N.; Saba, A.; Reza, M.T. Effect of hydrothermal carbonization temperature on pH, dissociation constants, and acidic functional groups on hydrochar from cellulose and wood. *J. Anal. Appl. Pyrolysis* **2019**, *137*, 138–145. [CrossRef]
43. Peterson, A.A.; Vogel, F.; Lachance, R.P.; Fröling, M.; Antal, J.M.J.; Tester, J.W. Thermochemical biofuel production in hydrothermal media: A review of sub- and supercritical water technologies. *Energy Environ. Sci.* **2008**, *1*, 32–65. [CrossRef]
44. Titirici, M.M.; Thomas, A.; Yu, S.-H.; Müller, A.J.-O.; Antonietti, M. A Direct Synthesis of Mesoporous Carbons with Bicontinuous Pore Morphology from Crude Plant Material by Hydrothermal Carbonization. *Chem. Mater.* **2007**, *19*, 4205–4212. [CrossRef]
45. Tumuluru, J.S. High moisture corn stover pelleting in a flat die pellet mill fitted with a 6 mm die: Physical properties and specific energy consumption. *Energy Sci. Eng.* **2015**, *3*, 327–341. [CrossRef]
46. Lucian, M.; Fiori, L. Hydrothermal Carbonization of Waste Biomass: Process Design, Modeling, Energy Efficiency and Cost Analysis. *Energies* **2017**, *10*, 211. [CrossRef]
47. Tumuluru, J.S. Specific energy consumption and quality of wood pellets produced using high-moisture lodgepole pine grind in a flat die pellet mill. *Chem. Eng. Res. Des.* **2016**, *110*, 82–97. [CrossRef]
48. Reza, M.T.; Lynam, J.G.; Vasquez, V.R.; Coronella, C.J. Pelletization of biochar from hydrothermally carbonized wood. *Environ. Prog. Sustain. Energy* **2012**, *31*, 225–234. [CrossRef]
49. Wang, T. An overview of IGCC systems. In *Integrated Gasification Combined Cycle (IGCC) Technologies*; Wang, T., Stiegel, G., Eds.; Woodhead Publishing: Cambridge, UK, 2017; pp. 1–80. [CrossRef]

Article

Sewage Sludge Treatment by Hydrothermal Carbonization: Feasibility Study for Sustainable Nutrient Recovery and Fuel Production

Gabriel Gerner [1,*], Luca Meyer [1], Rahel Wanner [1], Thomas Keller [2] and Rolf Krebs [1]

[1] Institute of Natural Resource Sciences, Campus Grüental, Zurich University of Applied Sciences (ZHAW), CH-8820 Wädenswil, Switzerland; luca.meyer@zhaw.ch (L.M.); rahel.wanner@zhaw.ch (R.W.); rolf.krebs@zhaw.ch (R.K.)

[2] Institute of Chemistry and Biotechnology, Campus Reidbach, Zurich University of Applied Sciences (ZHAW), CH-8820 Wädenswil, Switzerland; thomas.keller2@zhaw.ch

* Correspondence: gabriel.gerner@zhaw.ch; Tel.: +41-58-934-5588

Abstract: Phosphorus recovery from waste biomass is becoming increasingly important, given that phosphorus is an exhaustible non-renewable resource. For the recovery of plant nutrients and production of climate-neutral fuel from wet waste streams, hydrothermal carbonization (HTC) has been suggested as a promising technology. In this study, digested sewage sludge (DSS) was used as waste material for phosphorus and nitrogen recovery. HTC was conducted at 200 °C for 4 h, followed by phosphorus stripping (PS) or leaching (PL) at room temperature. The results showed that for PS and PL around 84% and 71% of phosphorus, as well as 53% and 54% of nitrogen, respectively, could be recovered in the liquid phase (process water and/or extract). Heavy metals were mainly transferred to the hydrochar and only <1 ppm of Cd and 21–43 ppm of Zn were found to be in the liquid phase of the acid treatments. According to the economic feasibility calculation, the HTC-treatment per dry ton DSS with an industrial-scale plant would cost around 608 USD. Between 349–406 kg of sulfuric acid are required per dry ton DSS to achieve a high yield in phosphorus recovery, which causes additional costs of 96–118 USD. Compared to current sewage sludge treatment costs in Switzerland, which range between 669 USD and 1173 USD, HTC can be an economically feasible process for DSS treatment and nutrient recovery.

Keywords: hydrothermal carbonization; phosphorus recovery; digested sewage sludge; hydrochar; nutrient recovery; climate-neutral fuel; energy efficiency; economic feasibility

1. Introduction

To meet global food demand, the utilization of mineral fertilizer in agriculture has become indispensable. While the production of phosphorus (P) and nitrogen (N) fertilizer cause a negative environmental impact, phosphate rock is also a limited resource and since 2014 it has been on the list of Critical Raw Materials for the European Union (EU) [1,2]. Phosphorus as an essential plant nutrient represents a considerable environmental burden due to its production from phosphate ores. According to Binder et al. [3] and Danesngar et al. [4], easily accessible phosphorus will only last for the next 300 years, while we could reach the phosphorus peak already by 2070 [5]. In addition to huge mining areas, the production of phosphoric acid from phosphate ore leaves behind hundreds of millions of tons of phosphogypsum, some of which is radioactive [6]. Radioactive material such as uranium as well as heavy metals (HM) like cadmium can be transferred from the ore to the mineral P fertilizer [7–12]. A study by the Braunschweig Federal Research Center for Cultivated Plants JKI (Julius Kühn-Institute) found in triple superphosphate 52–232 mg/kg of uranium [6]. Recycling of phosphorus from waste streams offers a sustainable alternative to conserve phosphorus deposits and reduce impurities in plant fertilizer. In contrast to phosphorus, which relies on a definite source, nitrogen fertilizer can be synthesized as ammonia from air by the Haber–Bosch process [13]. The negative aspect of this process is that to produce ammonia fertilizer, it consumes between 1–2% of global energy and produces around 1.4% of global CO_2 emissions [14]. Countries like Switzerland and Germany approved new regulations to reduce the dependency on phosphorus imports. It will be mandatory to recycle phosphorus from sewage sludge by 2026 in Switzerland and by 2029 in Germany [15]. This attempt to reduce mismanagement of this exhaustible resource and the need for high-quality fertilizer has increased the necessity of a phosphorus recycling process with high nutrient recovery and low environmental burden. Hydrothermal carbonization (HTC) is a thermochemical process allowing the direct usage of wet feedstock without drying them and converting it under high pressure and temperature to a coal slurry, which can be separated into an energy-rich solid phase (hydrochar) and nutrient-rich liquid phase (process water) [16]. The conversion leads to reduced NO emission in the hydrochar combustion [17] and simultaneously improves dewaterability of the carbonized sludge [18,19], which is crucial for an energy-efficient separation. In this process, phosphorus is mainly incorporated in the hydrochar and has to be removed by acid leaching [20–22]. Other processes use sewage sludge ash for P-leaching [23], where the sludge's fuel property is mainly used as process energy for the incineration. Therefore, it is lost to industrial processes like the cement industry or coal power plants as climate-neutral fuel.

Modern wastewater treatment plants (WWTP) employ biological and/or chemical phosphorus removal technologies to transfer around 90% of the input phosphorus load into the sewage sludge [24]. Larger WWTPs utilize the sludge as a feedstock for anaerobic digestion (AD) producing methane, which can be converted to electricity and heat to cover energy demands. Implementing an HTC plant on-site allows the methane yield to be increased by feeding the AD with HTC process water (PW), as a supplemental feedstock with high organic carbon content [20,25–30]. Recovered energy can at the same time be utilized for the HTC process and increase the energy efficiency.

Different studies show the potential of P-recovery by applying acids before or after the HTC treatment [20,22,31,32] and the need for using digested sewage sludge. For the acid application, digested sludge has a higher P availability compared to raw sludge, while with raw sludge only 50% of P can be recovered at pH 2 [23]. For a profitable application of the process a cost-effective usage of acid must be achieved.

The aim of this study was to investigate the recovery of plant nutrients (phosphorus and nitrogen) from digested sewage sludge for a sustainable fertilizer and fuel production with the main focus on minimizing the acid usage and increasing the PW utilization. Two paths of P-recovery are investigated in lab-scale experiments, with immediate acid application before and after liquid–solid separation and without prior drying to simulate industrial processing. Process liquids are analyzed for their nutrient and heavy metal content and hydrochars are examined for their fuel properties as possible substitutes for fossil fuels. Furthermore, the economic feasibility of the process in an industrial-scale HTC-plant is evaluated.

2. Materials and Methods
2.1. Digested Sewage Sludge

For this study anaerobically digested sewage sludge was used as a raw material. The digested sewage sludge (DSS) was collected at a WWTP in Switzerland (ARA Rietliau, Waedenswil, Switzerland), with an initial dry matter content of 22.30% (dried at 105 °C). The WWTP consists of four treatment trains, two with membrane bioreactors (MBR) and microfiltration (0.35 µm pore size) and two with conventional activated sludge (CAS) treatments. It was built for a population equivalent of 44,000 and treats communal and industrial wastewater. For chemical phosphorus precipitation, iron (III) chloride and poly aluminum chloride (PAC) are used by the WWTP. The collected DSS was immediately transferred to a convection oven and dried at 40 °C for 86 h (BINDER GmbH, Tuttlingen,

Germany). The slowly dried stock was homogenized by grinding to a fine powder (GM200, Retsch GmbH, Haan, Germany) and stored in airtight 1 L glass bottles (DURAN® GLS 80® laboratory wide mouth bottle, DURAN Group GmbH, Mainz, Germany) for HTC trials. These sludge pre-treatments are conducted to achieve a better reproducibility of the lab-scale experiments. In large-scale HTC-treatments sewage sludge will be treated directly after dewatering, without previous drying step.

2.2. Hydrothermal Carbonization

HTC lab experiments were carried out using a 1 L pressure vessel made from Hastelloy® C22® (Büchi AG, Uster, Switzerland). All trials were run in duplicates. The reactor vessel was loaded with 600 g of a mixture of dried DSS and deionized (DI) water. The input amount of DSS (dried at 40 °C) was adjusted by its remaining moisture content (measured at 105 °C) to obtain a final dry matter (DM) content of 20%. The reactor head space contained a small amount of air. Reaction time was set at 4 h after reaching a reaction temperature of 200 °C. The residence time was selected in accordance with the parameters from the industrial-scale reactor from GRegio Energie AG, Chur, Switzerland. During the reaction a maximum pressure of 18.8 bar (±0.2) was reached. The reactor jacket contained an electric heating mantle and water-cooling coil for short heating and cooling phases. Heating up from room to the set temperature took around 30 min and the cooling below 50 °C around 13 min. The mixture was stirred continuously at 200 rpm. All trials were monitored in situ for temperature and pressure. The resulting product of HTC is a mixture (HTC-slurry) of a solid (hydrochar) and liquid (process water) fraction. The HTC-slurry was transferred into a 1 L glass bottle (DURAN® GLS 80® laboratory wide mouth bottle) and afterwards vacuum filtered through a Buchner funnel with a Whatman filter paper (pore size 11 µm, Cat No 1001 125). The PW was stored until analysis at 6 °C and the hydrochar was, without a washing step, dried at 40 °C overnight. Dewaterability of the hydrochar was calculated based on the DM content (dried at 105 °C), achieved after simple vacuum filtration

2.3. Phosphorus Extraction

Figure 1 shows the flow charts of experimental set-ups with and without acid addition. To optimize the acid dose achieving maximum P recovery by minimizing the PW contamination with acid and heavy metals, the following three experimental set-ups were tested: (i) HTC-C (control): direct liquid–solid separation without any acid addition; (ii) HTC-PS (phosphorus stripping): acid addition to the slurry followed by L/S-separation; and (iii) HTC-PL (phosphorus leaching): Short-time sedimentation of hydrochar particles followed by decantation of surplus process water and acid addition to remaining hydrochar. The last set-up was conducted to reduce the amount of acidified process water, while concentrating the leached nutrients in the extract. Because of the hydrophobic characteristics of the hydrochar [18], the sedimentation took place immediately. Acid extractions (leaching or stripping) were carried out directly after HTC in the 1 L glass bottle, without any intermediate step. All slurries were stirred continuously at 600 rpm with a magnetic stirrer (Big Squid, IKA, Breisgau, Germany). Initial pH value was recorded, followed by a stepwise addition of 1 mL concentrated sulfuric acid (93–98%, Hiperpur, PanReac AppliChem, Darmstadt, Germany) to the mixture until a pH of 2 was reached. After reaching the desired pH value the reaction time of 2 h started [33]. During the reaction, the glass bottle was covered with aluminum foil to minimize the loss from evaporation and the pH value was measured every 15 min and adjusted with 0.5–1 mL acid addition. After 2 h the mixture was vacuum filtered through a Buchner funnel with a Whatman filter paper (pore size 11 µm, Cat No 1001 125). The liquid samples were stored at 6 °C and hydrochars were, without a washing step, directly transferred to the convection oven and dried at 40 °C over night. Remaining PW (incl. nutrients) on char was considered as a loss, to keep the treatment for industrial-scale applications simple and cost effective.

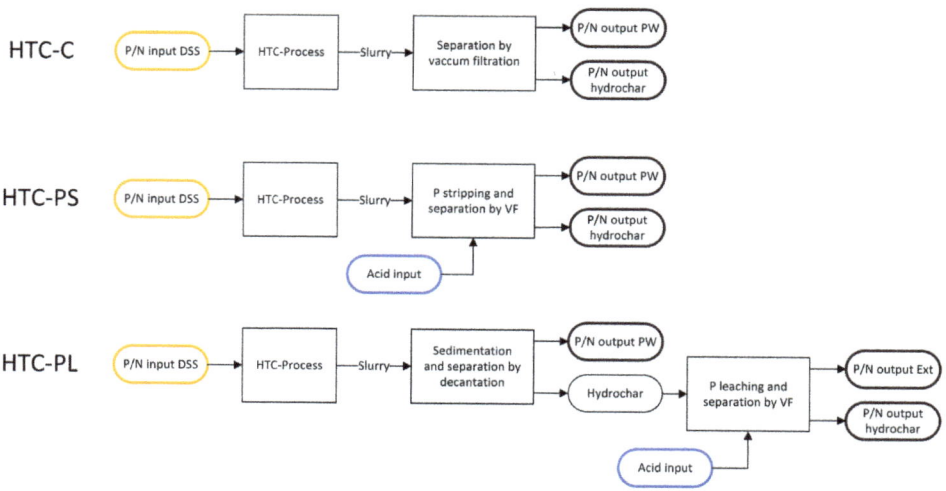

Figure 1. Flow charts of experimental set-ups with and without acid addition. (HTC = Hydrothermal carbonization, C = Control, PS = P-Stripping, PL = P-Leaching, DSS = Digested sewage sludge, PW = Process water, Ext = Leaching extract, VF = Vacuum filtration).

2.4. Characterization of Feedstock and HTC-Products

DSS and hydrochars were analyzed for DM and ash content. To avoid losses of mercury due to volatilization, all samples were previously dried at 40 °C [34] for further analyses and the initial DM content (% DM_{40C}) was recorded. As a second step, subsamples were taken and dried at 105 °C until constant weight was reached (% DM_{105C}). The effective DM content (% DM) was then calculated by Equation (1):

$$DM(wt.\%) = DM_{40C}(wt.\%) \times DM_{105C}(wt.\%) \div 100 \quad (1)$$

After determination of the DM_{105C}, samples were incinerated in a muffle furnace (L 40/11 BO, Nabertherm GmbH, Lilienthal, Germany) and the ash content was measured according to DIN EN 14775. Ultimate analyses for carbon (C), hydrogen (H) and N were undertaken with a CHN-analyzer (TruSpec Macro, LECO Instrumente GmbH, Mönchengladbach Germany) and sulfur was measured at an external lab with a CHNS-analyzer (vario EL cube, Elementar Analysensysteme GmbH, Langenselbold, Germany). All samples for ultimate analyses were dried at 105 °C. Oxygen content was calculated by difference (Equation (2)):

$$O(wt.\%) = 100 - C(wt.\%) - H(wt.\%) - N(wt.\%) - S(wt.\%) - Ash\ content(wt.\%) \quad (2)$$

Total content of HM (Cd, Cu, Ni, Pb and Zn) and phosphorus (P) was measured for liquid and solid samples spectroscopically by ICP-OES (Agilent 5100, Agilent Technologies, CA, USA). 2 mL of liquid sample or 0.35 g of solid sample (dried at 40 °C) was added to 8 mL of aqua regia and microwave digested (Speedwave Four, Berghof Products + Instruments GmbH, Eningen, Germany) for 35 min at 175 °C (in dependence on SN EN 13346, protocol C). After the acid digestion samples were transferred to a 25 mL volumetric flask and diluted with ultrapure water. Final results for solid samples were corrected by the remaining water content (% DM_{105C}). Acid digestion using aqua regia did not lead to a complete digestion for solid char samples. Therefore, measurement results were utilized for material balance and recovery efficiency was determined. The results showed good overall recovery with >83% for HM (excl. Pb, Cd and Hg) and nutrients. Dilution effects in liquid samples faced detection limitations, especially for elements with low concentrations as Pb, Cd and Hg. Phosphorus contents for hydrochars were calculated by difference. All samples were digested in triplicates and average results are reported. For the determination of mercury (Hg) samples dried at 40 °C were measured according to DIN EN 1483: 08.97. Mercury was analyzed using a cold vapor atomic absorption spectrometer (CV-AAS) (novAA® 350, Analytik Jena GmbH, Jena, Germany) equipped with a hydride generator (HS 60A, Analytik Jena GmbH, Jena, Germany).

Liquid samples (process water and leaching extract) were analyzed for total organic carbon (TOC), measured as non-purgeable organic carbon (NPOC), according to ASTM D7573 and total bound nitrogen (TNb) according to DIN EN 12260 with a TOC-L$_{CSH}$ analyzer equipped with a TNM-L unit (Shimadzu, Kyoto, Japan). For TOC measurements the samples were automatically acidified with HCl to pH < 3 and sparged with purified air to remove inorganic carbon. Purgeable organic carbon (POC) may also be lost during this sample treatment. The pH values of fresh process water and leachate were measured with a portable multi-parameter meter (HQ40d, Hach Lange, Düsseldorf, Germany).

Hydrochars were further characterized for its fuel properties. Regarding energy content, the higher heating value (HHV) of the raw material and the hydrochar was measured using a calorimeter (IKA C 200, Breisgau, Germany). The volumetric emissions of CO_2 and SO_2 were calculated according to Equations (3) and (4) adapted from Kaltschmitt [35], on the assumption of a complete combustion. Weight percentage of carbon (c) and sulfur (s) are given by the elemental analysis:

$$CO_2 \text{ emission } \left(\frac{m^3 \, CO_2}{kg \, DM}\right) = 22.41 \frac{c}{12} \quad (3)$$

$$SO_2 \text{ emission } \left(\frac{m^3 \, SO_2}{kg \, DM}\right) = 22.41 \frac{s}{32} \quad (4)$$

The hydrochar yield and energy efficiency (EE) were calculated by Equations (5) and (6), respectively [36–38].

$$\text{Hydrochar yield (\%)} = \left(\frac{DM_{hydrochar} \, (kg)}{DM_{raw \, material} \, (kg)}\right) \times 100 \quad (5)$$

$$\text{Energy efficiency (\%)} = (\text{Hydrochar yield}) \times \left(\frac{HHV_{hydrochar} \left(\frac{MJ}{kg}\right)}{HHV_{raw \, material} \left(\frac{MJ}{kg}\right)}\right) \quad (6)$$

2.5. Economic Feasabiliy of Hydrothermal Carbonization (HTC)

For the economic calculation, on-site treatment at a WWTP with anaerobic sludge digestion was presumed. Therefore, transport and PW treatment costs were neglected. The investment costs and annul fixed costs were obtained by the HTC-company GEegio Energie AG, Switzerland. All Swiss franc (CHF) amounts are reported in US dollars (USD) and were converted at the exchange rate of 1.1 USD/CHF. The costs include the HTC-process, liquid–solid separation with a hydraulic filter press (HPS, Bucher Unipektin AG, Switzerland), as well as drying and briquetting of the hydrochar. Annual fixed costs are based on Swiss rates and include electrical and thermal energy (electricity rate 0.22 USD/kWh; heating rate 0.066 USD/kWh), rent (198.00 USD/m^{2}a), labor, maintenance, provision and contingency insurance, as well as capital costs. Labor costs are based on a full time-position. Capital costs were calculated based on a payback period of 10 years with 2% annual interest. Annual provision and insurance costs are 1% and maintenance costs are 2.5% of the investment. Costs for the P-extraction were calculated based on the acid consumption in the lab experiments, and do not include industrial scale plant costs and/or fertilizer production. An average market price of 275 USD per ton of industrial grade H_2SO_4 98% was applied. Specific overall costs per dry ton DSS were determined by Equation (7).

$$\text{Specific overall costs (CHF per ton DSS)} = \left(\frac{\text{Annual fixed costs (CHF per year)}}{\text{Annual throughput (tons per year)}}\right) \quad (7)$$

The industrial scale HTC-plant from GRegio Energie AG is operating continuously with a reactor volume of 5.6 m³. The maximum flow rate of the current system is 2 m³/h and is limited by the heating system and heat recovery unit. The optimal DM content of the sludge input material is around 15%. For the calculation of the disposal fees the throughput was set at 11,200 tons dewatered sewage sludge (15%DM) per year. With a carbonization time of 4 h and a yearly operation time of 8000 h, the flow rate was set to 1.4 m³/h.

3. Results and Discussion

3.1. Carbonization and Separation

For the characterization of hydrochars and liquid phases extensive analysis were conducted. Higher heating values of the raw material and hydrochars are shown in Table 1. HTC of DSS causes a slight decrease in HHV for the control (HTC-C_HC) and hydrochar after P leaching (HTC-PL_HC). The HHV declined around 11% from 12.7 MJ/kg (DSS) to 11.3 MJ/kg (HTC-C_HC and HTC-PL_HC), which is comparable to reported heating values of other studies with DSS [8,14]. After P stripping the hydrochar (HTC-PS_HC) showed with 13.3 MJ/kg a slight increase of 4.5% compared to the starting material and was in average 17.7% above the HHV of HTC-C_HC and HTC-PL_HC.

Table 1. Ultimate analysis and higher heating values (HHV) of digested sewage sludge (DSS) and its derived hydrochars (HTC-xx). (C = Control; PS = P stripping; PL = P leaching; HC = Hydrochar) (n = 2).

	C [wt.%]	H [wt.%]	N [wt.%]	S [wt.%]	O_{diff} [wt.%]	Ash [wt.%]	H/C	O/C	HHV MJ/kg
DSS	30.2	4.4	4.2	1.0	17.3	42.8	1.7	0.4	12.7
HTC-C_HC	27.4	3.5	2.4	0.9	10.1	55.6	1.5	0.3	11.3
HTC-PS_HC	30.8	3.9	3.1	8.7	8.4	45.1	1.5	0.2	13.3
HTC-PL_HC	27.0	3.7	2.7	9.2	9.6	47.7	1.6	0.3	11.3

Table 1 summarizes the results from ultimate analyses, as well as gross calorific values on raw material and hydrochar samples. The ash content of the hydrochars decreased by 10% after the acid treatment, with a minimum ash content of 45.1% after P stripping. Volumetric carbon dioxide and sulfur dioxide emissions c are shown in Figure 2. Hydrochars after acid addition show an increase in sulfur content by 9–10 times, which has a direct influence on the SO_2 emissions. Hydrochars from HTC-PS and HTC-PL contained more than 6 wt.% sulfur which, according to the Swiss Clean Air Act (LRV, Annex 5 Number 2), is 3% above the recommended value for coal, coal briquettes and cokes [39]. The carbon content of the chars as well as the CO_2 emissions are very stable and did not show any major changes (Table 1, Figure 2).

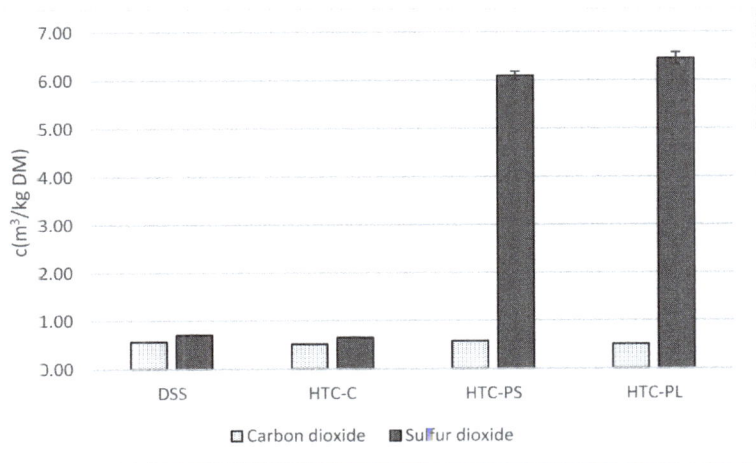

Figure 2. Volumetric carbon dioxide and sulfur dioxide emission per kg dry matter (DM) (Error bars show standard deviation of $n = 2$).

3.2. Nutrient Recovery

The aim of this study was to recover P and N from sewage sludge into the process water with minimal heavy metal contamination. Figure 3 shows the recovery of nutrients to the liquid phase. PS achieved with 84% a 13% higher P recovery than PL. The difference between the two methods in N recovery is even greater, because for PL it is divided into two liquid fractions. N recovery efficiencies of 53% for PS_PW and 22% for PL_Ext were reached. While for PL 32% of N were in the PW fraction. Compared to the control with 49% N recovery, even without additional treatment considerable amount of N could be recovered directly into the PW.

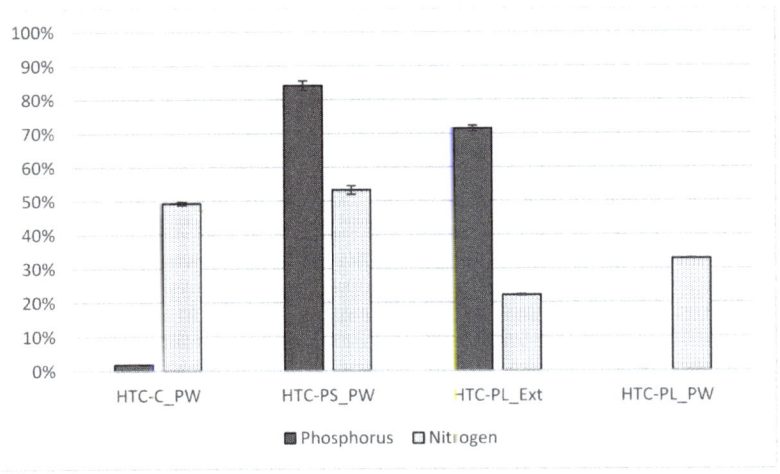

Figure 3. Recovery of main nutrients in the process water (PW) and leaching extract (Ext) (Error bars show standard deviation of $n = 2$).

Tables 2 and 3 show the nutrient and heavy metal contents in the raw material, hydrochars and liquid phase. Regarding heavy metal contaminations in the liquid phase, only cadmium and zinc could be detected and quantified. The appearance of Cd could not be explained by the authors, as it was not found in the raw material and must have accumulated in the PW. A release of Zn was also observed by Becker et al. [22]. Elemental analysis confirmed that most HM stayed in the hydrochar. An explanation for the high selectivity is the sorption property of the char [22] and the reduction of HM solubility after the HTC-treatment [40].

Table 2. Content analysis of raw material and hydrochars (HC). (C = Control; PS = P stripping; PL = P leaching; BLD = bellow limit of detection) (n = 2).

	Nutrient Content		Heavy Metal Content					
	P [mg kg^{-1}]	N [wt.%]	Pb [mg kg^{-1}]	Cd [mg kg^{-1}]	Cu [mg kg^{-1}]	Ni [mg kg^{-1}]	Zn [mg kg^{-1}]	Hg [mg kg^{-1}]
DSS	34,128	4.2	24.3	BLD	262.8	21.9	700.8	0.9
HTC-C_HC	49,505 *	2.4	20.8	BLD	337.5	33.2	900.5	0.5
HTC-PS_HC	9100 *	3.0	39.0	BLD	433.5	38.9	1031.9	1.0
HTC-PL_HC	14,745 *	2.7	25.9	BLD	367.8	43.5	894.0	1.0

* P content calculated by difference.

Table 3. Content analysis of process water (PW) and leaching extract (Ext). (C = Control; PS = P stripping; PL = P leaching; BLD = bellow limit of detection) (n = 2).

	Nutrient Content		Heavy Metal Content						pH
	P [mg L^{-1}]	N [mg L^{-1}]	Pb [mg L^{-1}]	Cd [mg L^{-1}]	Cu [mg L^{-1}]	Ni [mg L^{-1}]	Zn [mg L^{-1}]	Hg [mg L^{-1}]	
HTC-C_PW	208	6900	BLD	BLD	BLD	BLD	BLD	BLD	7.1
HTC-PS_PW	8113	6386	BLD	BLD	BLD	BLD	21.3	BLD	2.1
HTC-PL_Ext	16,050	6195	BLD	0.9	BLD	BLD	42.9	BLD	2.0
HTC-PL_PW	BLD	6960	BLD	BLD	BLD	BLD	BLD	BLD	7.3

Because of high sulfur and heavy metal content in the hydrochar, it can only be used for energy production in industrial-scale plants (e.g., cement industry, coal-fired power plant, mono-incineration) with appropriate fume cleaning systems. To decrease the sulfur content in the char, sulfuric acid could be substituted with nitric or organic acids (e.g., citric acid) for the P-recovery treatment. However, other mineral and organic acids can increase the leaching costs [22].

While the N concentrations were for all four liquid samples (PW and Ext) in a similar range (6195–6960 mg N/L), differences in P concentration could be observed. The PW after acid stripping contained 8113 mg P/L, whereas the highest P concentration was achieved in the leachate extract (HTC-PL_Ext) with 16,050 mg P/L. The acid consumption was for both methods (HTC-PS and HTC-PL) at around 14.2 kg (±0.3 kg) H_2SO_4 per kg P.

3.3. Feasability of Application

Figure 4 shows hydrochar dewaterability and yield, as well as energy efficiency. Normally, sewage sludge is difficult to separate to a high DM content, because of the microbial cell membranes strongly bound to water [41]. With conventional technologies like a decanter centrifuge or filter press, which are used in WWTPs, a DM content of 20–35% or 28–45%, respectively, can be obtained [42]. After HTC it was observed that the solid phase settles very quickly and with simple vacuum filtration a DM content of around 40% could be achieved. This is consistent with other findings, showing the positive effect on the dewaterability of biomass and sewage sludge [18,43]. The acid leaching treatment even increased the dewaterability slightly by 6%. All treatments included a single liquid–solid separation by vacuum filtration. It is assumed that the decantation, as an intermediate separation step, had no impact on the dewaterability of PL, as the process water was not removed completely. According to Chen et al. [44], an increase in acidity can improve the dewaterability of activated sludge. This would also explain the improvement in the PL treatment.

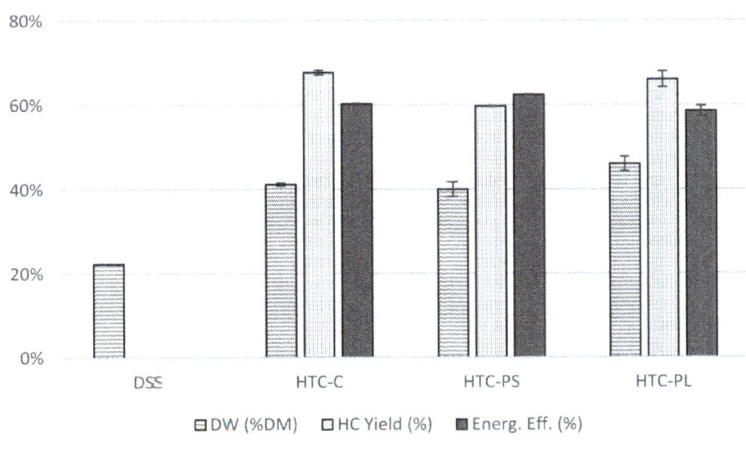

Figure 4. Dewaterability (DW), hydrochar (HC) yield and energy efficiency of the raw material (DSS as received), the control and acid treated hydrochars. Error bars show standard deviation of n = 2.

3.4. Profitability

Calculations for annual fixed costs are presented in Table 4. The specific costs per ton DSS for the HTC treatment depend on the DM content of the delivered sludge and can range between 91 USD and 183 USD with 15% and 30% DM content, respectively. On a dry basis, the specific overall costs are 608 USD per ton DM. In Switzerland average disposal costs (incl. transportation) per dry ton sewage sludge amounted in 2001 to: (i) 827 USD for sewage sludge incineration plant; (ii) 882 USD for municipal solid waste incineration plant; and (iii) 669 USD for cement plant (including drying) [45]. The WWTP Werdhölzli in Zurich is one of the biggest WWTPs in Switzerland and operates its own sludge incinerator. The sludge capacity of the plant is designed so that external WWTPs can also deliver their sewage sludge for disposal. The disposal of sludge at the WWTP Werdhölzli costs currently 1173 USD per dry ton [46]. Besides the disposal costs, by 2026 the acid treatment will be an important cost factor for WWTPs. In the current study the acid usage per dry ton DSS was for PS 406 kg and for PL 349 kg, which adds additional costs of 111 USD and 96 USD per dry ton DSS, respectively. Based on the acid usage per kg P recovered, both methods are equal with acid costs of 3.96 USD per kg P.

Table 4. Calculation of net costs for the HTC process from GRegio Energie AG (amortization time 10 years, interest rate 2%; db = dry basis; DSS = digested sewage sludge).

Annual Throughput of DSS	Tons (15%DM)	11,200
	Tons (db)	1680
Investment costs	USD	4,332,460.00
Annual fixed costs	USD/a	1,022,674.00
Capital costs	USD/a	482,317.00
Operation costs	USD/a	540,357.00
Electrical energy	USD/a	88,704.00
Heat energy	USD/a	72,441.00
Rent	USD/a	74,250.00
Labor costs	1 Person (USD)	110,000.00
Maintenance	2.5% of Invest. (USD)	108,312.00
Provision	1% of Invest. (USD)	43,325.00
Contingency insurance	1% of Invest. (USD)	43,325.00
Specific overall costs	USD per ton DSS (db)	608.00

4. Conclusions

Two methods for the nutrient recovery and fuel production were evaluated. For the treatments PS and PL the acid consumption per kg P recovered were, at around 14.2 kg (± 0.3 kg) H_2SO_4, equal, whereas for PL an advantage was the reduction of acidified PW and doubling the P-concentration in the extract. The additional treatment step with sedimentation and decantation and the reduced P-recovery efficiency compared to PS are the only drawbacks of this treatment. In comparison, PS produced a char with a slightly higher HHV, which could be favorable as a fuel substitute for industrial-scale plants. Both treatments exceed the regulatory requirements for P recovery and additionally transfer around 50% of N into the liquid phase, while leaving the main part of heavy metals in the hydrochar. A reduction in char-N content improves as well the combustion emissions regarding NO_x. Regarding the high char-S content, only industrial plants with appropriate exhaust gas treatments are applicable. Therefore, sewage sludge will be still available as a climate-neutral fuel for the cement industry and coal power plants.

HTC can compete with current costs for sewage sludge disposal and offers an advantageous alternative to mono-incineration as a sludge treatment, regarding fuel and nutrient recovery. For an implementation in current WWTP infrastructures plants with AD are favored because of their sludge property and the possibility of PW fermentation for higher methane recovery. To promote the P recovery from waste streams like sewage sludge, further investigations in a sustainable and economically feasible fertilizer production are needed.

Author Contributions: Conceptualization, G.G. and L.M.; methodology, G.G.; validation, G.G. and T.K.; formal analysis, G.G., L.M. and T.K.; investigation, G.G., L.M., R.W. and T.K.; resources, G.G., L.M., R.W. and T.K.; data curation, G.G. and L.M.; writing—original draft preparation, G.G.; writing—review and editing, G.G. and R.K.; visualization, G.G.; supervision, G.G. and R.K.; project administration, G.G.; funding acquisition, G.G. All authors have read and agreed to the published version of the manuscript.

Funding: This research was funded by cemsuisse, the Association of the Swiss Cement Industry. The article processing charges for the open access publication was funded by ZHAW Zurich University of Applied Sciences.

Institutional Review Board Statement: Not applicable.

Informed Consent Statement: Not applicable.

Data Availability Statement: Not applicable.

Acknowledgments: The authors would like to thank for the financial support received from cemsuisse, the Association of the Swiss Cement Industry and ZHAW Zurich University of Applied Sciences for funding the open access publication. The authors would also like to thank Samuel Solin and Simon Kaiser, Institute of Bioenergy and Resource Efficiency at the FHNW University of Applied Sciences and Arts Northwestern Switzerland, for the help with the elemental analysis.

Conflicts of Interest: The authors declare no conflict of interest. The funders had no role in the design of the study; in the collection, analyses, or interpretation of data; in the writing of the manuscript. The project funders agreed to publish the results.

References

1. Mathieux, F.; Ardente, F.; Bobba, S.; Nuss, P.; Blengini, G.; Alves Dias, P.; Blagoeva, D.; Torres De Matos, C.; Wittmer, D.; Pavel, C.; et al. *Critical Raw Materials and the Circular Economy—Background Report*; Publications Office of the European Union: Luxembourg, 2017.
2. European Commission. *Communication from the Commission to the European Parliament, the Council, the European Economic and Social Committee and the Committee of the Regions*; On the Review of the List of Critical Raw Materials for the EU and the Implementation of the Raw Materials Initiative; European Commission: Brussels, Belgium, 2014.
3. Binder, C.R.; de Baan, L.; Wittmer, D. *Phosphorflüsse in Der Schweiz. Stand, Risiken Und Handlungsoptionen*; Umwelt-Wissen; Bundesamt für Umwelt: Bern, Switzerland, 2009; p. 161.
4. Daneshgar, S.; Callegari, A.; Capodaglio, A.; Vaccari, D. The Potential Phosphorus Crisis: Resource Conservation and Possible Escape Technologies: A Review. *Resources* **2018**, *7*, 37. [CrossRef]

5. Cordell, D.; White, S. Life's Bottleneck: Sustaining the World's Phosphorus for a Food Secure Future. *Annu. Rev. Environ. Resour.* **2014**, *39*, 161–188. [CrossRef]
6. Hermann, L. *Rückgewinnung von Phosphor aus der Abwasserreinigung*; Umwelt-Wissen; Bundesamt für Umwelt: Bern, Switzerland, 2009; p. 196.
7. Bigalke, M.; Ulrich, A.; Rehmus, A.; Keller, A. Accumulation of Cadmium and Uranium in Arable Soils in Switzerland. *Environ. Pollut.* **2017**, *221*, 85–93. [CrossRef] [PubMed]
8. Bigalke, M.; Schwab, L.; Rehmus, A.; Tondo, P.; Flisch, M. Uranium in Agricultural Soils and Drinking Water Wells on the Swiss Plateau. *Environ. Pollut.* **2018**, *233*, 943–951. [CrossRef]
9. Bigalke, M.; Imseng, M.; Schneider, S.; Schwab, L.; Wiggenhauser, M.; Keller, A.; Müller, M.; Frossard, E.; Wilcke, W. Uranium Budget and Leaching in Swiss Agricultural Systems. *Front. Environ. Sci.* **2020**, *8*, 54. [CrossRef]
10. Schnug, E.; Haneklaus, N. Uranium in phosphate fertilizers—Review and outlook. In *Uranium—Past and Future Challenges*; Merkel, B.J., Arab, A., Eds.; Springer International Publishing: Cham, Germany, 2015; pp. 123–130. ISBN 978-3-319-11058-5.
11. Tulsidas, H.; Gabriel, S.; Kiegiel, K.; Haneklaus, N. Uranium Resources in EU Phosphate Rock Imports. *Resour. Policy* **2019** *61*, 151–156. [CrossRef]
12. Roth, N.; FitzGerald, R. *Human and Environmental Impact of Uranium Derived from Mineral Phosphate Fertilizers*; Swiss Centre for Applied Human Toxicology: Basel, Switzerland, 2015; p. 50.
13. Nutrient Source Specifics—Ammonia. Available online: www.ipni.net/specifics (accessed on 19 April 2021).
14. Kyriakou, V.; Garagounis, I.; Vourros, A.; Vasileiou, E.; Stoukides, M. An Electrochemical Haber-Bosch Process. *Joule* **2020**, *4*, 142–158. [CrossRef]
15. Krämer, J. *Phosphorrecycling: Wer, Wie, Was?—Umsetzung Einer Iterativen, Zielgruppenorientierten Kommunikationsstrategie*; Deutsche Phosphor Plattform: Frankfurt, Germany, 2019; p. 108.
16. Libra, J.A.; Ro, K.S.; Kammann, C. Funke, A.; Berge, N.D.; Neubauer, Y.; Titirici, M.-M.; Fühner, C.; Bens, O.; Kern, J.; et al. Hydrothermal Carbonization of Biomass Residuals: A Comparative Review of the Chemistry, Processes and Applications of Wet and Dry Pyrolysis. *Biofuels* **2011**, *2*, 71–106. [CrossRef]
17. Zhao, P.; Chen, H.; Ge, S.; Yoshikawa, K. Effect of the Hydrothermal Pretreatment for the Reduction of NO Emission from Sewage Sludge Combustion. *Appl. Energy* **2013**, *111*, 199–205. [CrossRef]
18. Escala, M.; Zumbühl, T.; Koller, C.; Junge, R.; Krebs, R. Hydrothermal Carbonization as an Energy-Efficient Alternative to Established Drying Technologies for Sewage Sludge: A Feasibility Study on a Laboratory Scale. *Energy Fuels* **2013**, *27*, 454–460. [CrossRef]
19. Ahmed, M.; Andreottola, G.; Elagroudy, S.; Negm, M.S.; Fiori, L. Coupling Hydrothermal Carbonization and Anaerobic Digestion for Sewage Digestate Management: Influence of Hydrothermal Treatment Time on Dewaterability and Bio-Methane Production. *J. Environ. Manag.* **2021**, *281*, 111910. [CrossRef] [PubMed]
20. Marin-Batista, J.D.; Mohedano, A.F.; Rodríguez, J.J.; de la Rubia, M.A. Energy and Phosphorous Recovery through Hydrothermal Carbonization of Digested Sewage Sludge. *Waste Manag.* **2020**, *105*, 566–574. [CrossRef]
21. Maurizio, V.; Luca, F.; Fabio, M.; Antonio, M. Andreottola Gianni Hydrothermal Carbonization as an Efficient Tool for Sewage Sludge Valorization and Phosphorous Recovery. *Chem. Eng. Trans.* **2020**, *80*, 199–204. [CrossRef]
22. Becker, G.C.; Wüst, D.; Köhler, H.; Lautenbach, A.; Kruse, A. Novel Approach of Phosphate-Reclamation as Struvite from Sewage Sludge by Utilising Hydrothermal Carbonization. *J. Environ. Manag.* **2019**, *238*, 119–125. [CrossRef]
23. Egle, L.; Rechberger, H.; Zessner, M. *Phosphorrückgewinnung aus dem Abwasser*; Bundesministerium für Land- und Forstwirtschaft, Umwelt und Wasserwirtschaft, Sektion VIIWasser: Vienna, Austria, 2014; p. 323.
24. Fux, C.; Theiler, M.; Irzan, T. *Studie Phosphorrückgewinnung aus Abwasser und Klärschlamm*; TBF+Partner AG: Zürich, Switzerland, 2015; p. 92.
25. Blöhse, D. Anaerobe Verwertung von HTC-Prozesswässern. In Proceedings of the Biokohle im Blick—Herstellung, Einsatz und Bewertung, 73. Symposium des ANS e.V., Berlin, Germany, 20 September 2012.
26. Ferrentino, R.; Merzari, F.; Fiori, L.; Andreottola, G. Coupling Hydrothermal Carbonization with Anaerobic Digestion for Sewage Sludge Treatment: Influence of HTC Liquor and Hydrochar on Biomethane Production. *Energies* **2020**, *13*, 6262. [CrossRef]
27. Zhao, K.; Li, Y.; Zhou, Y.; Guo, W.; Jiang, H.; Xu, Q. Characterization of Hydrothermal Carbonization Products (Hydrochars and Spent Liquor) and Their Biomethane Production Performance. *Bioresour. Technol.* **2018**, *267*, 9–16. [CrossRef] [PubMed]
28. Parmar, K.R.; Ross, A.B. Integration of Hydrothermal Carbonisation with Anaerobic Digestion; Opportunities for Valorisation of Digestate. *Energies* **2019**, *12*, 1586. [CrossRef]
29. Aragón-Briceño, C.; Ross, A.B.; Camargo-Valero, M.A. Evaluation and Comparison of Product Yields and Bio-Methane Potential in Sewage Digestate Following Hydrothermal Treatment. *Appl. Energy* **2017**, *208*, 1357–1369. [CrossRef]
30. Merzari, F.; Langone, M.; Andreottola, G.; Fiori, L. Methane Production from Process Water of Sewage Sludge Hydrothermal Carbonization. A Review. Valorising Sludge through Hydrothermal Carbonization. *Crit. Rev. Environ. Sci. Technol.* **2019**, *49*, 947–988. [CrossRef]
31. Shi, Y.; Luo, G.; Rao, Y.; Chen, H.; Zhang, S. Hydrothermal Conversion of Dewatered Sewage Sludge: Focusing on the Transformation Mechanism and Recovery of Phosphorus. *Chemosphere* **2019**, *228*, 619–628. [CrossRef]
32. Aragón-Briceño, C.I.; Pozarlik, A.K.; Bramer, E.A.; Niedzwiecki, L.; Pawlak-Kruczek, H.; Brem, G. Hydrothermal Carbonization of Wet Biomass from Nitrogen and Phosphorus Approach: A Review. *Renew. Energy* **2021**, *171*, 401–415. [CrossRef]

33. Biswas, B.K.; Inoue, K.; Harada, H.; Ohto, K.; Kawakita, H. Leaching of Phosphorus from Incinerated Sewage Sludge Ash by Means of Acid Extraction Followed by Adsorption on Orange Waste Gel. *J. Environ. Sci.* **2009**, *21*, 1753–1760. [CrossRef]
34. BBodSch, V. *Federal Soil Protection and Contaminated Sites Ordinance*; German Federal Council: Berlin, Germany, 1999; p. 1554.
35. Kaltschmitt, M.; Hartmann, H.; Hofbauer, H. (Eds.) *Energie aus Biomasse: Grundlagen, Techniken und Verfahren*; Springer: Berlin/Heidelberg, Germany, 2016; ISBN 978-3-662-47438-9.
36. Gao, Y.; Yu, B.; Wu, K.; Yuan, Q.; Wang, X.; Chen, H. Physicochemical, Pyrolytic, and Combustion Characteristics of Hydrochar Obtained by Hydrothermal Carbonization of Biomass. *BioResources* **2016**, *11*, 4113–4133. [CrossRef]
37. Li, Z.; Yi, W.; Li, Z.; Tian, C.; Fu, P.; Zhang, Y.; Zhou, L.; Teng, J. Preparation of Solid Fuel Hydrochar over Hydrothermal Carbonization of Red Jujube Branch. *Energies* **2020**, *13*, 480. [CrossRef]
38. Lucian, M.; Fiori, L. Hydrothermal Carbonization of Waste Biomass: Process Design, Modeling, Energy Efficiency and Cost Analysis. *Energies* **2017**, *10*, 211. [CrossRef]
39. *Luftreinhalte-Verordnung*; Swiss Federal Council: Bern, Switzerland, 1985; (Status as of 1 April 2020); p. 82.
40. Liu, M.; Duan, Y.; Bikane, K.; Zhao, L. The Migration and Transformation of Heavy Metals in Sewage Sludge during Hydrothermal Carbonization Combined with Combustion. *BioMed Res. Int.* **2018**, *2018*, 1913848. [CrossRef]
41. Wei, H.; Gao, B.; Ren, J.; Li, A.; Yang, H. Coagulation/Flocculation in Dewatering of Sludge: A Review. *Water Res.* **2018**, *143*, 608–631. [CrossRef]
42. Böhler, M.; Siegrist, H.; Pinnow, D.; Müller, D.; Krauss, W.; Brauchli, H. *Neue Presstechnologie zur verbesserten Klärschlammentwässerung*; EAWAG: Dübendorf, Switzerlnd, 2003; p. 42.
43. Funke, A.; Ziegler, F. Hydrothermal Carbonization of Biomass: A Summary and Discussion of Chemical Mechanisms for Process Engineering. *Biofuels Bioprod. Biorefin.* **2010**, *4*, 160–177. [CrossRef]
44. Chen, Y.; Yang, H.; Gu, G. Effect of Acid and Surfactant Treatment on Activated Sludge Dewatering and Settling. *Water Res.* **2001**, *35*, 2615–2620. [CrossRef]
45. Laube, A.; Vonplon, A. *Klärschlammentsorgung in Der Schweiz—Mengen- Und Kapazitätserhebung*; Umwelt-Materialien; BUWAL: Bern, Switzerland, 2004; p. 47.
46. Schlamm Aus Klärwerken—Preise Für Die Anlieferung von Schlamm Aus Klärwerken. Available online: www.stadt-zuerich.ch/ted/de/index/entsorgung_recycling/sauberes_wasser/entsorgen/schlamm_aus_klaerwerken.html (accessed on 21 January 2021).

Article

Thermal Analysis and Kinetic Modeling of Pyrolysis and Oxidation of Hydrochars

Gabriella Gonnella [1], Giulia Ischia [1], Luca Fambri [2] and Luca Fiori [1,*]

[1] Department of Civil, Environmental and Mechanical Engineering, University of Trento, 38123 Trento, Italy; ga.gonnella@gmail.com (G.G.); giulia.ischia-1@unitn.it (G.I.)
[2] Department of Industrial Engineering, University of Trento, Via Sommarive 9, 38123 Trento, Italy; luca.fambri@unitn.it
* Correspondence: luca.fiori@unitn.it

Abstract: This study examines the kinetics of pyrolysis and oxidation of hydrochars through thermal analysis. Thermogravimetric analysis (TGA) and differential scanning calorimetry (DSC) techniques were used to investigate the decomposition profiles and develop two distributed activation energy models (DAEM) of hydrochars derived from the hydrothermal carbonization of grape seeds produced at different temperatures (180, 220, and 250 °C). Data were collected at 1, 3, and 10 °C/min between 30 and 700 °C. TGA data highlighted a decomposition profile similar to that of the raw biomass for hydrochars obtained at 180 and 220 °C (with a clear distinction between oil, cellulosic, hemicellulosic, and lignin-like compounds), while presenting a more stable profile for the 250 °C hydrochar. DSC showed a certain exothermic behavior during pyrolysis of hydrochars, an aspect also investigated through thermodynamic simulations in Aspen Plus. Regarding the DAEM, according to a Gaussian model, the severity of the treatment slightly affects kinetic parameters, with average activation energies between 195 and 220 kJ/mol. Meanwhile, the Miura–Maki model highlights the distributions of the activation energy and the pre-exponential factor during the decomposition.

Keywords: hydrothermal carbonization; modeling; kinetics; thermal analysis; Aspen Plus; biomass; hydrochar

1. Introduction

The global increase in energy demand, driven mainly by the growing global population, urbanization, and economic growth, is depleting the world's reserves of fossil fuels, already strained by their high consumption rate, heightening environmental concerns, resource depletion, and rising costs. To face the problem, different renewable energy sources have been investigated, like wind, solar, hydropower, geothermal, and biomass. Among the different biomasses, those deriving from waste or agricultural residues, agro-industrial, horticulture, and wood processing, are cheap and convenient [1,2]. Besides, the use and valorizing of waste biomass are part of a circular economy context, where residues become the input for a new process [3]. At present, most of the waste produced in Europe is disposed of without any further treatment resulting in wastefulness of energy and resources and in serious environmental impact. A high-efficient use of waste not only increases its economic value but also results in environmental benefits [4].

In the last decades, biomass waste treatment and management have seen an increasing interest from the research community. Currently, biomass waste has characteristics that limit its use, including a low heating value, low fixed carbon content, high moisture content, low grindability, and non-uniformity. Studies on the conversion of biomass into high-quality biofuels have been performed to avoid the problems caused by direct combustion and find more effective methods of utilization, such as pyrolysis, gasification, torrefaction, and hydrothermal carbonization [5,6].

Despite pyrolysis, gasification, and torrefaction processes reducing moisture content and increasing fixed carbon content and heating value, all of them have a problem concerning high energy consumption and high pollutant emissions [6,7].

Hydrothermal carbonization (HTC) is a promising technique that can turn lignocellulose biomass into carbon-rich, solid, lignite-like fuel.

HTC is a thermochemical treatment able to recover energy from biomass wastes and eliminate organic contaminants and pollutants [8]. Similar to the natural coalification process, HTC lowers both the oxygen and hydrogen content of the feedstock, obtaining a final coal-like product referred to as hydrochar [9]. The process consists of subjecting the feedstock to mild temperatures (180–250 °C) and autogenous pressure in liquid water, ensuring a wide range of applicability for substrates with high moisture, up to 95 wt.% [10]. Hydrochar is endowed with higher calorific value and higher fixed carbon content compared to its precursor biomass [11]. Due to its potential, HTC is a mindful tool to treat biomass waste, recover materials, and provide an energy source. Hydrochar can be used as solid fuel and integrating HTC with other processes (like gasification, pyrolysis, anaerobic digestion [12], and composting [13]) can face shortcomings related to the single-stage, and improve the overall process efficiency. For example, HTC as a pre-treatment can affect the pyrolysis performance [1]. This last factor is important to control because it is the primary stage during the combustion and gasification processes (in view of using hydrochar as a fuel) and because it is a standalone process to produce biofuels (i.e., char, oil, syngas). In this framework, HTC can be used for the pre-conditioning of high moisture content biomass, making it hydrophobic (and thus allowing for a much cheaper dewatering/drying [2]), and then the biomass can undergo the dry pyrolysis treatment [14]. How much HTC could affect the pyrolysis rate depends on the type of feedstock involved [15,16].

In this context, understanding the kinetics of pyrolysis and oxidation of hydrochar could help in designing a thermochemical conversion process.

Even though the application of hydrochar is important in boilers and burners, the kinetics and combustion performance of hydrochar have been poorly studied. Sharma et al. [17] showed that adding hydrochar to pulverized coal helps to decrease pollutant gas emissions. Moreover, the study of process kinetics and the modeling of its control in the industrial field is weak. Different models (e.g., the Kissinger–Akahira–Sunose method and the Coats–Redfern method) have also been adopted to investigate the combustion kinetics of hydrochar [18,19], but the results obtained did not lead to quantitative data that fit well with the experimental values. Due to the importance of the kinetic behavior of hydrochar combustion, additional studies are needed. Actually, the distributed activation energy model (DAEM) is the most comprehensive model to investigate the thermal reaction kinetics of biomass. Bach et al. [20] employed DAEM to study the kinetic performance of combustion of wet-towered forest residues, while Chen et al. [6] applied DAEM to study the kinetic performance of pyrolysis of five lignocellulosic biomasses. Still, Zhang et al. [21] used a multi-Gaussian DAEM to investigate the decomposition characteristics of cellulose, hemicellulose, and lignin in different atmospheres. Therefore, the DAEM offers a higher quality of fit to the experimental data and can provide more information about the kinetics of biomass pyrolysis. DAEM was proposed originally by Vand [22] to model coal pyrolysis and is based on the hypothesis that the decomposition occurs through an infinite number of parallel reactions characterized by a certain distribution of activation energy. The energy distribution is often modeled through a continuous function, like the Gaussian, Weibull, or Boltzmann function [23]. Despite the diffusion of DAEMs to predict kinetics parameters of biomass and coal pyrolysis and oxidation [24–28], studies on the hydrochars are still very rare in the literature [20]. Therefore, this study used thermogravimetric analysis (TGA) to investigate the pyrolytic and oxidative behavior of hydrochars derived from an agro-industrial residue (grape seeds [29]).

In particular, TGA is a technique commonly adopted to analyze and examine biomass degradation in oxidation or pyrolysis patterns, and can be used to deduce kinetics parameters and mechanisms [30,31]. TGA data under inert and oxidative atmosphere were collected and used to determine kinetic parameters through two DAEMs (Gaussian and Miura–Maki DAEM). To do this, TGA data at different heating rates were collected, and activation energies and pre-exponential factors were computed. Analyses were also performed on the raw biomass and the oil contained inside the seeds. To investigate more in-depth the decomposition, derivative conversion curves were deduced, and the heat involved during the pyrolysis or oxidation was determined through differential scanning calorimetry (DSC). A simple thermodynamic model was also developed in Aspen Plus to compare experimental outputs with thermodynamic predictions. Through the software, pyrolysis and oxidation of the hydrochars were simulated and the outputs were compared with experimental measures from DSC to understand if this modeling approach can be used as a preliminary study of the heat of reactions.

2. Materials and Methods

2.1. Starting Samples: Grape Seeds and Hydrochars

The starting substrate consists of grape seeds and their corresponding hydrochars produced and characterized in previous work by our group [32]. Grape seeds are a residue of the winemaking industry and were provided by an alcohol producer company in the North of Italy (province of Trento). Hydrochars were obtained at 180, 220, and 250 °C through a solar HTC reactor, at a residence time of two hours, and a dry biomass to water weight ratio (B/W) of 0.3 [32]. Table 1 summarizes the main properties of the samples.

Table 1. Grape seeds and hydrochar (HC) properties, from the literature [32] (C: carbon, H: hydrogen, O: oxygen, N: nitrogen, S: sulfur, HHV: higher heating value).

Sample	Ultimate Analysis (wt.%)					Ash (wt.%)	HHV (MJ/kg)
	C	H	O	N	S		
Grape seeds	53.2	6.7	36.2	1.7	0.3	1.9	22.5
HC 180 °C	59.7	6.7	29.4	1.5	0.4	2.3	24.9
HC 220 °C	64.0	6.4	25.3	1.7	0.4	2.2	27.2
HC 250 °C	69.2	6.5	19.4	2.0	0.3	2.5	30.5

2.2. Sample Characterization

Grape seeds and hydrochars were characterized through thermogravimetric analysis (TGA), differential scanning calorimetry (DSC), and Fourier-transform infrared spectroscopy (FTIR).

TGA was performed in a thermobalance Mettler TG50 under non-isothermal conditions. 15 mg of whole dried samples (i.e., the entire raw and carbonized grape seeds) were placed into a 70 µL alumina crucible. Runs were performed between 40 and 700 °C at a heating rate of 1, 3, and 10 °C/min, under 100 mL/min of nitrogen or air (20.95% O_2 and 78.08% N_2). The thermobalance has a sensitivity of $\pm 1 \times 10^{-3}$ mg and 0.1 °C. The extent of conversion α at any time t was computed as:

$$\alpha(t) = \frac{m_0 - m(t)}{m_0 - m_f} = \frac{v(t)}{v_\infty} \quad (1)$$

where m_0, $m(t)$, and m_f are the mass at time zero (dry), at a given time t, and final time, respectively. This corresponds to the ratio between $v(t)$, which represents the amount of volatiles evolved up to a certain time t, and v_∞, which represents the amount of volatiles that evolved up to the end of the process.

DSC was performed in a Mettler DSC20 calorimeter between 40 and 600 °C, at a heating rate of 10 °C/min, under nitrogen or air at a flow rate of 100 mL/min. About 15 mg of material was placed into a 40 µL aluminum crucible. The heat released/absorbed was computed by integrating the measured heat flow over time in the 40–600 °C range, while the sample mass variation was recorded after every run.

FTIR spectra of samples were acquired using PerkinElmer FT-IR spectrometer Spectrum One. Spectra were obtained over a wavenumber range of 4000–650 cm^{-1}. Dried samples were ground to be compacted into a thin film and a flat top-plate Zinc Selenide crystal was used for the detection.

2.3. Distributed Activation Energy Models (DAEMs)

Two types of DAEMs (the Gaussian and the Miura–Maki [33]) were developed in software codes to model the decomposition of hydrochars, grape seeds, and grape seed oil.

In particular, when a first-order reaction is assumed to model biomass degradation, the variation of α vs. time can be expressed as the product between the reaction rate k and the conversion function f:

$$\frac{d\alpha}{dt} = k(T)f(\alpha) \tag{2}$$

By adopting the Arrhenius theory, k(T) can be expressed by Equation (3):

$$k(T) = k_0 \exp\left(-\frac{E}{RT}\right) \tag{3}$$

where k_0 represents the pre-exponential factor, E is the activation energy, and R is the universal gas constant.

To deduce the kinetic parameters, $f(\alpha)$ is usually assumed in advance (in the so called model-fitting methods) and then forced to fit experimental data [23]. In particular, non-isothermal models usually adopt a constant heating rate β equal to dT/dt, allowing the re-arrangement of Equation (3) into (4), which can be solved using thermogravimetric data [23].

$$\frac{d\alpha}{dT} \cong \frac{d\alpha}{dt}\frac{dt}{dT} = \frac{d\alpha}{dt}\frac{1}{\beta} = \frac{k(T)}{\beta}f(\alpha) = \frac{k_0}{\beta}\exp\left(-\frac{E}{RT}\right)f(\alpha) \tag{4}$$

Among model-fitting methods, DAEM assumes that pyrolysis can be described through an infinite number of irreversible first-order parallel reactions, with different rate parameters, which occur simultaneously [23,34]. In particular, the DAEM for a reaction order equal to 1 can be obtained by re-arranging Equations (2)–(4) as:

$$1 - \alpha = \int_0^\infty \exp\left(-\int_{T_0}^T \frac{k_0}{\beta}\exp\left(-\frac{E}{RT}\right)dT\right)f(E)dE \tag{5}$$

Equation (5) does not have an analytical solution and there are two families of methods to solve it. One is referred to as distribution-fitting method and the other as isoconversional methods. The distribution-fitting method assumes f(E) (like a Gaussian, Weibull, or Boltzmann distribution) and forces to fit the TGA data by applying a certain numerical method [23]. Meanwhile, isoconversional methods do not assume $f(\alpha)$, but adopt a series of thermogravimetric data at different heating rates to directly compute the activation energy. The reaction rate at a certain α is a function of temperature. Among these distribution-free methods, the Miura–Maki integral method is a common one [35].

2.3.1. Gaussian Model

In the Gaussian DAEM, the distribution function f(E) is assumed to be a Gaussian distribution with mean activation energy E_0 and standard deviation σ, which is:

$$f(E) = \frac{1}{\sigma\sqrt{2\pi}}\exp\frac{(E - E_0)^2}{2\sigma^2} \tag{6}$$

Applying the Coats–Redfern [36] and Fisher et al. [37] approximations, Equation (7) is obtained.

$$1 - \alpha = \int_0^\infty \left(\exp\left(-\frac{k_0 RT^2}{\beta E} \exp\left(-\frac{E}{RT} \right) \right) \right) \frac{1}{\sigma \sqrt{2\pi}} \exp \frac{(E - E_0)^2}{2\sigma^2} dE \qquad (7)$$

To solve Equation (7), the authors applied the simplification reported by Anthony and Howard [38], which consists in keeping k_0 constant and equal to 1.67×10^{13} s^{-1}, and the minimization procedure proposed by Güneş and Güneş [39] that solves the integral using the Simpson's 1/3 rule. To avoid oscillations in the results, dE was kept constant at 50 kJ/mol, while the extremes of integration were iteratively adjusted according to E_0. Then, the best solution was calculated by iterating the solution of the integral for several values of E_0 and σ and minimizing h, i.e., the root mean quadratic error between the model output and the experimental data:

$$h = \sqrt{\frac{\sum_{j=1}^n (\alpha_{TGA,j} - \alpha_{DAEM,j})^2}{n}} \qquad (8)$$

where $\alpha_{TGA,j}$ and $\alpha_{DAEM,j}$ are the experimental and calculated extent of conversion, respectively, and n is the number of available data points.

The code was developed in MATLAB and a direct search technique was adopted [39]. The code is reported in Supplementary Materials. Firstly, a larger grid procedure was used for both E_0 and σ, with a step size equal to 5. The first three values minimizing h were obtained. Then, for these last three values, an iteration process using $E_0 \pm 5$ and $\sigma \pm 5$ with a step equal to 1 was set. Finally, final values were used to solve the integral of Equation (7) and to calculate α_{DAEM}.

2.3.2. Miura–Maki Model

The Miura–Maki model adopts some assumptions to determine f(E) and k_0 without any a priori assumption on the energy distribution [33]. In particular, Miura and Maki approximated [33,35] the so called Φ function (Equation (9)) to a step-function E = E_s for a selected temperature T.

$$\Phi(E, T) = \exp\left(-k_0 \int_0^t \exp\left(-\frac{E}{RT} \right) dt \right) \cong \exp\left(-\frac{k_0}{\beta} \int_0^T \exp\left(-\frac{E}{RT} \right) dT \right) \qquad (9)$$

Thus, Equation (5) can be written as:

$$\alpha \cong 1 - \int_{E_s}^\infty f(E) dE = \int_0^{E_s} f(E) dE \qquad (10)$$

with E_s so that $\Phi(E_s,T) \cong 0.58$.

By the Miura and Maki approach, only a reaction having an activation energy E_s occurs at a given T and β—the model simplifies an actual reaction system by a set of N reactions, characterized by their own activation energy and pre-exponential factor.

Approximating the function $\Phi(E,T)$ as:

$$\Phi(E, T) \cong \exp\left(-\frac{k_0 RT^2}{\beta E} \exp\left(-\frac{E}{RT} \right) \right) \qquad (11)$$

the extent of conversion for the j-th reaction is expressed by Equation (12):

$$1 - \frac{\Delta v}{\Delta v_\infty} \cong \exp\left(-\frac{k_0 RT^2}{\beta E} \exp\left(-\frac{E}{RT} \right) \right) \qquad (12)$$

where Δv and Δv_∞ represent the amount of volatiles evolved and the effective volatile content for the j-th reaction.

After some mathematical steps [35], and after imposing $1 - \Delta v/(\Delta v_\infty) = \Phi(E,T) \cong 0.58$, the representative equation of the Miura–Maki model is given by Equation (13).

$$\ln\left(\frac{\beta}{T^2}\right) = \ln\left(\frac{k_0 R}{E}\right) + 0.6075 - \frac{E}{RT} \qquad (13)$$

Compared to the Gaussian DAEM, the approach of Miura [35] and Miura and Maki [33] relies on the dependence of k_0 on E. In particular, they proposed a simple procedure to estimate f(E) and k_0, consisting of the following steps:

(1) measurement of α vs. T relationship for at least three heating rates;
(2) calculation of β/T^2 at selected α values from the α vs. T relationship obtained in (1) for each heating rate;
(3) plotting of $\ln(\beta/T^2)$ vs. $-1/(RT)$ at the selected α values, and determination of E and k_0 from the Arrhenius plot at different α values using the relationship represented by Equation (13): E can be estimated by the slope, and k_0 by the intercept of each line with the y axis; and
(4) plotting the α value against the activation energy E, as calculated in (3), and differentiating the α vs. E relationship to obtain f(E).

2.4. Aspen Plus Model

The energy aspects of pyrolysis were thermodynamically investigated by performing some Aspen Plus simulations reproducing the pyrolysis of hydrochars. The model is based on thermodynamic equilibrium and the minimization of the Gibbs energy, without considering any kinetic phenomena. In particular, the thermodynamic properties of the conventional components were calculated by adopting the Peng–Robinson method.

Simulations were performed on grape seeds, hydrochars, and a commercial coal. Table 2 reports a summary of all input data, while Figure 1 shows the general Aspen scheme. The stream REACTANT represents the input feedstock entering the process, defined through a non-conventional stream characterized by its ultimate analysis and HHV (grape seeds and hydrochar in Table 1, coal Illinois No. 6 from literature [40]). Then, the RYIELD reactor, which is fictitious, splits the non-conventional stream into a conventional one (ELEMENTS) based on its elemental distribution. This stream enters the RGIBBS reactor, which computes the product distribution as a function of the operating conditions (i.e., temperature and pressure) by minimizing the Gibbs energy of the product mixture. The final products considered in the model are carbon graphite, carbon dioxide (CO_2), carbon monoxide (CO), water (H_2O), hydrogen (H_2), oxygen (O_2), methane (CH_4), acetylene (C_2H_2), ethylene (C_2H_4), and ethane (C_2H_6), to ensure a good model performance. Then, the net heat required/produced by the process is given by stream QOUT, while QIN represents the heat required to sustain RYIELD. Runs were performed between 300 and 900 °C (every block and stream were kept isotherm to avoid any sensible heat contribution) to deduce the amount of heat involved during the reaction. The product distribution at different temperatures was also investigated. Data were referred per gram of dry feedstock entering the process.

Table 2. Summary of parameters used in the Aspen Plus simulations.

Feedstock	Grape seeds, HC 180 °C, HC 220 °C, HC 250 °C, Coal
Products (at equilibrium)	C (graphite), CO_2, CO, H_2, O_2, H_2O, CH_4, C_2H_2, C_2H_4, C_2H_6
Equation of state	Peng–Robinson
Pyrolysis temperature	300–900 °C

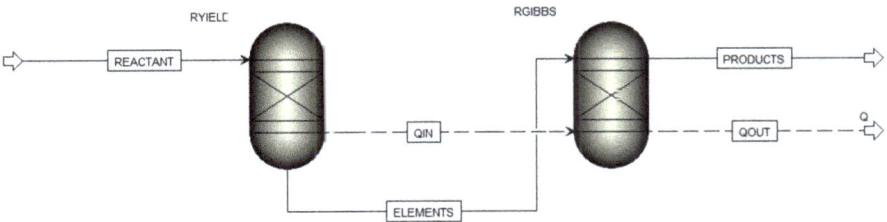

Figure 1. Aspen Plus scheme used to model pyrolysis.

3. Results and Discussion
3.1. Derivative Thermogravimetric Curves

Figure 2 shows the derivative thermogravimetric curves (DTGA) of grape seeds and hydrochars under pyrolytic and oxidative conditions, at different heating rates. Table 3 shows a sum of the position of local peaks, with their temperatures and intensities.

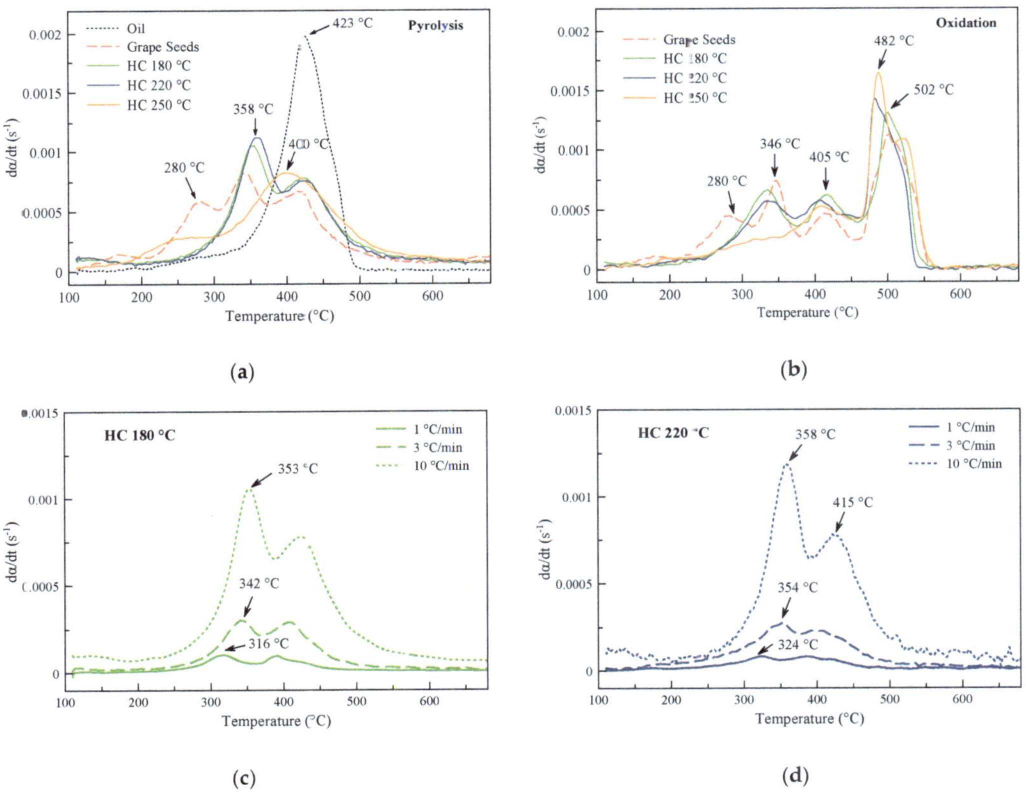

Figure 2. Cont.

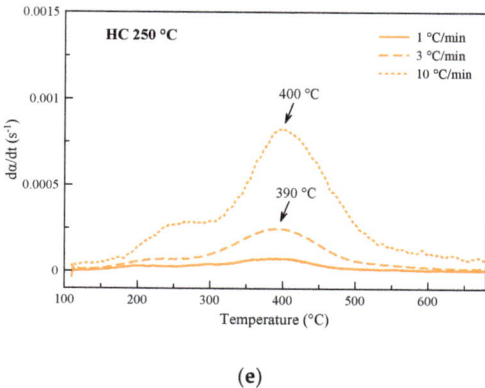

(e)

Figure 2. Derivative thermogravimetric (DTGA) curves: (**a**) during pyrolysis at 10 °C/min; (**b**) during oxidation at 10 °C/min; (**c**) effect of the heating rate during pyrolysis of hydrochar 180 °C; (**d**) effect of the heating rate during pyrolysis of hydrochar 220 °C; (**e**) effect of the heating rate during pyrolysis of hydrochar 250 °C. HC: hydrochars, dα/dt: derivative conversion curve of the extent of conversion α with respect to time t.

Table 3. DTGA peak temperature and decomposition rate of pyrolysis and oxidation (10 °C/min).

DTGA Peak		Pyrolysis					Oxidation			
		Oil	Grape Seeds	HC 180 °C	HC 220 °C	HC 250 °C	Grape Seeds	HC 180 °C	HC 220 °C	HC 250 °C
1	T (°C)	-	280	-	-	-	280	-	-	-
	dα/dt (s^{-1})	-	5.9×10^{-4}	-	-	-	4.4×10^{-4}	-	-	-
2	T (°C)	-	343	353	358	-	346	334	336	-
	dα/dt (s^{-1})	-	8.2×10^{-4}	1.0×10^{-3}	1.1×10^{-3}	-	7.4×10^{-4}	6.6×10^{-4}	5.7×10^{-4}	-
3	T (°C)	423	421	424	428	400	415	415	406	409
	dα/dt (s^{-1})	2.0×10^{-3}	6.6×10^{-4}	7.8×10^{-4}	7.7×10^{-4}	8.2×10^{-4}	4.6×10^{-4}	6.2×10^{-4}	5.8×10^{-4}	5.3×10^{-4}
4	T (°C)	-	-	-	-	-	502	502	482	480–523
	dα/dt (s^{-1})	-	-	-	-	-	1.1×10^{-3}	1.3×10^{-3}	1.4×10^{-3}	1.6–1.1×10^{-3}

Overall, both grape seeds and hydrochars mostly decompose between 200 and 500 °C, indicating that most of the volatile matter is removed in this range. Grape seeds present three main decomposition peaks, occurring at 280, 343, and 421 °C under a pyrolytic atmosphere. These peaks can be attributed to the overlapped decomposition of hemicellulose, lignin plus cellulose, and oil, which are the main constituents of grape seeds (around 7% cellulose, 31% hemicellulose, 44% lignin, and 10–15% oil [41,42]). Indeed, it is well known that under pyrolytic conditions, due to its amorphous and little polymerized structure, hemicellulose is very reactive at low temperatures (in the range of 250–300 °C [21,43]). In the DTGA curves, this reactivity translates into the first decomposition peak. Meanwhile, the second peak can be attributed to the overlapped decomposition of lignin and cellulose, which both generally decompose in the 300–350 °C range [43]. Considering the low content of cellulose (around 7% [41,42]), lignin clearly dominates at this temperature. Beyond this decomposition, lignin also contributes by forming a decomposition "baseline" to the profile, overlapping the other components. Indeed, due to its high polymerized structure and higher heterogeneity, it decomposes also (less intensively than at 300–350 °C) in a broad range of temperatures, from around 230° up to 600 °C [43]. Therefore, during the pyrolysis of grape seeds, its decomposition could overlap to that of the other constituents, forming a "baseline" to the overall profile. Meanwhile, comparing the measured curve with that of grape seed oil from previous work by our group [28], it is clear that the peak occurring at 423 °C can be attributed to the decomposition of the oil. Regarding the hydrochars, through HTC, grape seeds undergo carbonization, decreasing their atomic O/C and H/C ratios with a natural reduction of the volatile matter as the harshness of the process increases [32]. In DTGA curves, this translates in flattening the first decomposition peak at around 280 °C, which is indeed absent for all the hydrochars. Therefore, the authors can affirm that hemicellulosic compounds are mostly degraded during HTC. Interestingly, the 250 °C hydrochar exhibits a broad and not intense decomposition between 200 and 300 °C. Since hemicellulose degrades during HTC, this reactivity can be explained by the presence of re-polymerized compounds from the aqueous phase. Indeed, during HTC, sugar-derived compounds dissolved in the aqueous phase (like 5-HMF) can undergo condensation and re-polymerization, forming a solid phase called "secondary char" [44]. Beyond this, fatty acids formed during the hydrolysis of the oil could also be embedded in the solid phase, conferring to the sample an extra-reactivity at low temperatures. Both grape seeds and 180 and 220 °C hydrochars show their highest reactivity at around 350 °C, attributable to the degradation of lignin and cellulosic derived compounds. Differences can be due to both the heterogeneity of the feedstock and the synergistic behavior among the constituents, which are present in different ratios in the various samples. Conversely, the 250 °C hydrochar, after the initial slow decomposition at 200–300 °C, shows only one main decomposition peak at 400 °C. Therefore, while the 180 and 220 °C hydrochars tend to follow the feedstock profile (2nd and 3rd peaks), the 250 °C hydrochar highly deviates from that, highlighting the impact of the HTC operating temperature on the devolatilization profile. Overall, it is important to highlight to positive effect of the HTC treatment: after HTC, curves tend to shift towards the right region of the graph, demonstrating a higher thermal stability during the degradation process.

Regarding oxidative conditions (Figure 2b), as expected by typical combustion profiles of biomass fuels [20], all the profiles can be divided into two regions: a first phase during which the feedstock devolatilizes to char, and a second phase where char oxidation occurs. The first phase, between 250 and 450 °C, resembles the pyrolytic behavior: hemicellulose, lignin plus cellulose, and oil volatilize in the same temperature ranges as during pyrolysis. Then, the second phase, 480–520 °C, corresponds to the oxidation of the char formed during the first phase. This phase is predominant with respect to the previous phase. As for pyrolysis, the 180 and 220 °C hydrochars resemble the behavior of grape seeds, exhibiting three main DTGA peaks. Meanwhile, the 250 °C hydrochar presents only two DTGA peaks, demonstrating the harshness of severe HTC conditions on the feedstock structure and composition. Oxidation and pyrolysis TGA curves where (1-α) is plotted vs. T are reported in Supplementary Materials Figure S1.

Figure 2c–e show the effect of the heating rate on DTGA curves. The heating rate does not affect the shape of the decomposition profile. Meanwhile, the DTGA peak moves towards higher temperatures at increasing heating rates, which can be due to heat and mass transfer phenomena and intrinsic devolatilization kinetics. Indeed, at higher rates, the mismatch between the temperature in the furnace and the particle is higher, and therefore there is a delay between the temperature measured and the decomposition stage [45]. In addition, the $d\alpha/dt$ is higher at higher rates because, at a fixed α, dt is smaller. In addition, higher rates cause a bigger gradient between the surface and core temperature of the sample [45].

3.2. DSC Curves

Figure 3 shows DSC curves under nitrogen and oxidative atmosphere. Under an inert atmosphere, both grape seeds and hydrochars show an initial endothermic phase followed by an exothermic one at higher temperatures. The transition from an endo- to an exothermic behavior occurs in a temperature range between 200 and 290 °C. The heat request comprises both the heat necessary to evaporate water and that to trigger endothermic reactions. Water derives from both some residual moisture and some produced during condensation reactions occurring at the beginning of pyrolysis. Meanwhile, after the transition temperature, heat is released during pyrolysis. This behavior was already observed in literature with hydrochars derived from other feedstock [40,46] and can be attributed to the presence of non-oxidized oxygen on the hydrochar, like in the form of carboxylic groups. Indeed, this oxygen is enough to oxidize the elemental carbon, leading to a heat release. As shown in Table 4, the net energy balance along the temperature spectrum 40–600 °C is positive for all the samples, highlighting that overall pyrolysis releases thermal energy. Not surprisingly and consistently with elemental analysis (Table 1), a higher heat release is associated with a higher volatile matter loss (Table 4). Details on the progressive integration of DSC data are given later in the paper (namely, in Section 3.5).

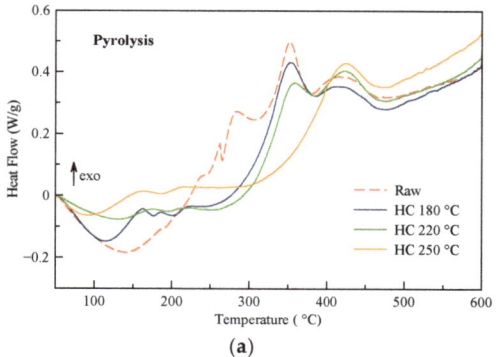 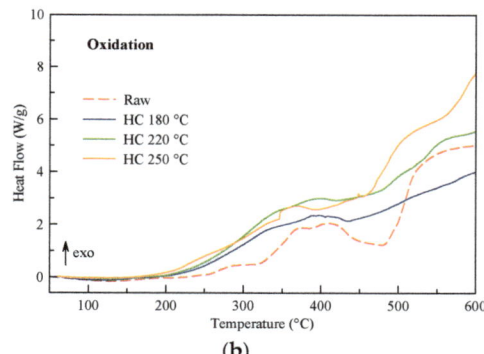

Figure 3. Differential scanning calorimetry (DSC) curves of grape seeds and hydrochars (HC) at 10 °C/min: (**a**) under inert and (**b**) oxidative atmosphere (> 0 exothermic, < 0 endothermic); per grams of dry feedstock.

Table 4. Details of DSC curves: integrals, mass loss, and transition point from endothermic to exothermic behavior; per kilograms of dry feedstock. The positive values of "Net energy" correspond to exothermal behavior during thermal treatments.

	Pyrolysis				Oxidation			
	Raw	HC 180 °C	HC 220 °C	HC 250 °C	Raw	HC 180 °C	HC 220 °C	HC 250 °C
Endothermic [1] (J/g)	118	90	67	23	98	54	31	19
Exothermic [1] (J/g)	705	587	585	598	4406	4928	6678	7533
Net energy [1] (J/g)	587	497	518	575	4309	4874	6647	7514
Mass loss [1] (%)	61.2	56.2	55.4	43.7	73.7	69.0	73.3	69.0

[1] Integral and mass losses computed/measured in the range 40–600 °C.

As for DTGA curves, samples show heat profiles with local peaks. Indeed, grape seeds show three main peaks, at 280, 352, and 410–425 °C, in correspondence with the degradation of hemicellulose, lignin plus cellulose, and oil. The 180 and 220 °C hydrochars show two peaks, while the 250 °C shows only one at 420 °C. A comparison between DTGA and DSC curves under pyrolytic atmosphere is reported in Supplementary Materials Figure S2.

As expected from the nature of the process, under oxidative conditions (Figure 3b), the amount of heat released increases as the carbonization degree of the sample increases. All the samples show a slight endothermic behavior at low temperature, justified by the absence or low presence of oxidative reactions at this condition. Overall, the heat release profile is broad and rises to the final observed temperature. Indeed, at 600 °C, a certain quantity of matter that undergoes oxidation is still present. Even if less sharp than under an inert atmosphere, all the samples show a peak at 350–450 °C that can be associated with the oxidation of lignin-like compounds. Grape seeds also show a small inflection at 280 °C, probably due to the oxidation of hemicellulosic compounds.

3.3. FTIR

Figure 4 reports the FTIR spectra of the samples. The deflection at 3300 cm^{-1} indicates the stretching of the hydroxyl group, occurring generally in the range 3700–3200 cm^{-1} [47]. Increasing HTC temperature results in a decreasing intensity in O-H stretching, because dehydration phenomena occur during HTC treatment. The peak at 3010 cm^{-1} represents =C-H group stretch, which tends to gradually disappear in the hydrochars. This fact can be related to double bonds present in the oil, which degrades, and it is even not present in HC 250 °C. Purnomo et al. [48] identified hydrogen bonded to the unsaturated carbon chain (C-H), whereas peaks at lower wavelength represent attachment to the saturated carbon chain. The peak at 1744 cm^{-1} indicates a typical bending of carbonyl group >C=O, possibly in the form of aldehyde [5], and/or the ester group in glyceride compounds. It tends to attenuate by increasing HTC treatment severity. Whereas grape seeds and HC 180 °C still present a peak, the HC 220 °C spectrum moves toward a lower wavelength. In the HC 250 °C spectrum the peak disappeared, confirming degradation and formation of new compounds during the hydrothermal treatment. Furthermore, Diaz et al. [41] state that peaks around 1030 cm^{-1} indicate C-O and C-O-C stretching, confirming that both decarboxylation and dehydration occurred. The grape seed spectrum exhibits a shoulder at a wavelength equal to 1655 cm^{-1}, attributed to stretching of double bond >C=C< [47], that tends to flatten out in hydrochars. Unsaturated double-bound >C=C< could be related to the presence of oil [6]—grape seed oil is mainly composed of linoleic acid [48,49], having two double bonds C=C in position 9 and 12. In general, functional groups act a stretch toward the region at a shorter wavelength.

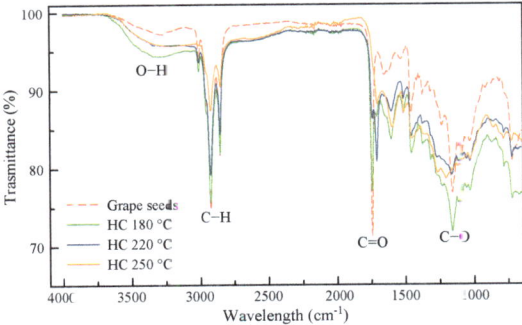

Figure 4. Fourier transform infrared spectroscopy (FTIR) curves of grape seeds and hydrochars (HC).

3.4. Kinetic Models

3.4.1. Gaussian Model

Table 5 shows E_0, σ, and h computed through the Gaussian model at different heating rates for the various substrates, while Figure 5 shows the experimental data compared with the modeling curves for pyrolysis (data and curves relevant to oxidation are available in Supplementary Materials Figure S3). Experimental data are the starting point to compute DTGA curves reported in Figure 2.

Table 5. Values of E_0 and σ for the Gaussian model for the different substrates and heating rates; per moles of dry feedstock. k_0 is let constant and equal to 1.67×10^{13} s^{-1}.

			Grape Seeds	Oil	HC 180 °C	HC 220 °C	HC 250 °C
Pyrolysis	E_0 (kJ/mol)	1 °C/min	-	-	205	207	209
		3 °C/min	-	-	203	202	206
		10 °C/min	193	218	200	200	204
	σ (kJ/mol)	1 °C/min	-	-	15	16	17
		3 °C/min	-	-	14	15	16
		10 °C/min	16	11	15	15	16
	h (-)	1 °C/min	-	-	0.083	0.090	0.089
		3 °C/min	-	-	0.061	0.072	0.073
		10 °C/min	0.084	0.022	0.064	0.062	0.076
Oxidation	E_0 (kJ/mol)	1 °C/min	-	-	211	214	219
		3 °C/min	-	-	215	215	220
		10 °C/min	213	-	214	213	221
	σ (kJ/mol)	1 °C/min	-	-	14	15	15
		3 °C/min	-	-	15	14	15
		10 °C/min	17	-	16	15	15
	h (-)	1 °C/min	-	-	0.057	0.065	0.065
		3 °C/min	-	-	0.071	0.066	0.066
		10 °C/min	0.102	-	0.084	0.077	0.077

Data demonstrate that for both pyrolysis and oxidation, the effect of the heating rate on E_0 and σ is negligible. Therefore, every substrate can be associated with one single pair of E_0 and σ, regardless of the heating rate, which agrees with previous studies performed on other biomass substrates [28]. Under both inert and oxidative atmospheres, the Gaussian DAEM adopted does not highlight the effects of both HTC and its severity on the starting feedstock (their maximum relative difference is 3.7%). This is not surprising: the model basis on the assumption of approximating the entire decomposition profile, which includes more stages, through a one-step profile characterized by a Gaussian distribution. Therefore, the final values of E_0 and σ can be considered as a sort of weighted average among the values of the single components (biomass constituents), i.e., hemicellulosic, lignin-like plus cellulosic compounds, and oil. Probably, the difference in the relative composition of the hydrochars is not enough to affect the final values. In particular, under an inert atmosphere, E_0 passes from 193 kJ/mol of grape seeds to 200–209 kJ/mol of hydrochars, while σ ranges between 14 and 17 kJ/mol. Values agree with those reported in the literature on the single components. For example, during pyrolysis, the activation energies usually range in 80–116 kJ/mol for hemicellulose, 195–286 kJ/mol for cellulose, and 176–300 kJ/mol for lignin [35,42]. The pyrolysis of oil alone has an E_0 of 218 kJ/mol. With values between 211 and 215 kJ/mol, the activation energy of oxidation is the same for grape seeds and hydrochars obtained at 180 and 220 °C. Meanwhile, it slightly increases at 220 kJ/mol for the 250 °C hydrochar.

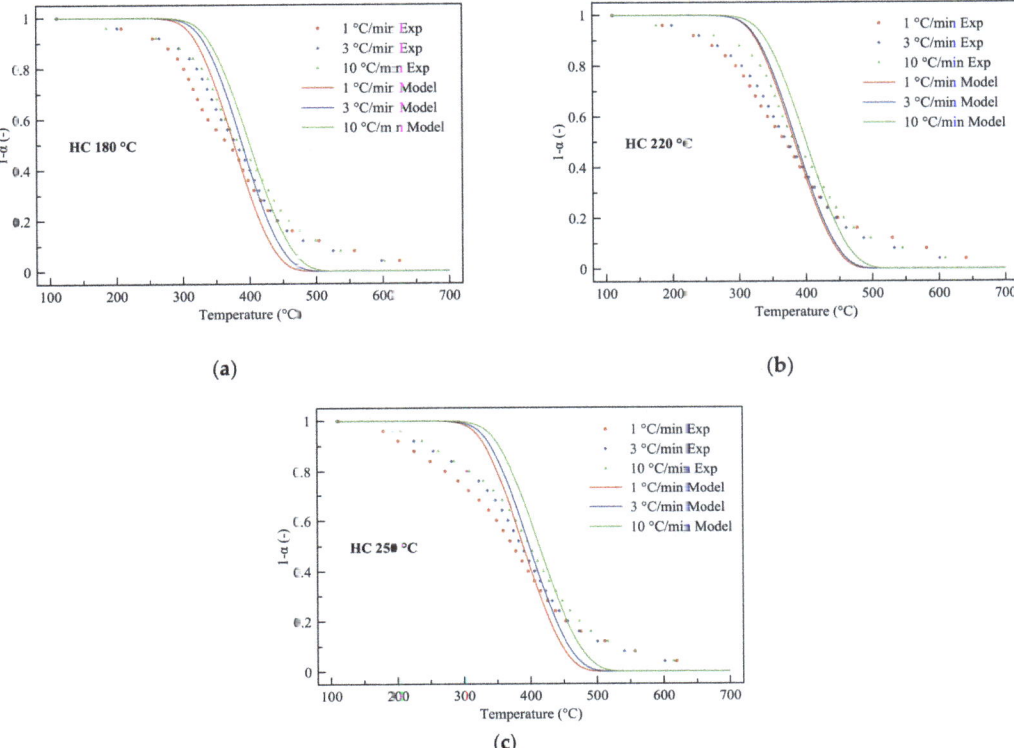

Figure 5. 1-α vs. temperature: comparison between experimental data (dots) and predicted data (lines) computed through the Gaussian model, during pyrolysis of different samples. (**a**) hydrochar 180 °C; (**b**) hydrochar 220 °C; (**c**) hydrochar 250 °C.

3.4.2. Miura–Maki Model

Figure 6 shows the values of E, while Tables S1 and S2 (Supplementary Materials) report the values of both E and k_0 with the correlation coefficient R^2. Figure 7 shows the plot of $\ln(\beta/T^2)$ vs. $-1/RT$ at different values of α. Each line corresponds to a certain α, while the three points on each line correspond to different heating rates. Each α corresponds to a certain temperature and the sequence $T_{\beta=1\,°C/min} < T_{\beta=3\,°C/min}\, T_{\beta=10\,°C/min}$ has to be respected. The maximum value of conversion for which the correct sequence of temperature was respected is defined with $α_{max} < 1$. This was found in the range of α 0.1–0.8 for pyrolysis and 0.1–0.9 for oxidation. Then, for the valid range of α, E and k_0 were determined graphically using the schemes in Figure 7; E is given by the slope of each line for each α, while k_0 is obtained from the intercept and by applying Equation (13).

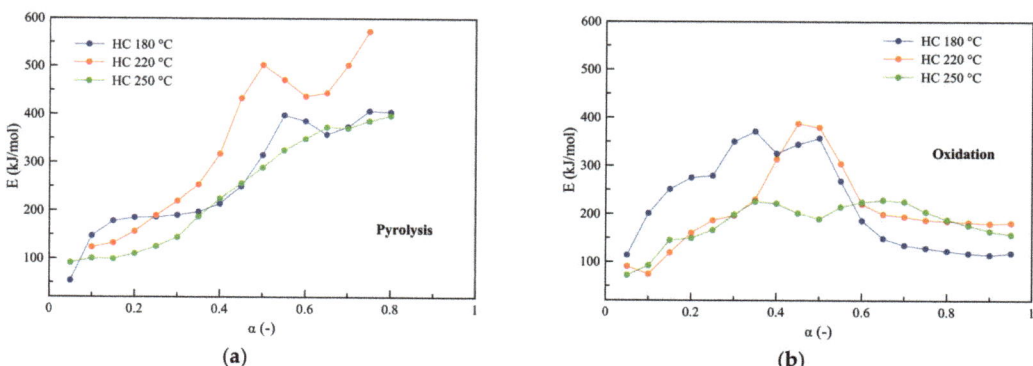

Figure 6. Distribution of E at different α of hydrochars, computed through the Miura–Maki model during (**a**) pyrolysis and (**b**) oxidation; per mol of dry feedstock.

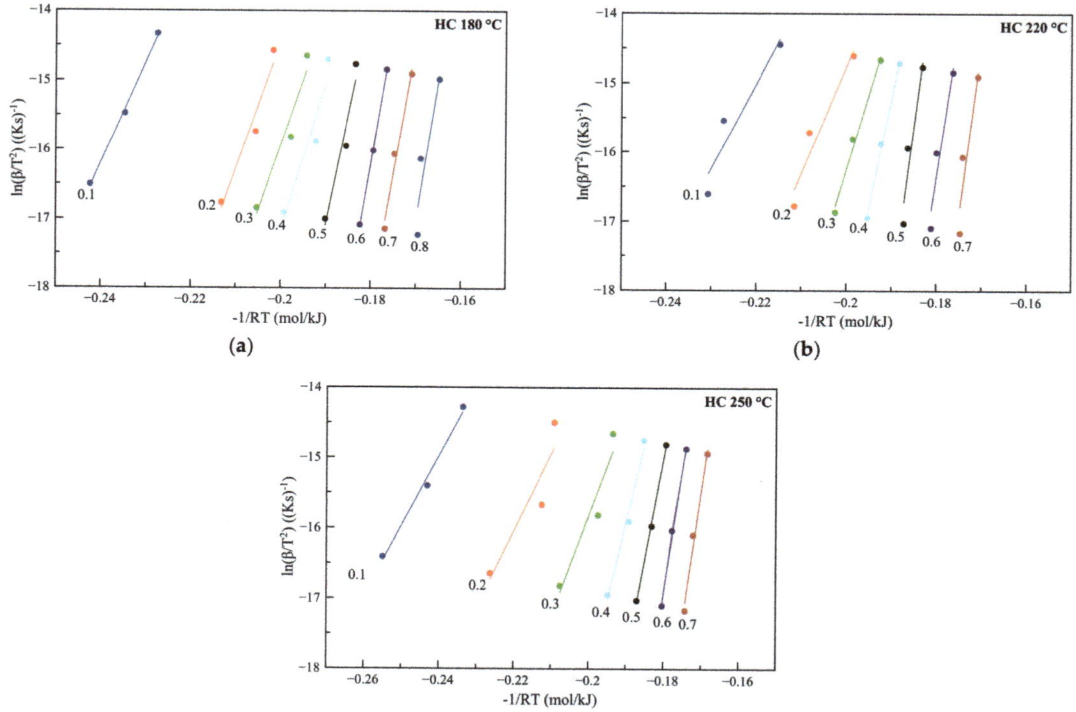

Figure 7. Miura–Maki diagrams of pyrolysis of: (**a**) hydrochar 180 °C; (**b**) hydrochar 220 °C; (**c**) hydrochar 250 °C.

Under pyrolysis, the fitting is satisfactory, with a R^2 always higher than 0.95, except for one outlier at α = 0.7 for the 220 °C hydrochar. Results show that α hugely affects the kinetic parameters. All the hydrochars present an E that increases with the degree of conversion: it rapidly passes from 100–147 kJ/mol (at α = 0.1) to 371–386 kJ/mol in a range of α 0.6–0.7. This phenomenon can be explained by the progressive increase of the degree of carbonization with temperature. As pyrolysis proceeds, the biomass undergoes volatilization and is converted into a high-carbon content matrix. This char is more difficult to degrade than the volatile matter and therefore causes a higher E. Generally, values are considerably higher than those commonly found in literature for biomass substrates (for example, for lignin it is 237–267 kJ/mol, and for pine wood it is 186–271 kJ/mol [27]), which can be explained by the nature of the substrate. Meanwhile, k_0 reaches its maximum at an α of 0.5–0.6, with maximum values of 10^{24-27} for the 180 and 250 °C hydrochars and 10^{38} for the 220 °C one.

Figure 6b shows how E of different hydrochars vary during oxidation. For the 180 and 220 °C hydrochars, E increases up to α = 0.3–0.5 to values of 357–379 kJ/mol. Meanwhile, the 250 °C hydrochar shows a very similar trend to the 220 °C one up to an α of 0.35, and then stabilizes at much lower values of 221–225 kJ/mol. The 180 °C hydrochar shows higher values of E at the beginning of the conversion. Therefore, the HTC severity decreases the kinetic parameters of both volatile matter and char produced during the process. Similar trends were observed by Bach et al. [20] on hydrochars produced from wood residues. Meanwhile, pre-exponential factors k_0 show a very similar trend to E (they reach their maxima in correspondence of the maximum of α, arriving up to 10^{24-27} for the 180 and 220 °C hydrochars, and 10^{14} for the 250 °C one).

3.4.3. Comparison and Suggestions for Future Work

Table 6 reports a comparison table with the average values of E, σ, and k_0, computed through the Gaussian and Miura–Maki models.

Table 6. Average values of E, σ, and k_0 computed through the Gaussian and Miura–Maki models (k_0 constant and equal to 1.67×10^{13} s^{-1} for the Gaussian model).

Model	Atmosphere	Parameter	Grape Seeds	Oil	Hydrochar 180 °C	Hydrochar 220 °C	Hydrochar 250 °C
Gaussian	Pyrolysis	E (kJ/mol)	193	218	203	203	206
		σ (kJ/mol)	16	11	15	15	16
	Oxidation	E (kJ/mol)	213	-	213	214	220
		σ (kJ/mol)	17	-	15	15	15
Miura-Maki	Pyrolysis	E (kJ/mol)	-	-	265	339	239
		k_0 (s^{-1})	-	-	7.6×10^{27}	6.0×10^{38}	9.8×10^{24}
	Oxidation	E (kJ/mol)	-	-	222	215	181
		k_0 (s^{-1})	-	-	3.4×10^{26}	1.0×10^{26}	2.9×10^{13}

Due to its starting hypothesis, i.e., approximating the profile with a single Gaussian curve, the Gaussian model does not highlight the effect of the severity of the HTC process on E, which ranges between 193 and 206 kJ/mol for pyrolysis. This does not occur for the Miura–Maki model, which assumes a distribution of E that varies on the entire range leading to more diverse average values. Indeed, the Gaussian approach models a decomposition involving several stages through a single decomposition stage characterized by a Gaussian profile of the activation energy. Therefore, this assumption "averages" the profiles, canceling the differences among the samples obtained at different severities. Meanwhile, the Miura–Maki model avoids this averaging since it does not impose a similar strong assumption on the distribution.

Since the literature lacks the modeling of the kinetics of pyrolysis of hydrochar derived from grape seeds, the comparison of the results obtained was possible considering typical values of activation energy for decomposition of the main constituents of lignocellulosic materials. In particular, cellulose decomposition activation energy ranges between 175 and 279 kJ/mol [21,27,50] hemicellulose is in the range 132–186 kJ/mol [27,51,52], whereas lignin ranges from 62 to 271 kJ/mol [27,50–52], with the broadest range. These values seem to prove the consistency of the results obtained in the present work.

In general, the Gaussian model has the advantage of being valid over the entire range of conversion, while the Miura–Maki model requires always that the temperature sequence must be satisfied, an aspect that is not obvious (in this study this occurs only for α in the range of 0.1–0.8). Overall, both the models suffer from a strong starting hypothesis: k_0 fixed for the Gaussian model and a strong integral simplification in the Miura–Maki model that can lead to an overestimation of the average values of the kinetics parameters [53].

To improve the prediction, the authors suggest extending the single Gaussian model to a Multi Gaussian, in which the entire decomposition profile is divided into single decomposition peaks, each one approximated by a Gaussian curve [21]. Indeed, the single approach suits well homogeneous substrates (like the oil) characterized by a single decomposition peak, but seems too simple for complex substrates like biomasses. As a result, future works should include a comparison with other kinetics models.

Then, applying other DAEMs (like Kissinger–Akahira–Sunose method and the Coats–Redfern method) on the same feedstock could help to validate the results. Apart from contributing to understanding the mechanisms behind the decomposition, kinetic parameters can help in technological design and optimization. Moving to a larger scale, integration with other modeling techniques is necessary. Among these, it is worth mentioning models that consider the effects of particle distribution and geometry on heat transfer phenomena (like fractal models [54,55]), statistical models [56], and comprehensive computational models [57].

3.5. Aspen Plus Model

Figure 8a,c show the trend of heat produced/required as computed by Aspen Plus under pyrolytic and oxidative atmosphere.

Regarding pyrolysis, all samples follow a similar trend, characterized by an exothermic behavior at low temperatures and an endothermic one at high temperatures. Figure 9 reports the trends of the products at different pyrolysis temperatures resulting from the 180 °C hydrochar (the other samples present the same qualitative trends) and can help to interpret the heat trends. At low temperatures, thermodynamics favors exothermic reactions like char oxidation to CO_2 and CO, oxidation of hydrogen, and methanation. Conversely, higher temperatures enhance endothermic reactions, like the Boudouard reaction (and therefore the conversion of CO_2 into CO), and methane steam reforming (thus the conversion of CH_4 and H_2O to CO and H_2). At low temperatures, more carbonized substrates show a higher exothermic behavior due to a higher carbon partial oxidation and methane production (they present a higher C/H ratio than less carbonized substrates). Conversely, at higher temperatures, less carbonized samples show a higher endothermic behavior, probably due to the higher production of CO. As expected from its high C/O ratio and low volatile matter content, the profile of coal is flatter than hydrochar, and it is always endothermic. The different starting points can be explained by the higher ash content of coal than hydrochars (13.7% vs. 1.9–2.5%): ash, being not reactive, lowers the energy input/output per unit of mass involved. The slight difference in the coal profile with the previous work of our group [40] is due to the different inlet temperatures (isotherm to the reactors here vs. at ambient temperature in [31]).

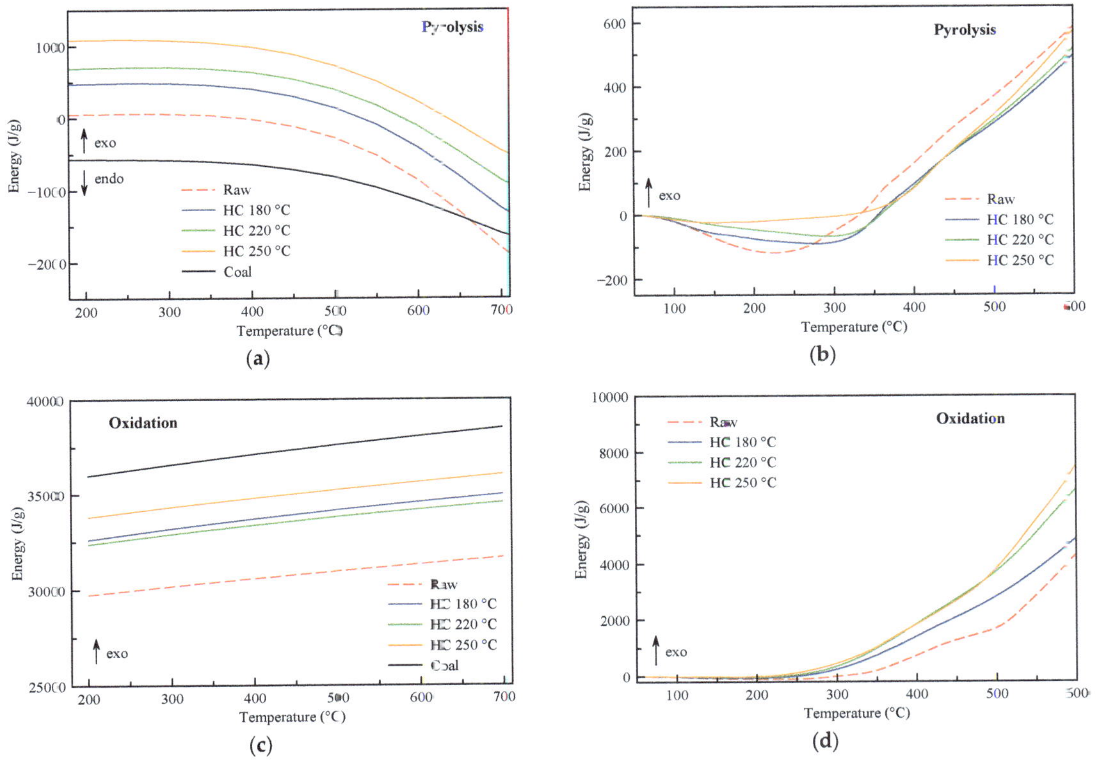

Figure 8. Energy released (> 0) or absorbed (< 0) by the samples computed from: (**a**) Aspen Plus during pyrolysis; (**b**) DSC data during pyrolysis at 10 °C/min; (**c**) Aspen Plus during oxidation; (**d**) DSC data during oxidation at 10 °C/min; per grams of dry feedstock.

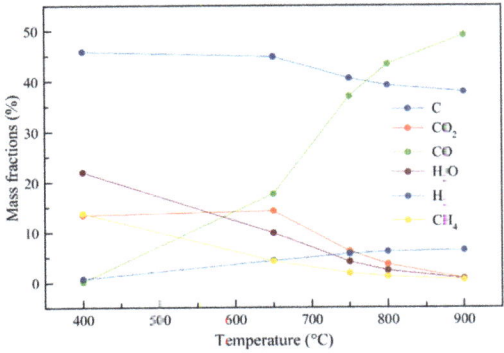

Figure 9. Aspen Plus results: main products from pyrolysis of hydrochar 180 °C.

Besides, Figure 8b shows the progressive integration of pyrolytic DSC data; computing the energy integral is necessary because DSC measures the instantaneous heat flow absorbed or released (which depends on the "history" of the sample due to the thermal program), while thermodynamics considers the energy absorbed or released when converting all the reactants into products at that determined temperature (and therefore the values obtained from thermodynamics do not depend on the thermal history of the sample). Overall, DSC data show a trend towards a cumulative exothermic behavior, whose final value corresponds to the integral reported in Table 4. The different behavior concerning Aspen data can be due to the nature of the thermodynamic approach. Indeed, real systems do not behave as predicted by thermodynamics at low temperatures because of reaction kinetics constraints. Moreover, as previously mentioned, thermodynamics does not consider the "history" of what happened before a certain condition. Indeed, inside the DSC, the starting material progressively undergoes pyrolysis, and the material at every time is the result of what happened before (i.e., the thermal program). In addition, thermodynamics does not consider typical "out of equilibrium" products, like tarry compounds, always present in real systems. Therefore, the thermodynamic approach requires particular attention and criticism before being used.

Regarding oxidation, both thermodynamic and DSC curves are clearly exothermic, with values that progressively increase with temperature. Under both atmospheres, absolute values differ between the Aspen and DSC approaches. This is due to the mass involved: DSC curves measure the energy absorbed/released by an amount of material that progressively decreases with time (the process is occurring), while Aspen predicts the energy absorbed/released by the same mass at every temperature. Since the energy is normalized by the amount of starting feedstock (and not the mass measured at every temperature), the amount of heat computed through the DSC approach is smaller than the Aspen one. Finally, it is interesting to note that in an oxidative environment, and for the whole range of temperature investigated, the thermodynamic approach foresees the complete oxidation of the biomass with the production essentially of CO_2, H_2O, and residual ash (equal to about 2% of the biomass fed to the reactor, Table 1). Conversely, in an oxidative environment in DSC, at the maximum temperature reached of 600 °C, there is still a substantial mass of non-oxidized carbon: the residual mass at 600 °C is of the order of 30% (Table 4). Through this critical comparison, the two approaches, thermodynamic and DSC, turn out to actually be very distant from each other, which is an aspect worthy of interest.

4. Conclusions

This work investigated the kinetic behavior during pyrolysis and oxidation of hydrochars derived from an agro-industrial residue (grape seeds). The topic was approached through experimental techniques (TGA and DSC) that were used to develop two different DAEMs for biomass decomposition. Indeed, there is a certain lack in the literature regarding hydrochar decomposition, especially for the computation of the kinetic parameters. Therefore, this work covers both the experimental aspects and the comparison of two DAEMs using a critical approach. Thermogravimetric analysis and differential scanning calorimetry highlighted the importance of the HTC severity on the decomposition profiles. Indeed, hydrochars obtained at 180 and 220 °C still decompose with two/three stages (pyrolysis/oxidation) due to the presence of lignocellulosic constituents and oil that only partially degraded during HTC. Meanwhile, the 250 °C hydrochar shows a more stable profile, with one/two decomposition peaks, highlighting that HTC temperature hugely affected the structure of the biomass. Interestingly, all hydrochars show a certain exothermic behavior during pyrolysis.

DAEMs predicted the kinetic parameters involved during the pyrolysis and oxidation processes, i.e., the activation energy and the pre-exponential factor. The single-stage Gaussian model does not highlight differences among the various hydrochars obtained at different HTC temperatures, with average values of activation energy in the range 203–206 kJ/mol for pyrolysis and 213–220 kJ/mol for oxidation. Actually, the Gaussian DAEM turned out to be unsatisfactory to model complex feedstock such as hydrochars characterized by multi-decomposition peaks. Meanwhile, the Miura–Maki model, even if not applicable over the entire decomposition region, enabled the determination of the distribution of activation energies: the more severe the HTC process, the lower the activation energy values.

Supplementary Materials: The following supporting information can be downloaded at: https://www.mdpi.com/article/10.3390/en15030950/s1, Figure S1: (1-α) at different temperatures of feedstock during: (a) pyrolysis; (b) oxidation. Heating rate: 10 °C/min.; Figure S2: Comparison between DTGA and DSC curves during pyrolysis at 10°C/min of: (a) grape seeds and hydrochar 180 °C; (b) hydrochars 220 and 250 °C; Figure S3: Comparison between experimental and predicted data (1-α vs temperature) computed through the Gaussian model, during oxidation of different samples (a) hydrochar 180 °C; (b) hydrochar 220 °C; (c) hydrochar 250 °C; Table S1: Details of Miura-Maki model, pyrolysis (on a dry basis); Table S2: Details of Miura-Maki model, oxidation (on a dry basis); Gaussian DAEM – MATLAB code.

Author Contributions: G.G.: methodology, software, data curation, writing—original draft preparation; G.I.: data curation, visualization, writing—original draft preparation; L.F. (Luca Fambri): experimental planning, writing—review and editing, supervision, project administration; L.F. (Luca Fiori): writing—review and editing, supervision, project administration. All authors have read and agreed to the published version of the manuscript.

Funding: This research received no external funding.

Institutional Review Board Statement: Not applicable.

Informed Consent Statement: Informed consent was obtained from all subjects involved in the study.

Data Availability Statement: Not applicable.

Acknowledgments: All the authors acknowledge Claudia Gavazza for thermal analysis. G.I. acknowledges the financial support provided by the ERICSOL project of the University of Trento, which partially covered her Ph.D. scholarship. All individuals included in this section have consented to the acknowledgement.

Conflicts of Interest: The authors declare no conflict of interest.

References

1. Li, J.; Pan, L.; Suvarna, M.; Tong, Y.W.; Wang, X. Fuel Properties of Hydrochar and Pyrochar: Prediction and Exploration with Machine Learning. *Appl. Energy* **2020**, *269*, 115166. [CrossRef]
2. Ahmed, M.; Andreottola, G.; Elagroudy, S.; Negm, M.S.; Fiori, L. Coupling Hydrothermal Carbonization and Anaerobic Digestion for Sewage Digestate Management: Influence of Hydrothermal Treatment Time on Dewaterability and Bio-Methane Production. *J. Environ. Manage.* **2021**, *281*, 111910. [CrossRef] [PubMed]
3. De Medeiros, A.D.M.; Da Silva Junior, C.J.G.; De Amorim, J.D.P.; Do Nascimento, H.A.; Converti, A.; De Santana Costa, A.F; Sarubbo, L.A. Biocellulose for Treatment of Wastewaters Generated by Energy Consuming Industries: A Review. *Energies* **2021**, *14*, 1–18. [CrossRef]
4. Yu, C.; Ren, S.; Wang, G.; Xu, J.; Teng, H.; Li, T.; Huang, C.; Wang, C.; Yu, C.; Ren, S.; et al. Kinetic Analysis and Modeling of Maize Straw Hydrochar Combustion Using a Multi-Gaussian-Distributed Activation Energy Model Kinetic Analysis and Modeling of Maize Straw Hydrochar Combustion Using a Multi-Gaussian-Distributed Activation. *Energy Model.* **2021**, *28*, 611–620.
5. Zhang, N.; Wang, G.; Zhang, J.; Ning, X.; Li, Y.; Liang, W.; Wang, C. Study on Co-Combustion Characteristics of Hydrochar and Anthracite Coal. *J. Energy Inst.* **2020**, *93*, 1125–1137. [CrossRef]
6. Chen, Z.; Hu, M.; Zhu, X.; Guo, D.; Liu, S.; Hu, Z.; Xiao, B.; Wang, J.; Laghari, M. Characteristics and Kinetic Study on Pyrolysis of Five Lignocellulosic Biomass via Thermogravimetric Analysis. *Bioresour. Technol.* **2015**, *192*, 441–450. [CrossRef]
7. Chen, J.; Wang, Y.; Lang, X.; Ren, X.; Fan, S. Evaluation of Agricultural Residues Pyrolysis under Non-Isothermal Conditions: Thermal Behaviors, Kinetics, and Thermodynamics. *Bioresour. Technol.* **2017**, *241*, 340–348. [CrossRef]
8. Ferrentino, R.; Merzari, F.; Fiori, L.; Andreottola, G. Coupling Hydrothermal Carbonization with Anaerobic Digestion for Sewage Sludge Treatment: Influence of HTC Liquor and Hydrochar on Biomethane Production. *Energies* **2020**, *13*, 6262. [CrossRef]
9. Funke, A.; Ziegler, F. Hydrothermal Carbonization of Biomass: A Summary and Discussion of Chemical Mechanisms for Process Engineering. *Biofuels, Bioprod. Biorefining* **2010**, *4*, 160–177. [CrossRef]
10. Olszewski, M.P.; Arauzo, P.J.; Wądrzyk, M.; Kruse, A. Py-GC-MS of Hydrochars Produced from Brewer's Spent Grains. *J. Ancl. Appl. Pyrolysis* **2019**, *140*, 255–263. [CrossRef]
11. Ischia, G.; Cazzanelli, M.; Fiori, L.; Orlandi, M.; Miotello, A. Exothermicity of Hydrothermal Carbonization: Determination of Heat Profile and Enthalpy of Reaction via High-Pressure Differential Scanning Calorimetry. *Fuel* **2022**, *310*, 122312. [CrossRef]
12. Ischia, G.; Fiori, L. Hydrothermal Carbonization of Organic Waste and Biomass: A Review on Process, Reactor, and Plant Modeling. *Waste and Biomass Valorization* **2021**, *12*, 2797–2824. [CrossRef]
13. Scrinzi, D.; Andreottola, G.; Fiori, L. Composting Hydrochar-OFMSW Digestate Mixtures: Design of Bioreactors and Preliminary Experimental Results. *Appl. Sci.* **2021**, *11*, 1–15. [CrossRef]

14. Lin, J.C.; Mariuzza, D.; Volpe, M.; Fiori, L.; Ceylan, S.; Goldfarb, J.L. Integrated Thermochemical Conversion Process for Valorizing Mixed Agricultural and Dairy Waste to Nutrient-Enriched Biochars and Biofuels. *Bioresour. Technol.* **2021**, *328*, 124765. [CrossRef]
15. Hu, Q.; Yang, H.; Xu, H.; Wu, Z.; Lim, C.J.; Bi, X.T.; Chen, H. Thermal Behavior and Reaction Kinetics Analysis of Pyrolysis and Subsequent In-Situ Gasification of Torrefied Biomass Pellets. *Energy Convers. Manag.* **2018**, *161*, 205–214. [CrossRef]
16. Román, S.; Libra, J.; Berge, N.; Sabio, E.; Ro, K.; Li, L.; Ledesma, B.; Alvarez, A.; Bae, S. Hydrothermal Carbonization: Modeling, Final Properties Design and Applications: A Review. *Energies* **2018**, *11*, 216. [CrossRef]
17. Sharma, H.B.; Panigrahi, S.; Dubey, B.K. Hydrothermal Carbonization of Yard Waste for Solid Bio-Fuel Production: Study on Combustion Kinetic, Energy Properties, Grindability and Flowability of Hydrochar. *Waste Manag.* **2019**, *91*, 108–119. [CrossRef]
18. Islam, M.A.; Kabir, G.; Asif, M.; Hameed, B.H. Combustion Kinetics of Hydrochar Produced from Hydrothermal Carbonisation of Karanj (Pongamia Pinnata) Fruit Hulls via Thermogravimetric Analysis. *Bioresour. Technol.* **2015**, *194*, 14–20. [CrossRef]
19. He, C.; Giannis, A.; Wang, J.-Y. Conversion of Sewage Sludge to Clean Solid Fuel Using Hydrothermal Carbonization: Hydrochar Fuel Characteristics and Combustion Behavior. *Appl. Energy* **2013**, *111*, 257–266. [CrossRef]
20. Bach, Q.V.; Tran, K.Q.; Skreiberg, Ø. Combustion Kinetics of Wet-Torrefied Forest Residues Using the Distributed Activation Energy Model (DAEM). *Appl. Energy* **2017**, *185*, 1059–1066. [CrossRef]
21. Zhang, J.; Chen, T.; Wu, J.; Wu, J. Multi-Gaussian-DAEM-Reaction Model for Thermal Decompositions of Cellulose, Hemicellulose and Lignin: Comparison of N2 and CO2 Atmosphere. *Bioresour. Technol.* **2014**, *166*, 87–95. [CrossRef]
22. Vand, V. A Theory of the Irreversible Electrical Resistance Changes of Metallic Films Evaporated in Vacuum. *Proc. Phys. Soc.* **1943**, *55*, 222–246. [CrossRef]
23. Wang, S.; Dai, G.; Yang, H.; Luo, Z. Lignocellulosic Biomass Pyrolysis Mechanism: A State-of-the-Art Review. *Prog. Energy Combust. Sci.* **2017**, *62*, 33–86. [CrossRef]
24. Radojević, M.; Janković, B.; Jovanović, V.; Stojiljković, D.; Manić, N. Comparative Pyrolysis Kinetics of Various Biomasses Based on Model-Free and DAEM Approaches Improved with Numerical Optimization Procedure. *PLoS ONE* **2018**, *13*, 1–25. [CrossRef]
25. Hameed, S.; Sharma, A.; Pareek, V.; Wu, H.; Yu, Y. A Review on Biomass Pyrolysis Models: Kinetic, Network and Mechanistic Models. *Biomass Bioenergy* **2019**, *123*, 104–122. [CrossRef]
26. White, J.E.; Catallo, W.J.; Legendre, B.L. Biomass Pyrolysis Kinetics: A Comparative Critical Review with Relevant Agricultural Residue Case Studies. *J. Anal. Appl. Pyrolysis* **2011**, *91*, 1–33. [CrossRef]
27. Cai, J.; Wu, W.; Liu, R.; Huber, G.W. A Distributed Activation Energy Model for the Pyrolysis of Lignocellulosic Biomass. *Green Chem.* **2013**, *15*, 1331–1340. [CrossRef]
28. Fiori, L.; Valbusa, M.; Lorenzi, D.; Fambri, L. Modeling of the Devolatilization Kinetics during Pyrolysis of Grape Residues. *Bioresour. Technol.* **2012**, *103*, 389–397. [CrossRef]
29. Basso, D.; Weiss-Hortala, E.; Patuzzi, F.; Baratieri, M.; Fiori, L. In Deep Analysis on the Behavior of Grape Marc Constituents during Hydrothermal Carbonization. *Energies* **2018**, *11*, 1379. [CrossRef]
30. Cai, J.M.; Bi, L.S. Kinetic Analysis of Wheat Straw Pyrolysis Using Isoconversional Methods. *J. Therm. Anal. Calorim.* **2009**, *98*, 325. [CrossRef]
31. Branca, C.; Albano, A.; Di Blasi, C. Critical Evaluation of Global Mechanisms of Wood Devolatilization. *Thermochim. Acta* **2005**, *429*, 133–141. [CrossRef]
32. Ischia, G.; Orlandi, M.; Fendrich, M.A.; Bettonte, M.; Merzari, F.; Miotello, A.; Fiori, L. Realization of a Solar Hydrothermal Carbonization Reactor: A Zero-Energy Technology for Waste Biomass Valorization. *J. Environ. Manage.* **2020**. [CrossRef] [PubMed]
33. Miura, K.; Maki, T. A Simple Method for Estimating f(E) and K0(E) in the Distributed Activation Energy Model. *Energy and Fuels* **1998**, *12*, 864–869. [CrossRef]
34. Gao, L.; Volpe, M.; Lucian, M.; Fiori, L.; Goldfarb, J.L. Does Hydrothermal Carbonization as a Biomass Pretreatment Reduce Fuel Segregation of Coal-Biomass Blends during Oxidation? *Energy Convers. Manag.* **2019**, *181*, 93–104. [CrossRef]
35. Miura, K. A New and Simple Method to Estimate f(E) and K0(E) in the Distributed Activation Energy Model from Three Sets of Experimental Data. *Energy and Fuels* **1995**, *9*, 302–307. [CrossRef]
36. Coats, A.W.; Redfern, J.P. Kinetic Parameters from Thermogravimetric Data. *Nature* **1964**, *201*, 68–69. [CrossRef]
37. Fischer, P.E.; Jou, C.S.; Gokalgandhi, S.S. Obtaining the Kinetic Parameters from Thermogravimetry Using a Modified Coats and Redfern Technique. *Ind. Eng. Chem. Res.* **1987**, *26*, 1037–1040. [CrossRef]
38. Anthony, D.B.; Howard, J.B. Coal Devolatilization and Hydrogasification. *AIChE J.* **1976**, *22*, 625–656. [CrossRef]
39. Güneş, M.; Güneş, S. A Direct Search Method for Determination of DAEM Kinetic Parameters from Nonisothermal TGA Data (Note). *Appl. Math. Comput.* **2002**, *130*, 619–628. [CrossRef]
40. Ischia, G.; Fiori, L.; Gao, L.; Goldfarb, J.L. Valorizing Municipal Solid Waste via Integrating Hydrothermal Carbonization and Downstream Extraction for Biofuel Production. *J. Clean. Prod.* **2021**, *289*, 125781. [CrossRef]
41. Spanghero, M.; Salem, A.Z.M.; Robinson, P.H. Chemical Composition, Including Secondary Metabolites, and Rumen Fermentability of Seeds and Pulp of Californian (USA) and Italian Grape Pomaces. *Anim. Feed Sci. Technol.* **2009**, *152*, 243–255. [CrossRef]
42. Moldes, D.; Gallego, P.P.; Rodríguez Couto, S.; Sanromán, A. Grape Seeds: The Best Lignocellulosic Waste to Produce Laccase by Solid State Cultures of Trametes Hirsuta. *Biotechnol. Lett.* **2003**, *25*, 491–495. [CrossRef]
43. Hubble, A.H.; Goldfarb, J.L. Synergistic Effects of Biomass Building Blocks on Pyrolysis Gas and Bio-Oil Formation. *J. Anal. Appl. Pyrolysis* **2021**, *156*, 105100. [CrossRef]

44. Volpe, M.; Fiori, L. From Olive Waste to Solid Biofuel through Hydrothermal Carbonisation: The Role of Temperature and Solid Load on Secondary Char Formation and Hydrochar Energy Properties. *J. Anal. Appl. Pyrolysis* **2017**, *124*, 63–72. [CrossRef]
45. Mackuľak, T.; Prousek, J.; Olejníková, P.; Bodík, I. Slovak Society of Chemical Engineering Institute of Chemical and Environmental Engineering Slovak University of Technology in Bratislava. *Direct* **2010**, 1407–1412.
46. Volpe, M.; Goldfarb, J.L.; Fiori, L. Hydrothermal Carbonization of Opuntia Ficus-Indica Cladodes: Role of Process Parameters on Hydrochar Properties. *Bioresour. Technol.* **2018**, *247*, 310–318. [CrossRef]
47. Guo, S.; Dong, X.; Wu, T.; Shi, F.; Zhu, C. Characteristic Evolution of Hydrochar from Hydrothermal Carbonization of Corn Stalk. *J. Anal. Appl. Pyrolysis* **2015**, *116*, 1–9. [CrossRef]
48. Purnomo, C.W.; Castello, D.; Fiori, L. Granular Activated Carbon from Grape Seeds Hydrothermal Char. *Appl. Sci.* **2018**, *8*, 1–16. [CrossRef]
49. Diaz, E.; Manzano, F.; Villamil, J.A.; Rodriguez, J.; Mohedano, A. Low-Cost Activated Grape Seed-Derived Hydrochar through Hydrothermal Carbonization and Chemical Activation for Sulfamethoxazole Adsorption. *Appl. Sci.* **2019**, *9*, 5127. [CrossRef]
50. Jiang, G.; Nowakowski, D.J.; Bridgwater, A.V. A Systematic Study of the Kinetics of Lignin Pyrolysis. *Thermochim. Acta* **2010**, *498*, 61–66. [CrossRef]
51. Wang, S.; Ru, B.; Lin, H.; Sun, W.; Luo, Z. Pyrolysis Behaviors of Four Lignin Polymers Isolated from the Same Pine Wood. *Bioresour. Technol.* **2015**, *182*, 120–127. [CrossRef] [PubMed]
52. Gomez, I.J.; Arnaiz, B.; Cacioppo, M.; Arcudi, F.; Prato, M. Nitrogen-Doped Carbon Nanodots for Bioimaging and Delivery of Paclitaxel. *J. Mater. Chem. B* **2018**, *6*, 5540–5548. [CrossRef] [PubMed]
53. Cai, J.; Li, T.; Liu, R. A Critical Study of the Miura-Maki Integral Method for the Estimation of the Kinetic Parameters of the Distributed Activation Energy Model. *Bioresour. Technol.* **2011**, *102*, 3894–3899. [CrossRef] [PubMed]
54. Xu, P.; Yu, B. The Scaling Laws of Transport Properties for Fractal-like Tree Networks. *J. Appl. Phys.* **2006**, *100*, 104906. [CrossRef]
55. Liang, M.; Fu, C.; Xiao, B.; Luo, L.; Wang, Z. A Fractal Study for the Effective Electrolyte Diffusion through Charged Porous Media. *Int. J. Heat Mass Transf.* **2019**, *137*, 365–371. [CrossRef]
56. Mäkelä, M.; Yoshikawa, K. Simulating Hydrothermal Treatment of Sludge within a Pulp and Paper Mill. *Appl. Energy* **2016**, *173*, 177–183. [CrossRef]
57. Álvarez-Murillo, A.; Sabio, E.; Ledesma, B.; Román, S.; González-García, C.M. Generation of Biofuel from Hydrothermal Carbonization of Cellulose. Kinetics Modelling. *Energy* **2016**, *94*, 600–608. [CrossRef]

MDPI
St. Alban-Anlage 66
4052 Basel
Switzerland
www.mdpi.com

Energies Editorial Office
E-mail: energies@mdpi.com
www.mdpi.com/journal/energies

Disclaimer/Publisher's Note: The statements, opinions and data contained in all publications are solely those of the individual author(s) and contributor(s) and not of MDPI and/or the editor(s). MDPI and/or the editor(s) disclaim responsibility for any injury to people or property resulting from any ideas, methods, instructions or products referred to in the content.